T0206162

# Biophysics and Nanotechnology of Ion Channels

# Biophysics and Nanotechnology of Ion Channels

Mohammad Ashrafuzzaman

**CRC Press**
Taylor & Francis Group
Boca Raton  London  New York

CRC Press is an imprint of the
Taylor & Francis Group, an **informa** business

First edition published 2022
by CRC Press
6000 Broken Sound Parkway NW, Suite 300, Boca Raton, FL 33487-2742

and by CRC Press
2 Park Square, Milton Park, Abingdon, Oxon, OX14 4RN

ISBN: 978-0-367-44545-4 (hbk)
ISBN: 978-1-032-07374-3 (pbk)
ISBN: 978-1-003-01065-4 (ebk)

DOI: 10.1201/9781003010654

Typeset in Times
by codeMantra

# Contents

# Preface

Ion channels of the cellular membranes are molecular machines made up of membrane proteins. The channel machinery is electromechanical because the mechanical structural components, channel subunits, contain electrical charge properties (functional charge groups) due to which they are found to respond to the localized electric fields. The building blocks of this electromechanical system respond, move, and collectively function under the influence of naturally inbuilt membrane potentials. The working region of the ion channel machines is across lipid membranes, so their functional dimension falls within low nanometer lengths.

An electrochemical gradient (a gradient of electrochemical potential) exists across cell membranes. Ion channels bypass the hydrophobic core of the membrane and transport materials and information between two hydrophilic regions which contain varying concentrations of a large variety of ions. Because ion channels are known to regulate the electrical activities in living cells, they are always at the forefront of scientific interests while it comes to understanding the fundamentals of cells and biological life. Ion channel function in biological cells is a biophysical process that is considered central to life at many levels. Ion channels of plasma, mitochondrial, and nuclear membranes play crucial physiological roles that together work to maintain cell health. Any alterations in structures and functions of channels of any cellular boundaries may lead to the rise of serious physiological conditions. As a result, certain types of cell-originated diseases may emerge. Among them, cancer, neurological disorders, and Alzheimer's are especially mentionable.

Cellular membrane constituents including various lipids and cholesterols usually associate with specific membrane proteins to geometrically accommodate across membrane's thickness (hydrophilic/hydrophobic/hydrophilic) to construct the ion channels. Therefore, the membrane's physical and chemical properties are found to play crucial roles to let the membrane proteins physically penetrate across the bilayer thickness. Thus, the potency of channel formation and stability does not depend only on the properties of channel-forming membrane proteins, but also on the channels' surrounding creating membrane environment to which they are physically coupled. Both mechanical properties of the bilayer and its surrounding electrical environments cause the regulation of the channel conformations.

The development of electrical excitability or the expression of various voltage-dependent ion channels is one of the fundamental aspects of neural differentiation. Ion channels are found to get expressed at certain stages of cellular differentiation in various muscular cells, endocrine cells, and nonexcitable glial cells.

Researchers have been working on the understanding of the electrical properties of biological cells for over centuries. Cellular electrical properties are largely scored due to that of ion channel hosting cellular membranes and the physiological environments thereof. Ion channel understanding has historically been made utilizing physical techniques that help detect or address mainly the physical properties of the membranes including the localized electrical conditions and membranes' mechanical integrity, both influencing the integral ion channel energetics and kinetics. The technological aspects of these ion channel nanomachines have recently been taken into light through in-depth analysis of both naturally inbuilt nanotechnology in the channel systems and their possible interventions using artificial means. Even channels are now constructed using synthesized materials.

All of the biophysics and nanotechnology approaches to addressing ion channels are naturally explored using interdisciplinary scientific initiatives. The intervention in biological systems by physicists started many centuries ago. Eminent physicist Robert Hooke's discovery of the cell in the 17th century was perhaps the first-ever interdisciplinary demonstration of a biologically super-important system using an established physics technique. This biophysics discovery of cell opened up avenues for physicists, chemists, engineers, technologists, pharmacologists, and medical scientists to invest their talents and efforts to understand the crucial aspects of cells. During the last century, relevant translational research has ensured medical benefits available for human health care. The issue of ion channel research appeared in the forefront among the most exciting fields of the cell. A lot of groundbreaking discoveries, both fundamental and applied types, helped us achieve knowledge on many crucial features about ion channels, including their normal functioning, abnormal behaving physical states as well as mutations in their structural building blocks. As a result, we now know about a lot of cellular diseases that perhaps originate or (at least their physiological status quo) linked to ion channel aberrations, alterations, mutations, etc. A huge amount of literature in the form of peer-reviewed scientific publications is now available. Processing them to creating a ground for understanding ion channels is a timely due. In this book "Biophysics and Nanotechnology of Ion Channels", I have taken an initiative.

Ten chapters of the whole book can be classified into several groups addressing various areas of the subject. The first three chapters are dedicated to explaining the fundamentals of various types of ion channels in general. Besides explaining ion channels of the plasma membrane, I have distinguishably presented the channels of the mitochondrial and nuclear membranes. I then brought in information related to the progress made on the artificial construction of channels in Chapter 4. This will briefly address how specific nanotechnology and general engineering techniques have started revolutionizing human intervention in cellular ion channel functions. There are cases where we see gating mechanisms happening in the membranes which do not fall in traditional norm of classified channels. The delivery mechanisms through non-channel route(s) across the cell membrane are largely unaddressed. Ion channel transports mostly refer to the controlled transport phenomena, while non-channel transports may be both controlled and uncontrolled. Chapter 5 is constructed to summarize important findings in this area. Channel gating mechanisms, related energetics, and principles active behind are then explored in Chapter 6. This chapter may be found important to help understand the channels using especially a few fundamental physics laws. Chapter 7 has been dedicated to explaining how we can intervene in ion channel structures and thus achieve controlled transport across the channel hosting membranes. This chapter is more about the engineering of the channel features. The medically relevant most important information will be found in Chapter 8. Understanding ion channel features including their natural norms and mutated ones is important in applied medical sciences, because a lot of diseases are linked to the physiological states of respective ion channels. Understanding them would aid in drug discovery. Bioinformatics techniques are now so relevant in exploring cellular systems including ion channels that a monogram without having a dedicated chapter would be considered missing something. The readers will be guided toward applying a series of artificial intelligence techniques including machine learning and deep learning techniques. This alternative in computer exploration of ion channels helps heavily to cover so many aspects within so limited means and costs. Chapter 9 has a brief report on them. Quantum mechanics is an important physics subject that has entered into exploring ion channels quite recently. Super-fast features in super-small dimensions within ion channel structures may be explained easily utilizing the powerful theoretical formalisms of quantum mechanics. I, therefore, wished to familiarize the readers with the latest progress in this area in Chapter 10. This book will certainly be helpful for individuals who are open to developing further.

I hope that this monograph will be found of use as a source of valuable information and conceptual inspiration to both students of biophysics, biology, medical science and expert researchers including faculties. I have avoided going deep into calculations and presented only the necessary forms of physics formulas to make the presentations quite understandable for all. Nanotechnology of ion channels has also been briefly explained. Similarly, a bulk of experimental data are also avoided from presentation and only the summary of them are mentioned. Discussion on biophysics perspectives of ion channels could be extended to covering many other areas as this is a large field now, but I have limited myself within most important aspects that we usually find necessary for addressing the important physics active in cell biology at low dimensions across the lipid bilayers where ion channels are accommodated.

# Epilogue

Ion channels are low-nanometer-scale geometric structures that are accommodated across cellular membranes. Strong mechanical and electrical interactions among channel subunits and hosting membrane subunits help ion channels to be stable inside the lipid bilayers. Any distortion in the localized membrane environment and channel structures leads to destabilization in the channel–bilayer coupling. Channel subunit structural alterations due to environmental damages or permanent mutations may raise issues responsible for variety of disease conditions. Application of biophysics principles in addressing the structures and functions of cell's nanometer dimension sections, e.g., ion channels, started not too long ago. But advances in this area over the last half a century are not limited too. Nanotechnology of ion channels has also been addressed quite rigorously during the last couple of decades. It has promoted the possibility of not only synthesizing channel subunits of specific interests, but also incorporating them in biological cellular membranes using engineering techniques. Thus, certain disease conditions may even be altered applying technological interventions that may use artificial means. Considering all these progresses, I took an initiative with Taylor and Francis Group-CRC Press to publish a book with the title "Biophysics and Nanotechnology of Ion Channels". In this book, I have presented quite a lot of information about biological and biophysical aspects of ion channel structures and functions. To be in line with the title of the book, my focus has always been to limit the discussion in the specific nanoscale dimension of the ion channels and their surroundings by presenting the analysis of existing and novel experimental and theoretical data along with the techniques of their production, methods of interpretations, and understanding of their applications. This book addresses the structural and functional mechanism perspectives of both normal cell conditions and perturbed conditions under the influence of diseases. Special attention has been given to explaining the causes of diseases that originate in ion channel structural subunits including their mutated conditions.

This book starts strategically with the search of a descriptive definition of ion channels and their classification. I then provide topical explanations on various ion channel perspectives including explaining their physiological origins, environmental adaptations, and structural mutations. I also explain about artificial means including technology-based interventions to create synthetic materials that can work alongside biological ion channels having channel-like properties. Besides using biophysics and nanotechnology explanations, I also use the power of bioinformatics and quantum mechanics principles to analyze ion channel features.

The book is organized in such a way that it will help the readers to get a summary of huge information on all of ion channel aspects that are available to date. I hope that this monograph will be a catalyst for future studies of ion channels. It is my great hope that this book will become a source of empirical information and conceptual inspiration to students of biophysics, biology, biomedical sciences, and engineering disciplines, and expert researchers and faculties including those involved in pharmaceutical sciences and industries.

# Acknowledgment

I am thankful to Professors Jack Tuszynski and Michael Houghton (Nobel Laureate in Medicine or Physiology, 2020) of the University of Alberta, Professor Olaf Sparre Andersen of Weill Medical College of Cornell University, Professor Marco Colombini of the University of Maryland, and Dr. Chih-Yuan Tseng of MDT Canada Inc. for many insightful discussions on various aspects that helped develop many of the ideas incorporated in this book. Writing of this book would be impossible without using a bulk of experimental and theoretical data from many publications. All of these articles have been quoted in references. It is a great pleasure to thank all of them for their contributions in the field. Hundreds of discussions with colleagues, academic friends and research group members, and students helped me to shape ideas while writing this book during last four years. Editorial assistance and encouragements provided by the staff members at CRC Press, Taylor & Francis Group, especially Dr. Rebecca Davies and Dr. Kirsten Barr, are thankfully acknowledged. I am grateful to my parents whose blessings have always been with me. I especially value the emotional support given by my beloved wife Anwara, sons Imtihan Ahmed and Yakin Ahmed, and daughter-in-law Emma Mannan. Anwara has always been on trips between my residence in Canada and Saudi workstation to provide much needed social and emotional supports in my academic activities.

**Mohammad Ashrafuzzaman**
*King Saud University, Riyadh, Saudi Arabia, March 2021*

# Author

**Mohammad Ashrafuzzaman,** a biophysicist and condensed matter scientist, is passionate about investigating biological and biochemical processes utilizing principles and techniques of physics. His theoretical and experimental works have created a generalized template to address the membrane interactions of cell-targeted agents, including drugs and nanoparticles. He has generalized the screened Coulomb interaction method to apply in biological systems and has developed a set of techniques (US patented) to help design novel drug molecules, including especially aptamers, validate the drugs' biological target binding potency, and help address the drug efficacy against specific diseases. Before joining (current affiliation) the King Saud University's Biochemistry Department of College of Science, Riyadh, Saudi Arabia, he held academic ranks at Bangladesh University of Engineering and Technology, Neuchatel University (Switzerland), Helsinki University of Technology (Finland), Cornell University (United States), and Alberta University (Canada). He is the co-founder of MDT Canada Inc., and the founder of Child Life Development Institute, Edmonton, Canada. He also authored *Nanoscale Biophysics of the Cell and Membrane Biophysics.*

# 1 Ion Channels – Physical Structures and Gating Mechanisms

Ion channels are transporters of materials and electrolytes across membranes. The channels also act as communicators between regions on both sides of the membrane. Generally, they play crucial roles in modestly compromising to breaking the membrane's physical barriers against materials and information. The exchange of electrolytes, water molecules, nutrients, proteins, genetic materials, etc. across semi-permeable membranes of biological cells is a dynamic, yet active or passive transport phenomenon. Ion channels are membrane proteins that have specific and/or nonspecific participation in maintaining the membrane exchange phenomena following physical energetics principles. Statistical mechanics principles consider energetics of the structural channel integrity, fluctuations, and transitions among various physical states inside the lipid membrane, and thus channel functions are determined. Ion channels have mainly two physical states, namely, closed or inactive and open or active state. The distinction between these two states is clear considering their ability to transport electrolytes across the membrane. However, therte are additional physical states that might be considered from the viewpoint of crucial energetics of the channel conformations. This chapter aims to explain all the structural, energetics, and phenomenological aspects of various types of channels falling under different categories. We shall try to generalize their structural and functional similarities and dissimilarities. Biological and biophysical inspections will guide us to addressing the channels regarding their natural physiological states and roles, mutations due to disease or genetic conditions, and possible natural and artificial intervention strategies.

## 1.1 INTRODUCTION TO GENERAL ASPECTS OF CHANNELS

Ion channels are cell membrane-hosted physical structures that help the membrane to be selectively permeable to certain ions. Channels are constructed because of interactions among membrane proteins and membrane constituents, especially lipids. Local hydrophobic and hydrophilic environments of the membrane and the transmembrane electrical and thermodynamic conditions have leading roles in helping the channels to get constructed and remain stable and undergo transitions between different energy states. The channel constituents and channels themselves are dynamic in membrane environment following statistical mechanical principles to transit from one energy state to another led by energetics. Ion channels are considered to fluctuate between stable, quasi-stable, and unstable states. During the stable open state (OS) of ion channels, a specific amount of ions are allowed to cross through the channels under a transmembrane potential, determining the conductance of the channel. As the channel structure transfers to a different open or close state, the conductance gets altered to another value (corresponding to open state) or zero (corresponding to the close state). This is the simplistic phenomenological interpretation based on the electrophysiological (EP) measurements of channel currents. Figure 1.1 depicts the scenario for two simple channels (Ashrafuzzaman et al., 2008).

Ion channel currents are determined by mainly the ion concentration difference across the membrane hosting the channels and the diffusive transport of the ions toward the channels and away from them. Calculating the ion channel currents requires solving the diffusion equation around the channels. Bentele and Falcke (2007) provided a quasi-steady approximation for the channel current

DOI: 10.1201/9781003010654-1

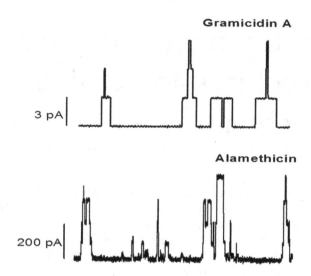

**FIGURE 1.1** Current fluctuations (in EP records) through two simple channels: β–helical gramicidin A (gA) and barrel-save alamethicin (Alm) channels/pores in model membrane. A 10 s gA current trace across phosphatidylcholine bilayer under the influence of 200 mV transbilayer potential and a 0.26 s Alm channel current trace across identical bilayer under influence of 150 mV transbilayer potential are presented here. For details, see Ashrafuzzaman et al. (2008).

and the local concentrations at the channel together with formulas linking the channel current and local concentrations at the channel to bulk concentrations and diffusion properties of the compartments. The quasi-steady approximation correlates currents with the bulk concentrations and provides formulas for the local concentrations. Thus, it provides a tool for the experimental analysis of *in vivo* currents and concentrations if the values of the relevant parameters in the formulas are known. For details on the models and associated equations, readers should consult the original article. Their hypothesis was that the knowledge about the concentration gradients around transport molecules may be required for modeling by the presence of the regulatory binding sites for conducted ions within the range of large gradients. That should apply to all ion channels regulated by the ions they conduct and to communicate with other compartments, channels, or chemical species. However, the *in vivo* ion channel current measurements started long before this kind of pinpointed theoretical understanding. Almost a century ago, voltage-clamp techniques were developed and applied in understanding membrane potentials. This technique allows the membrane voltage to be manipulated independent of the ionic currents, which allows us to study the current-voltage relationships of membrane-hosted channels.

During the 1940s, Kenneth Cole and George Mormont started to develop the voltage-clamp technique, where the large cell membrane potential could be measured and controlled. This led to the earliest descriptions of the membrane electrical properties and specifically the conductance that underlie neuronal action potentials. Within two decades, Alan Hodgkin and Andrew Huxley started creating grounds and finally refined the technique to only discover that the action potential was not simply a relaxation of the membrane potential to zero, but constituted an overshoot of the membrane potential to positive potentials. They also discovered that the action potential depolarizing phase was due to sodium ($Na^+$) flux into the cell and the repolarization back to the resting membrane potential was due to potassium ($K^+$) efflux. Thus, the concept of channels (sodium and potassium channels) emerged with a considerable level of understanding. This is one of the early breakthroughs modern medical science celebrates and will never forget in the coming future.

The two electrodes that Cole, Hodgkin, and Huxley used were fine wires that could be inserted only into extremely large cells, for example, the squid giant axon approximately 1 mm in diameter.

Scientists were not able to stop there as they had to inspect at smaller resolutions, for example, at the cell membrane scale which is of the order of low nanometer (~5 nm) thickness.

In the 1980s, new cell-based electrical conductance were discovered, including voltage- and ligand-gated conductance. The confirmation of voltage- and ligand-gated pores constructed by distinct proteins naturally required another technological breakthrough, which came on the shoulders of Bert Sakmann, Erwin Neher, and colleagues who developed the patch-clamp technique. They demonstrated that a very high resistance, of the order of Giga ohm (GΩ), a seal could be formed between a glass micropipette and cell's plasma membrane. Thus, the voltage-clamp technique could be applied to much smaller cells and even to small patches of the membrane, occasionally containing a single-channel protein. Now electrophysiology recording (EPR) of single ion channel currents is obvious research using the patch-clamp technique. Biophysics of ion channels has emerged as a unique subject based on mainly a bulk amount of research and discoveries using especially the experimental voltage-clamp and patch-clamp techniques. The function of channel protein molecules can now be monitored in real time. Thus, the statistical nature of the channel is observed.

During the 1980s and 1990s, the theoretical and computational research to explore the kinematics of ion channels in dynamic membrane systems started gaining popularity. *In silico* modeling alongside general theoretical and experimental analysis of ion channel functions helped understand channel conformational energetics and energetics of the flow of charges through channels. Martin Karplus, Arieh Warshel, and Michael Levitt are among the best modern-day scientists who revolutionized areas like multiscale methods for various biological complex systems. They developed methods and found their applications *in silico* simulating the behavior of molecules at various biologically relevant scales, ranging from single molecules to proteins. These simulations helped address the behavior of the channels in plasma membrane systems whose structures and functions were already known during the 1970s through mainly the groundbreaking discovery of the fluid mosaic cell membrane model by Singer and Nicolson. To date, we do know even about mutations in ion channel subunits, so have early-stage understanding of the roles of channels in various cellular diseases. In subsequent sections, we will cover a lot of these aspects in detail and create a background for topics covered in all chapters of this book.

Let us now address some fundamentals of ion channels that we may draw from all the developments achieved so far. As seen in Figure 1.1, ion channels switch between close state (CS) conducting no ion currents and OS conducting a certain amount of ion currents. In both states, ion channels may or may not maintain stable structural integrity. We know that both close and open states are nothing but complex physical structural states of the channels. If in OS condition the channel structure stays stable, we usually observe a lifetime of the channel (the duration of time the channel conducts current) that can be measured using popular EP records. In Figure 1.1, all lines with nonzero current values that are parallel to X axis represent stable conductance states of channels. Transitions within and between complex physical structural states are among the important hallmarks of ion channel functions. Every state and transitions between states are represented by physical energy conditions. Transitions between different states require a change in energy conditions often referred to as energetics.

The general ion channel energetics may be hypothetically represented by a curve in Figure 1.2. This energy diagram considers the ion channel as a single entity.

According to Figure 1.1, as a channel undergoes a transition between closed or nonconducting state and open or conducting state (e.g., gA and Alm channels) or between various conducting states within conducting state (e.g., Alm channels), the channel itself goes through different energy conditions. Degeneracy for the channels is perhaps not possible or observed. However, strong statistical mechanical behavior in the phenomenology of ion channel conformations is evident. Therefore, it is expected that as the channel changes its state, its energy condition changes, which happen with time. Figure 1.2 presents such change of energies with time as a channel undergoes through transitions between different states. We also emphasize in this diagram that although drastic changes in the energy values associated with the channel may not happen as long as it stays within a state, with

**FIGURE 1.2** Energy (vertical direction) versus time (horizontal progression). It shows that channel energy changes with time.

time the energy may also change modestly. There lies the possibility of observing any semi-classical or even quantum states (if there is any at all). We shall analyze this issue in a couple of chapters in this book.

As an ion channel undergoes a CS ↔ OS transition, we may apply the laws of physics to address these phenomenological channel states and transitions thereof. The ion channel opening probability relative to the closing probability (p) may be presented as:

$$p = P_O/P_C = \text{Exp}\{-\Delta G/k_B T\} \tag{1.1}$$

$P_O$ and $P_C$ are ion channel opening and closing probabilities, respectively. $\Delta G$ is free energy of transition, meaning this much energetic changes the complex channel structure must experience for the phenomenological changes described by probability in equation (1.1). $k_B$ is the Boltzmann's constant, and T is absolute temperature.

According to the EP records, CS ↔ OS transitions occur without spending time. This is perhaps due to some sort of lacking (ignorable though in classical many body systems) in the EP record techniques as it depends on certain frequency to filter the current. This frequency at very high value gets confused with background noise. The time t (s)-dependent ion current I in ampere (A) follows a transition function, which is either of the following two types:

  i. a step function (for two sharp transitions: lower current value to higher and vice versa), as shown here (current along y axis, time along x axis, Figure 1.3),
  ii. or a function as follows (Figure 1.4):

The time-dependent part of channel current I(t) varies with time, I(t), during both channel opening time $\Delta t^{CO}$ (the time taken by the channel to transform from CS to OS: a slow or nontransient channel opening process) and channel closing time $\Delta t^{OC}$ (the time taken by the channel to transform from OS to CS: a slow or nontransient channel closing process), respectively. $\Delta t^O$ is the time when the channel remains fully open and energetic in a stable state with a fixed channel current $I_0$. I(t) is perhaps a function of a few parameters, such as the number of charges ($n_e{}^c$) the channel may accumulate/conduct during CS → OS transition, the maximum time ($t_c$) a charge can stay inside a

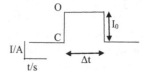

**FIGURE 1.3** In this case, the channel opening time ($\Delta t$ s) is deterministic for the channel open state with a constant current through the channel $I_0$ ampere (A).

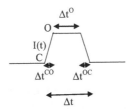

**FIGURE 1.4** In this case, the channel opening time ($\Delta t$ s) is deterministic, but it is the sum of three times representing three different phenomenological states, corresponding to three different energy states (one stable and two changing) of the channel, corresponding to three different current conduction conditions of the channel. Therefore, $\Delta t = \Delta t^{CO} + \Delta t^{O} + \Delta t^{OC}$.

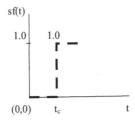

**FIGURE 1.5** Plot of sf(t) (dashed line). For simplicity, we consider the journey of charges starts at 0th s ($t = 0.0$). Sf (t) = 0.0 for $t < t_c$, and 1.0 for $t \geq t_c$.

channel before getting released, etc. Here, $n_e^c = 0, 1, 2, \ldots, n_{max}$. Superscript or subscript c is added to let $n_e^c$ or $t_c$ be channel structure-specific, meaning every channel allows certain maximum number of charges to accumulate inside the channel complex before fully opening its current conduction state represented by the flat region in current versus time plot (Figure 1.4). $t \geq t_c$.

Here, we may consider (e.g., as a gross fit for CS $\rightarrow$ OS transition), $I(t) = f(n_e^c, (t/t_c)^c)$, see Figure 1.4.

For electrophysiology recording (EPR), as we lose information on this time-dependent current change for both CS $\rightarrow$ OS and OS $\rightarrow$ CS transitions to avoid considering the current change (due to accumulation of charges) inside the closed channel structure (Figure 1.3), we indirectly use this function I(t) as $I^{EPR}(t)$ in built, where (e.g., for CS $\rightarrow$ OS), $I^{EPR}(t) = I(t) sf(t)$. Here, sf(t) is a step function (see Figure 1.5), defined in plot (plot of sf(t) (y axis) as a function of t (x axis)):

The above explanation predicts that we cannot measure any current, even for the conduction of any number of charges through the channel before the lapse of time $t_c$ in EPR. Therefore, $t_c$ is the microscopic time naturally lost in any EPR before getting any current measured as conducted through the channel. It is also the minimum time the channel structural changes require for preparing the channel to conduct any measurable amount of current, irrespective of how small that may be.

For simplicity, we may assume, $\Delta t^{CO} = \Delta t^{OC}$, so

$$\Delta t = \Delta t^{O} + 2\Delta t^{CO} = \Delta t^{O} + 2\Delta t^{OC} \qquad (1.2)$$

Electrophysiological experiments cannot detect the values of $\Delta t^{CO}$ or $\Delta t^{OC}$ and consider both to be 0, that is, the transition happens instantaneously, as seen in case (i), see Figure 1.3. From a macroscopic viewpoint, this consideration is valid. However, as the channel's CS $\leftrightarrow$ OS transitions correspond to actual changes in protein arrangements/compositions in channel structures, despite happening fast, these structural transitions (to either state) require nonzero time (at least on a microscopic viewpoint): $\Delta t^{CO} > 0$ s, $\Delta t^{OC} > 0$ s, so the case explained in Figure 1.4 is realistic. Moreover, given individual channel-specific gating mechanisms, if we consider that the lowest possible theoretical value of I(t) corresponds to the flow of an electron or proton charge, we can theoretically calculate the time taken by this single and lowest possible charge considering the channel's current conduction length $l_{ch}$ (this is apparently of the order of hydrophobic membrane thickness, the length associated to any channel that needs to be traveled by channel conducting charges bypassing the membrane's electric insulation barrier).

Theoretically, both $\Delta t^{CO}$ and $\Delta t^{OC}$ can be (approximately) deductible. We shall show this here. Let us consider a channel to conduct the smallest amount of current due to the flow of 1 electron charge (e) or 1 proton charge. Using the definition of electric current (I), $I \times t = e$, we may consider the following: $\Delta t^{CO} (e) = \Delta t^{OC} (e) = e/\Delta I(t)$. For a specific channel, for example, gramicidin A (gA) channel, we may deduce the value of $\Delta I(t)$ for the flow of 1e charge, considering 2 pA current is conducted through a gA channel under 200 mV membrane potential during a channel lifetime of 10 ms (assume) (Ashrafuzzaman et al., 2006). The parameters explained here may vary considerably depending on the channel type and the electrochemical environment of the channel hosting membranes.

There are three major types of ion channels: voltage-gated, extracellular ligand-gated, and intracellular ligand-gated along with two additional groups of miscellaneous ion channels. Based on the stimulus (mainly physical) to which they respond, ion channels are divided into three super families:

- voltage-gated ion channels (VGICs),
- ligand-gated ion channels (LGICs), and
- mechanosensitive ion channels (MSICs).

We shall explain them individually later in this chapter and in various subsequent chapters. First, let us discuss a rather crucial issue – the energetics related to channel conformational states and the transitions thereof.

Ion channels are mostly the geometrically clustered form of proteins hosted in the hydrophobic regions of cell membranes. Thus, the channels help the membrane to build a connection between two hydrophilic regions bypassing the hydrophobic core. However, the channels, including the participating proteins or peptides, stay energetically at different states at different time points on a molecular reaction and/or microscopic time scale. This confirms that any channel hosted in a hydrophobic core behaves as a dynamic entity following universal kinetics and energetics. Before going into the details on various channels, I wish to propose a model to explain the ion channel gating scenario which may lead to developing necessary theories (following classical mechanical approaches) and finding related analytical expressions. We may also find some experimental facts already addressed for different channels that might fit in the proposed model presented here.

Three major structural compositions and transitions thereof may be relevant while addressing the general aspect of channel gating mechanism, namely, the CS (having no conduction) structure, widely OS (having full conduction) structure, and lastly the back-and-forth structural transitions between CS $\leftrightarrow$ OS structures. There are two additional states that may also be considered to fully understand the structural moiety of a channel, namely, the preopen state (PrOS) and postclosed state (PoCS). These four phenomenological structural states (OS, CS, PrOS, and PoCS) are general participants in a channel gating mechanism, meaning the channel must move (energetic transitions) through energy landscapes corresponding to these four phenomenological states. However, there is one more state that falls in a category of nonparticipants to the gating mechanism. This may be called the tightly closed state (TCS), negating the gating mechanism. From the channel current conduction perspective, this state is yet to undergo substantial energetic changes to even fall in the energetic regime following the gating mechanisms. All these phenomenological geometric states and stepwise transitions among or between them are energy and time-dependent dynamic phenomena. We may consider OS, CS, PrOS, and PoCS to appear with individual time and energy coordinates (t, E) $(t_O, E_O)$, $(t_C, E_C)$, $(t_{PrO}, E_{PrO})$, $(t_{PoC}, E_{PoC})$, respectively. We may also refer TCS with identical energy/time coordinate, $(t_{TC}, E_{TC})$. To fully understand channel kinematics, we need to consider all these parameters, explained here, as illustrated in Figure 1.6.

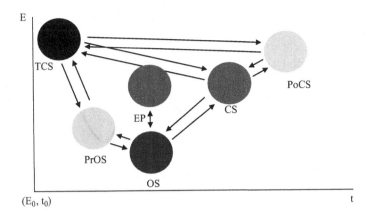

**FIGURE 1.6** OS, CS, PrOS, and PoCS appear with individual time and energy coordinates. TCS is drawn at different times (ith and jth), for example, let's consider them as ($t^i_{TCS}$, $E_{TCS}$) and ($t^j_{TCS}$, $E_{TCS}$) to show that it has equal possibility to exist before the initiation of channel gating and after the closing of the channel gating with practically (assumed) same energy $E_{TCS}$. Actual time change in any dashed arrow represents minus (−) the time change in this plot, which means dashed arrows represent transitions with increase of time similar to non-dashed arrows (we chose this presentation method to avoid repeated drawing of states at two different times for each). Our model suggests that the channel (energy) states may repeatedly appear with the progression of time. Only bidirectional vertical arrow connecting OS and CS (additionally drawn for EP purpose only) is relevant if we consider EP record of ion channel currents can correctly detect the OS ↔ CS transitions taking no time, but changing the energetics only.

As shown in Figure 1.6, a bidirectional vertical arrow connecting OS and CS is relevant if we consider that EP record of ion channel currents can correctly detect the OS ↔ CS transitions taking no time, but changing the energetics only. However, earlier in this chapter (see Figures 1.3–1.5), I argued on back-and-forth transitions hypothesizing that the OS ↔ CS transitions take nonzero time (so the EP detected bidirectional arrow showing transitions taking no time in Figure 1.6 should be considered imaginary), which is of the order of perhaps some molecular time scale (the time scale on which molecular movement occurs in ion channel gating). We may assume this time scale to be at best as low as of the order of time followed by the elementary dynamics of hydrogen bonds and related proton transfer reactions, both occurring in the ultra-fast time domain between $10^{-14}$ and $10^{-11}$s (Elsässer and Bakker, 2002). This is the classical limit. Beyond this perhaps exists the quantum mechanical treatment of the movement in which the time scale might follow the so-called quantum time scale (Li, 2019). Quantum time scale may also refer to the Planck time of approximately $5.4 \times 10^{-44}$s (Lieu and Hillman, 2003). However, this scale, if accepted, may theoretically be valid in a vacuum, not in a biological system in a condensed phase like the channels are in. We know in quantum mechanics things may move at no time which in the classical system (like the channel movement) should not or does not happen. The electron speed in the condensed phase or through conducting ion pores is certainly order(s) of magnitude less than the speed of light. Considering all these it is perhaps an acceptable assumption if we measure the time for the OS ↔ CS transitions of the order not more than the ultra-fast time domain, that is, less than a femtosecond ($10^{-15}$s) for dynamics of hydrogen bonds (Elsässer and Bakker, 2002).

We recently addressed the channel gating issue using energetics related to the chemical kinetics, see Figure 1.7 (Ashrafuzzaman and Tuszynski, 2012a; 2012b). The energetics of the OS ↔ CS transitions follows a theoretical formalism that we developed in light of the charge-based screened Coulomb interactions among the charges in the complex of proteins or peptides and lipids participating in a channel hosted inside the hydrophobic region of a cell membrane bilayer. As this energetics issue has been rigorously addressed in our earlier publications, I invite readers to go through these studies (Ashrafuzzaman and Tuszynski, 2012a; 2012b).

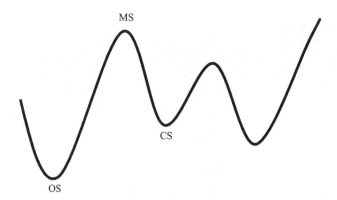

**FIGURE 1.7** Energy (vertical axis) plot (chemical kinetics) against reaction coordinates (horizontal axis), representing different channel (energy) states and transitions between. The higher energy landscape/unstable energy state between successive stable energies representing OS and CS is metastable (energy) state (MS). For details, see Ashrafuzzaman and Tuszynski (2012a, 2012b).

## 1.2 VOLTAGE-GATED ION CHANNELS

VGICs are multisubunit protein complexes in cell membranes that alter conformations in response to changes in channel hosting membrane potential and that the conformational alterations may lead to gating or opening-closing transitions of a transmembrane ion-selective pore. The VGIC proteins have three properties enabling them to participate in the channel gating process, namely, opening in response to the membrane potential changes "voltage gating," subsequent closing and inactivation of channels, and similar to other ion channels, ion specificity that allows for those ions that will permeate and others not.

### 1.2.1 VOLTAGE-GATED ION CHANNEL FAMILIES

Three major VGIC families distinguished by the cations conducted through them are as follows:

- Sodium channels
- Potassium channels
- Calcium channels

Voltage-gated sodium channels (VGSCs) are large integral membrane proteins that are encoded by at least 10 genes in mammals. Various sodium channels have remarkable functional similarities. Although modest changes in sodium channel function due to any possible physiological or locally active biophysical reason are biologically relevant, these are underscored by multitude of mutations in sodium channel proteins that may cause human diseases.

Voltage-gated potassium channels (VGPCs) form a large and diverse family that is evolutionarily conserved. There are 40 VGPC genes belonging to 12 subfamilies in humans. Each potassium channel contains four pore-forming $\alpha$-subunits as well as auxiliary $\beta$-subunits possibly affecting the channel function and/or localization potency. These channels display broad distributions in the nervous system and tissues. They also regulate the waveform and firing pattern of action potentials in excitable cells such as neurons, cardiomyocytes, and muscles. They may also regulate the cell volume, proliferation, and migration.

The voltage-gated calcium channel (VGCC) family has 10 members in mammals, which play distinct roles in cellular signal transduction. Contraction, secretion, regulation of the gene expression, integration of synaptic inputs in neurons, and synaptic transmissions at ribbon synapses in specialized sensory cells, initiation of synaptic transmission at fast synapses, repetitive firing of action potentials in rhythmically firing cells such as cardiac myocytes and thalamic neurons, etc. are among the broad spectrum of cellular processes in which VGCCs are engaged.

### 1.2.2 VOLTAGE-GATED SODIUM CHANNEL STRUCTURE AND FUNCTION

Proteins of sodium channels in the mammalian brain are composed of a complex containing a 260 kDa $\alpha$ subunit being associated with one or more auxiliary 33–36 kDa $\beta$ subunits ($\beta$1, $\beta$2, and/or $\beta$3). All nine $\alpha$ subunits ($Na_v$1.1–$Na_v$1.9) have already been functionally characterized. There is a tenth related isoform ($Na_x$), predicted to also function as a sodium channel. The sodium channel $\alpha$ subunit is predicted to fold into four domains (I–IV), each containing six $\alpha$-helical transmembrane segments (S1–S6). The voltage sensor in each of the domains is located in the S4 segments containing positively charged amino acid residues at every third position. The detailed architecture of the channel structure including the transmembrane region of the channel, ion-selective filter at the extracellular end of the pore, etc. are schematized, see Figure 1.8 (Yu and Catterall, 2003).

2.45 Å resolution crystal structure of the complete $Na_v$Ms prokaryotic sodium channel in a fully open conformation has been revealed (Sula et al., 2017). A canonical conformation on activation of

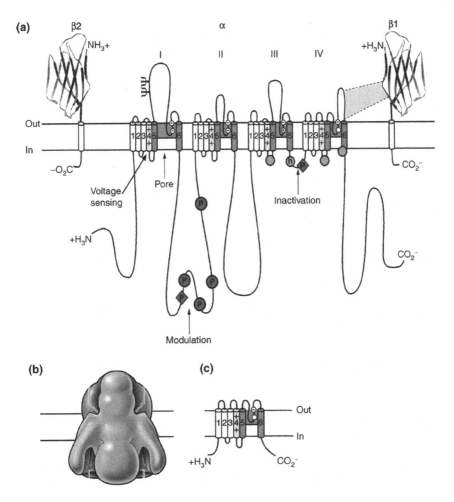

**FIGURE 1.8** VGSC structure – schematic diagram (Yu and Catterall, 2003). (a) Schematic representation of the sodium channel subunits. (b) Three-dimensional structure of the $Na_v$ channel $\alpha$-subunit at 20 Å resolution compiled from electron micrograph reconstructions. (c) A schematic representation of NaChBac, the bacterial VGSC.

the voltage sensor S4 helix and an open selectivity filter (SF) leading to an open activation gate on the intracellular membrane surface and the intracellular C-terminal domain are found in the structure. A heretofore unseen interaction motif between W77 of S3, the S4–S5 interdomain linker, and the C-terminus which is associated with regulation of opening and closing of the intracellular gate are included. Figure 1.9 illustrates this structure.

Bacterial VGSCs serve as models of their vertebrate counterparts because they have similar functional components in a simpler structure. High-resolution structures of both tightly closed and open states of VGSC have been recently investigated (Lenaeus et al., 2017). In the CS, the activation gate fully occludes the conduction pathway. The intracellular C-terminal domain is observed as a long four-helix bundle. On the other hand, in the OS, the activation gate is found with an orifice of ~10 Å.

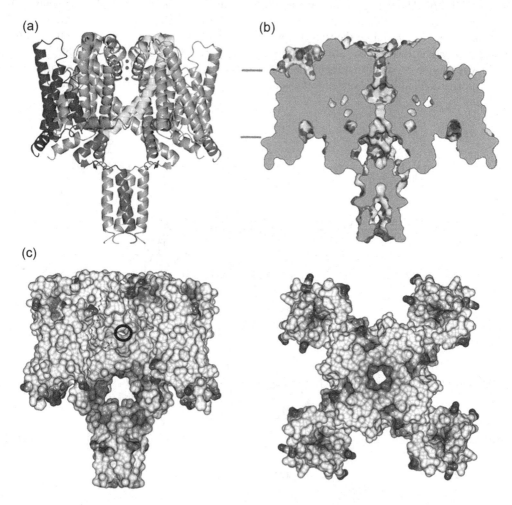

**FIGURE 1.9** NavMs channel structure. (a) Cartoonrepresentation of the NavMs tetrameric channel. One of the monomers is colored as follows: voltage sensor (blue), S4–S5 linker (green), pore helices (yellow), C-terminal domain (red), and sodium ions (purple). The other three monomers are depicted in gray for ease of viewing. (b) Cross-sectional view showing the open SF, hydrophobic cavity, and open pore gate. The opening does not extend into the stabilizing coiled-coil bundle at the end of the CTD, but provides an egress via the bundle linker region. The approximate locations of the membrane surfaces are indicated by the cyan lines. (c) Space-filling model (colored according to electrostatics). The locations of detergent and polyethylene glycol molecules are shown on the surface in green and red stick format. Left: view from the membrane normal. The location of the fenestration leading to the hydrophobic cavity is circled in black. Right: view from the extracellular surface, extending from the N-termini to the ends of the S6 helices. (For color figure see eBook figure.)

Here, structures of two Na$_V$Ab mutants that capture tightly closed and open states at a resolution of 2.8–3.2 Å are presented. The introduction of two humanizing mutations in the S6 segment (Na$_V$Ab/FY: T206F and V213Y) generates a persistently closed form of the activation gate in which the intracellular ends of the four S6 segments are drawn tightly together to block ion permeation completely. This construct reveals the complete structure of the four-helix bundle that forms the C-terminal domain. Truncation of the C-terminal 40 residues in Na$_V$Ab/1–226, in contrast, captures the activation gate in an open conformation, revealing the OS of a bacterial VGSC with intact voltage sensors (Lenaeus et al., 2017). Here, we present a more interesting set of data plots on molecular dynamics (MD) simulation of the system. MD simulations have confirmed that this conformation allows the permeation of hydrated Na$^+$, see Figures 1.10 and 1.11, and presents the condition of

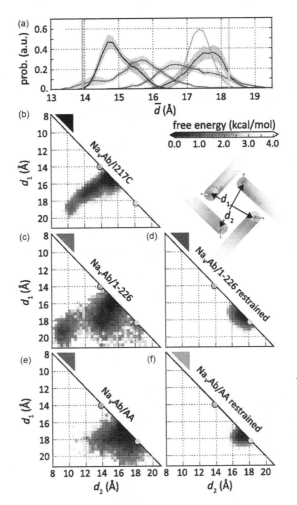

**FIGURE 1.10** Structural fluctuations of the activation gate. (a) Probability distribution of the mean diameters of the activation gate. Na$_V$Ab/I217C, black; Na$_V$Ab/1–226 restrained open, blue; Na$_V$Ab/1–226, unrestrained, red; Na$_V$Ab/AA restrained open, tan; Na$_V$Ab/AA unrestrained. a.u., arbitrary units; prob., probablility. (b) Free energy of the pairwise distances of residue 215–218 Cα atom center of mass in diagonal subunits, d$_1$ and d$_2$, from molecular simulations with several channel models. (b–f) Free energy profiles are shown for: top row, Na$_V$Ab/I217C preopen state without TM restraints (black; b); middle row, Na$_V$Ab/1–226 without (red; c) and with (blue; d) TM restraints; and bottom row, Na$_V$Ab/V213A/C217A (Na$_V$Ab/AA) without (e) and with (f) TM restraints. The average pairwise distance of diagonal subunits in the open- and closed-state crystal structures are shown as pink and blue circles, respectively. *Inset*, a schematic representation of distances d$_1$ and d$_2$. (For color figure see eBook figure.)

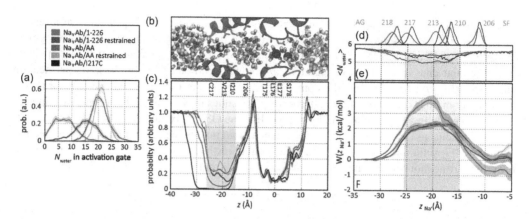

**FIGURE 1.11** Hydration and free energy of Na⁺ conduction through the activation gate. (a) Distribution of the number of water molecules, $N_{water}$, in the activation gate. Color coding is indicated in the key. (b) Snapshot of the unconstrained Na$_V$Ab/1–226 I217C channel with one Na⁺ cation in the activation gate and two Na⁺ cations in the selectivity filter shown as blue spheres. Two pore domain monomers are shown as red ribbons with selected side chains indicated below. Water molecules are shown as red and white licorice. (c) Distribution of water molecules along the pore axis. Scale and alignment are identical to b. The activation gate is highlighted in gray shading. (d) Axial distribution of Cα atoms of selected side chains lining the activation gate and the CC of the channel from restrained (blue) and unrestrained (red) simulations of Na$_V$Ab/1–226 I217C. AG, activation gate; SF, selectivity filter. (e) Average number of water molecules in the first hydration shell of Na⁺ from bulk water (left) to the CC (right). (f) PMF profiles $W(z_{Na}{}^+)$ for movement of a Na⁺ ion along the pore axis. Differences in the free energy of Na⁺ in the CC ($z > -15$ Å) are due to fluctuations in the number of cations present in the selectivity filter. In (c–f), the 11-Å-wide barrier region is highlighted in gray shading. Data for Na$_V$Ab/I217C are only shown in (c) because the long, narrow dehydrated intracellular activation gate precludes the passage of Na⁺ ions. (For color figure see eBook figure.)

hydration and free energy of Na⁺ conduction through the activation gate. This study provides a complete closed–open–inactivated conformational cycle in a single VGSC.

For understanding the pattern of actual gating currents of sodium channels, EP records of channel currents were made. Figure 1.12 summarizes the results showing the currents in the OS relative to the CS. Na$_V$Ab/FY was found to conduct gating currents, but not ionic currents. This is consistent with MD data showing the activation pore with the ability to conduct currents, as seen in EP records. Although the Na$_V$Ab/FY protein was expressed at higher levels in insect cells, it did not show any inward Na⁺ currents under standard experimental protocols. Because of the unusually negative voltage-dependent activation of Na$_V$Ab, which might also be further negatively shifted by the FY mutation, the N49K mutation was introduced to shift the voltage dependence of gating ~75 mV in the positive direction. Recordings of this triple-mutant construct reveal no ionic current, but a gating current. This represents the outward movement of the gating charges in the S4 segments, with the conformational changes leading to channel activation (Bezanilla, 2000). The Q/V curve of Na$_V$Ab/FY/N49K was observed to overlap with that of Na$_V$Ab/N49K, indicating that the FY mutation does not actually alter the voltage dependence of voltage sensor function, although it prevents ionic conductance.

The current state of our knowledge on VGSCs is based on continued progress over many decades. We cannot cover even a fraction of them due to lack of space. Here, we just presented the latest biophysical and physiological status of the channels. However, for understanding the stepwise historical progress, readers may refer to hundreds of published studies, including the early ones (Bezanilla and Armstrong, 1977; Meves, 1979; Fenwick et al., 1982; Horn and Vandenberg, 1984; Aldrich and Stevens, 1987).

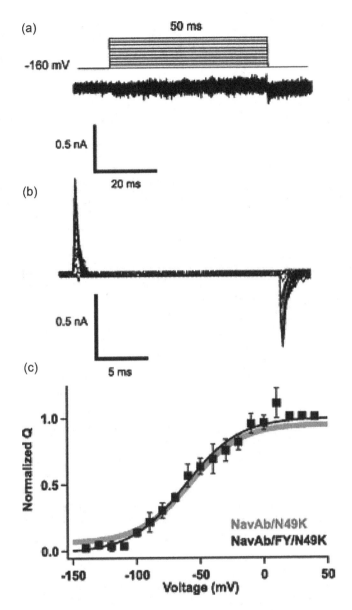

**FIGURE 1.12** Na$_v$Ab/FY conducts gating currents but not ionic currents. (a) No ionic currents were detected from Na$_v$Ab FY construct. (b) At very high expression of Na$_v$Ab FY construct, gating currents were elicited using a voltage-clamp protocol, where depolarizing pulses were applied for 50 ms from −160 to +60 mV in 10-mV increments. (c) Q–V curve of Na$_v$Ab/FY. For Na$_v$Ab/FY/N49K, $V_{1/2}=-62.2\pm3$ (n=7); for Na$_v$Ab/ N49K, $V_{1/2}=-65\pm2.2$. Details in Lenaeus et al. (2017).

### 1.2.3 VOLTAGE-GATED POTASSIUM CHANNEL STRUCTURE AND FUNCTION

VGPCs play crucial roles in generating and propagating the membrane action potential. They are transmembrane channels sensitive to cell membrane voltage changes and specific to potassium transport. Their roles are also vital in helping depolarized cells to return to their resting states.

The discovery of VGPCs began more than a century ago. Bernstein first hypothesized that cells at rest were permeable exclusively to K+ ions, but were permeable to other ions only during excitation

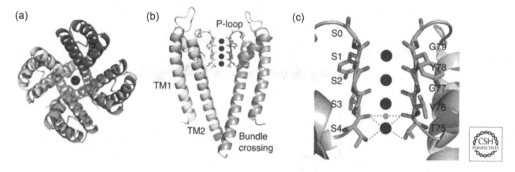

**FIGURE 1.13** Structural details of KcsA (K channel of streptomyces A). (a) KcsA tetramer as viewed from the top of the membrane (PDBID:14KC). Each subunit is uniquely colored. The central ion conduction pore is shown with a $K^+$ ion (blue sphere). (b) KcsA as viewed from the side with two opposing subunits removed for clarity. The P-loop contains the signature sequence TVGYG (shown as sticks) and forms the SF. Four potassium ions are shown in the selectivity filter. Below the selectivity filter is the aqueous cavity formed by the TM2 helices that form the bundle crossing. (c) A detailed view of the SF of KcsA. Each binding site, S0–S4, is formed by the oxygen cages originating from backbone carbonyl and side-chain hydroxyls of the selectivity filter signature sequence TVGYG. Dashed lines depict coordination of the $K^+$ ion in S4 and the oxygens. Sodium (orange sphere) binds in the plane between sites S3 and S4. Dashed lines represent its coodination with carbonyl oxygens. Details in Kim and Nimigean (2016). (For color figure see eBook figure.)

(Bernstein, 1902). Later came the breakthrough discoveries of Hodgkin and Huxley and others in their research performed on the squid giant axon. Their understanding of the action potentials and the coordinated changes in the cell membrane permeability to $Na^+$ and $K^+$ ions helped them develop a model directly correlating these fluxes with excitation and electrical conduction (Goldman, 1943; Hodgkin and Huxley, 1945, Hodgkin and Keynes, 1955).

Besides being highly selective for $K^+$ ions, VGPCs conduct ions at an extremely fast rate, almost close to the diffusion rate (Hille, 2001). This is largely possible owing to the naturally inbuilt architecture of the SF inside the channel. Structural and functional studies (Nobel prize-winning works) of potassium channels conclude that the placement of ions in the SF leads to the so-called "knock-on" mechanism, see Figure 1.13 (Kim and Nimigean, 2016), ensuring the charge-charge repulsion of $K^+$ ions that march in a single file through the SF (Hodgkin and Keynes, 1955; Doyle et al., 1998).

An easy to understand illustration is presented here showing models for coupling of voltage sensing to channel opening in potassium channels, see Figure 1.14 (Kim and Nimigean, 2016). Detailed analysis is discussed in the figure caption.

The membrane, hosting potassium channels, modulates the integral channel functions due to the regulation in lipid compositions and integrity, confirmed in MD simulations (Kasimova et al., 2014). Experimental and MD studies have revealed a direct interaction between the lipid headgroups and the ion channel residues, suggesting an influence on the ion channel function. The alteration of the lipids may, in principle, modify the overall electrostatic environment of the channel, and hence the transmembrane potential, leading to an indirect modulation, that is, a global effect. The structural and dynamical properties of the VGPC Kv1.2 embedded in bilayers were investigated using modified upper or lower leaflet compositions corresponding to realistic biological scenarios. Kv1.2 embedded in different bilayers shows different quantitative properties, summarized in Table 1.1. The states of the channel structures in different lipid bilayer environments have been presented in Figure 1.15.

The membrane upper leaflet modification is predicted to alter the channel's global electrostatic environment. Moreover, the modification of local salt bridge pairings was proposed as a key element of the channel modulation. It has been shown that local interactions between the ion channel and

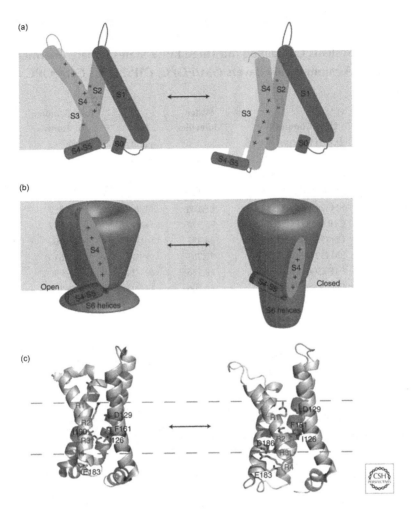

**FIGURE 1.14** A model for coupling of voltage sensing to channel opening in Kv channels. (a) A cartoon depiction of the paddle model (Long et al., 2007) voltage-sensing mechanism. Helices S1–S4 and the S4–S5 linker are shown as cylinders of different colors. Gating charges are located on the S4 helix and represented by a black+sign. Countercharges on S0 (a short α-helical region before S1), S1, and S2 are represented by a red dash. In the up or open conformation (left), S3 and S4 are located within the membrane and the gating charges are closer to the extracellular side, interacting with the external cluster of countercharges. The hydrophobic plug is represented by an orange circle on the S2 helix. A change in membrane potential would cause the S4 helix to move into the down conformation (right) with the gating charges now closer to the intracellular side and interacting with the internal cluster of countercharges. This displacement of S4 pushes down on the S4, S5 linker, tilting it toward the intracellular side, poising it to interact with the S6 helices to close the pore. (b) The proposed mechanism for coupling of voltage sensing to gating in Kv channels (Long et al., 2007). The channel pore and bundle crossing are represented in blue. S4 and the S4, S5 helix of the voltage-sensing domain are depicted in green and red, respectively. Based on the crystal structure, S4 is in the up conformation (left), and the S4, S5 linker rests on the S6 helices in the bundle crossing in the open state. Transition to the hypothetical CS of the channel requires movement of S4 into the down conformation (right). This downward movement of the S4 helix pushes on the amino-terminal end of the S4, S5 helix, which tilts toward the intracellular side and pushes the S6 helices down into the CS. (c) A comparison of the conformational states of the voltage-sensing phosphatase Ci-VSP. R217E Ci-VSP (left, PDBID:4G7V) shows the voltage sensor in the up conformation, whereas wild-type Ci-VSP (right, PDBID:4G80) shows the voltage sensor in the down conformation. The structures were aligned using the S1 helix for reference. Critical Arg residues R1–R4 are shown as green sticks. Countercharges are shown along with residues comprising the hydrophobic plug region. The gray dotted lines denote the position of R1 (top) and R4 (bottom) on the up conformation of R217E Ci-VSP for comparison to their positions on the down conformation of wild type. Details in Kim and Nimigean (2016). (For color figure see eBook figure.)

**TABLE 1.1**

**Summary of the 11 Systems Containing the Three Kv1.2 States (Open, Intermediate, and Closed) in the Four Asymmetrical Bilayers (SM/POPC, C1P/POPC, Cer/POPC, and POPC/PIP2)**

| | Channel State | Lipid Composition | Water Molecules | Ions | Total Number of Atoms | Size (Å³) |
|---|---|---|---|---|---|---|
| 1 | Open | SM (319) POPC (331) | 84,909 | 141/97 | 368,254 | 153×159×146 |
| 2 | Open | C1P (297) POPC (318) | 78,345 | 392/51 | 338,665 | 165×141×142 |
| 3 | Open | Cer (320) POPC (323) | 86,143 | 143/99 | 364,923 | 157×148×153 |
| 4 | Open | POPC (368) PIP2 (65) | 86,116 | 613/224 | 343,436 | 145×145×159 |
| 5 | Intermediate | SM (315) POPC (323) | 85,106 | 142/98 | 367,227 | 150×152×156 |
| 6 | Intermediate | C1P (313) POPC (320) | 75,374 | 534/177 | 332,192 | 157×142×155 |
| 7 | Intermediate | Cer (320) POPC (326) | 84,482 | 141/97 | 359,400 | 152×146×158 |
| 8 | Closed | SM (323) POPC (333) | 97,757 | 159/115 | 407,650 | 153×151×171 |
| 9 | Closed | C1P (306) POPC (316) | 88,136 | 424/74 | 368,896 | 162×132×167 |
| 10 | Closed | Cer (326) POPC (323) | 84,506 | 141/97 | 359,760 | 155×143×158 |
| 11 | Closed | POPC (376) PIP2 (54) | 87,465 | 562/248 | 346,869 | 144×158×156 |

*Source:* Details in Kasimova et al. (2014).

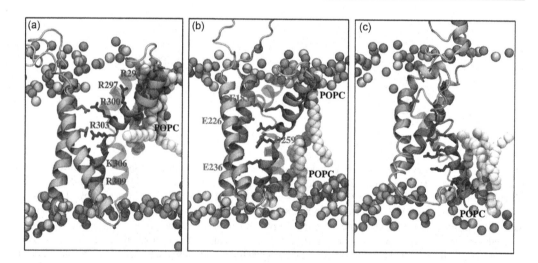

**FIGURE 1.15**  Kv1.2 embedded in a POPC bilayer (only one VSD is shown). A: Kv1.2 is in the active state. The upper residues of S4, R293, and R297 form salt bridges with the adjacent phosphate groups of the lipid heads from the upper bilayer leaflet. R300, R303, K306, and R309 interact with the negatively charged residues of S1–S3 helices (E183, E226, E236, and D259). B: Kv1.2 is in an intermediate state. The upper and bottom positive residues of S4, R293, and R309 interact with both bilayer leaflets. C: Kv1.2 is in the resting state. R303, K306, and R309 form salt bridges with the lipid phosphate groups from the bottom bilayer leaflet. S1, S2, and S3 helices are shown in orange, and S4 is shown in cyan. The lipid nitrogen and phosphorus atoms are shown as purple and yellow spheres respectively. Details in Kasimova et al. (2014). (For color figure see eBook figure.)

the lipid headgroups are key elements of the modulation, which is consistent with our own studies of model membrane-hosted peptide channels, namely, gramicidin A and alamethicin channels (Ashrafuzzaman and Tuszynski, 2012a; 2012b).

Ion channel currents talk a lot about the overall properties of the channels. Among many available studies, we pick the EP recording of VGPCs in primary rat peritoneal macrophages to demonstrate

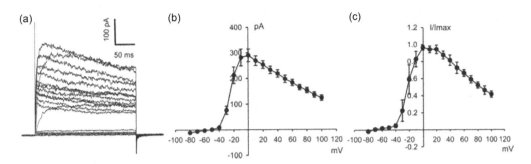

**FIGURE 1.16** (a) Example raw traces showing voltage-dependent outward currents recorded in a rat peritoneal macrophage after 12–24 h in culture. Holding potential was −60 mV and testing potentials were between −80 and 100 mV with a 10 mV increment. (b) Average current-voltage relationship (I–V curve) calculated from a total of nine cells. C. I/Imax versus Vm plot, which was well fit by the Boltzmann function with a half maximal activation potential V1/2 of −25.69±2.23 mV. Details in Wu et al. (2013). (For color figure see eBook figure.)

it here (Wu et al., 2013). With intracellular solution containing K$^+$ as the main charge carrier, all cells showed outward currents in response to the membrane depolarization. See Figure 1.16.

The currents were found to be inhibited by TEA (10 mM), a nonselective blocker for voltage-gated K$^+$ channels, and attenuated when intracellular K$^+$ was substituted with Cs$^+$. The outward currents were inhibited by changing the holding potential from −80 to −30 mV or −10 mV. In contrast, ATP concentration increase in the intracellular solution decreased the amplitude of the outward currents. Thus, rat peritoneal macrophages express several types of functional VGPCs.

### 1.2.4 VOLTAGE-GATED CALCIUM CHANNEL STRUCTURE AND FUNCTION

VGCCs in mammals have 10 members with distinct roles in cellular signal transduction. The Ca$_V$1 subfamily is known to initiate the contraction, secretion, regulation of gene expression integration of synaptic input in neurons; and synaptic transmission at ribbon synapses in specialized sensory cells. The primary role of the Ca$_V$2 subfamily is the initiation of synaptic transmission at fast synapses. The function of Ca$_V$3 subfamily is vital for the repetitive firing of action potentials in rhythmically firing cells, such as cardiac myocytes and thalamic neurons.

Calcium channels in different cell types activate on membrane depolarization and mediate Ca$^{2+}$ influx in response to action potentials, as well as subthreshold depolarizing signals. Cell entering Ca$^{2+}$ through VGCCs serves as the second messenger of electrical signaling, initiating many different cellular events, see Figure 1.17 (Catterall, 2011). In cardiac and smooth muscle cells, activation of Ca$^{2+}$ channels is observed to initiate contraction directly by increasing the cytosolic concentration of Ca$^{2+}$ and indirectly by activating calcium-dependent calcium release by ryanodine-sensitive Ca$^{2+}$ release channels in the sarcoplasmic reticulum (Reuter, 1979).

The pore-forming α1 subunit and the auxiliary β subunits of skeletal muscle Ca$_V$1.1 channels (Curtis and Catterall, 1985) and cardiac Ca$_V$1.2 channels (Hell et al., 1993) are phosphorylated by PKA. These α1 subunits are also truncated by proteolytic processing of the carboxy-terminal domain, see Figure 1.18 (Jongh et al., 1989). The voltage-dependent potentiation of Ca$_V$1.1 channels on the 50 ms time scale requires PKA phosphorylation. This suggests a close association between PKA and Ca$^{2+}$ channels.

Some of the early crystallographic structural studies revealed the functional domains and predicted specific information on channel gating sites (Opatowsky et al., 2004; Petegem et al., 2004). The crystal structure of a voltage-gated channel in a mutant form of a bacterial sodium channel homolog, Na$_V$Ab, which is a single domain channel forming homo-tetramers, revealed crucial relevant information (Payandeh et al., 2011). The mutation confirmed the pore to be Ca$^{2+}$-selective

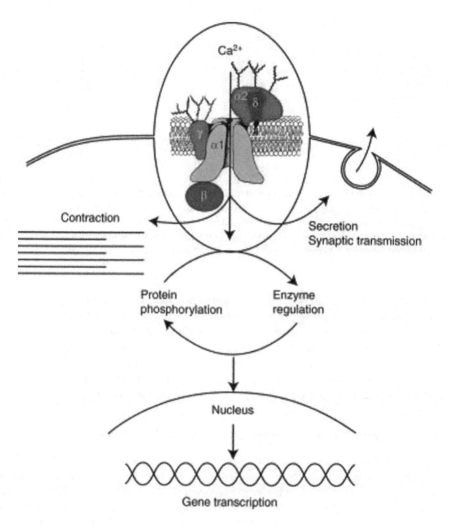

**FIGURE 1.17** Signal transduction by VGCCs. Cells entering $Ca^{2+}$ initiate numerous intracellular events, including contraction, secretion, synaptic transmission, enzyme regulation, protein phosphorylation/dephosphorylation, and gene transcription. (Inset) Subunit structure of VGCCs. The five-subunit complex that forms high-voltage-activated $Ca^{2+}$ channels is illustrated with a central pore-forming α1 subunit, a disulfide-linked glycoprotein dimer of α2 and δ subunits, an intracellular β subunit, and a transmembrane glycoprotein γ subunit (in some $Ca^{2+}$ channel subtypes). Details in Catterall (2011).

and forming $Ca_VAb$. This structure has provided multiple insights, including confirming the $Ca^{2+}$ permeation process (Tang et al., 2013), providing insights into the binding and mechanism of drug actions (Tang et al., 2016), and other details on subunit interaction sites (Wu et al., 2016; Tang, 2019). Detailed analysis on the current status of VGCC regarding structure, function, interior gating mechanisms, drug binding sites, etc. may be found in Dolphin (2018). Figure 1.19 representing both recorded currents and structure of the VGCCs demonstrates the structural basis, gating mechanisms, etc. (Tang et al., 2013).

Using MD simulations VGCC structure and function was addressed two decades ago, and the three-dimensional (3D) structure and behavior of a molecular model of the L-type calcium channel pore were proposed (Barreiro et al., 2002). Four glutamic acid residues, the EEEE locus, located at highly conserved P loops (also called SS1–SS2 segments) of the α1 subunit, molecularly express the calcium channel selectivity. The proposed α-helix structure for the SS1 segment, analyzed in simulations

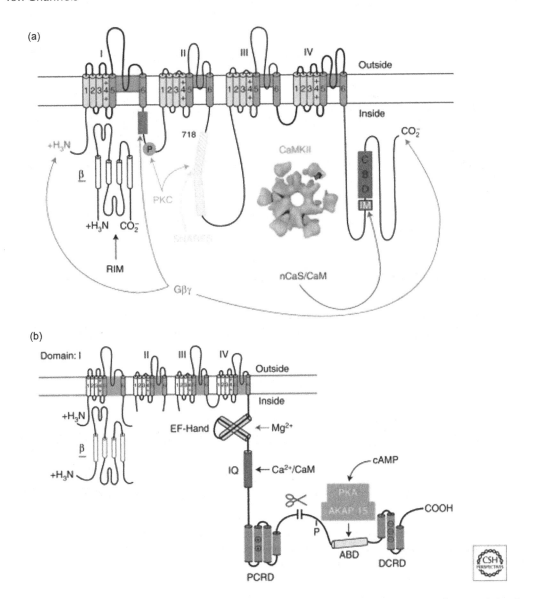

**FIGURE 1.18** Signaling complexes of calcium channels. (a) This represents a presynaptic Ca²⁺ channel signaling complex. A presynaptic Ca²⁺ channel α1 subunit is illustrated as a transmembrane folding diagram. Sites of interaction of SNARE proteins (synprint site), Gβγ subunits, protein kinase C (PKC), CaMKII, and CaM and CaS proteins are illustrated. IM, IQ-like motif; CBD, CaM-binding domain. (b) The cardiac Ca²⁺ channel signaling complex. The carboxy-terminal domain of the cardiac Ca²⁺ channels is shown in expanded presentation to illustrate the regulatory interactions clearly. ABD, AKAP15 binding domain; DCRD, distal carboxy-terminal regulatory domain; PCRD, proximal carboxy-terminal regulatory domain; scissors, site of proteolytic processing. The DCRD binds to the PCRD through a modified leucine zipper interaction. Details in Catterall (2011).

in aqueous phase, was validated by plotting Ramachandran diagrams for the average structures and by analyzing $i$ and $i+4$ helical hydrogen bonding between amino acid residues. These results suggest that the Ca²⁺ permeation through the channel can be derived from a competition between two ions for the only high-affinity binding site. Figure 1.20 demonstrates the summary of the results.

In a recent study (by a former colleague Barakat and his colleagues), efforts have been made to build an atomistic model for a calcium-selective VGCC, human Ca$_V$1.2 (hCa$_V$1.2), by studying

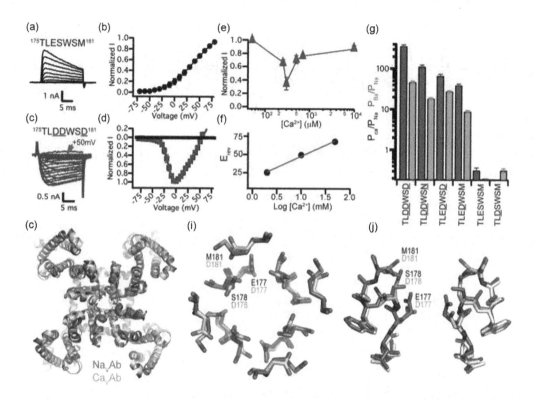

**FIGURE 1.19** Structure and function of the Ca$_V$Ab channel. (a, b) Outward Na$^+$ current conducted by Na$_V$Ab with 10 mM extracellular Ca$^{2+}$ and 140 mM intracellular Na$^+$. Holding potential, −100 mV; 20 ms, 10 mV step depolarizations. (c, d) Voltage-dependent conductance of inward Ca$^{2+}$ current by Ca$_V$Ab under the same conditions, 20 ms, 5 mV step depolarizations. (e) Biphasic anomalous mole fraction effect of increasing Ca$^{2+}$, as indicated, with Ba$^{2+}$ as the balancing divalent cation: 10 mM Ba$^{2+}$ with 0–0.5 mM Ca$^{2+}$, 9.3 mM Ba$^{2+}$ with 0.7 mM Ca$^{2+}$, and 0 mM Ba$^{2+}$ with 10 mM Ca$^{2+}$ (n=4–10). (f) Reversal potential (E$_{rev}$) versus Ca$^{2+}$ concentration. (g) Relative permeability of Ca$_V$Ab and its derivatives as measured from bi-ionic reversal potentials. P$_{Ca}$/P$_{Na}$, blue; P$_{Ba}$/P$_{Na}$, green (n=5–22). (h) Cartoon representation of the overall structure of Ca$_V$Ab (yellow) superimposed with Na$_V$Ab (slate). (i, j) Top and side views of the superimposed selectivity filters of Ca$_V$Ab (yellow) and Na$_V$Ab (slate) in stick representation. The three original Na$_V$Ab residues (black) and substituted Ca$_V$Ab residues (orange) are indicated. Errors bars in (a)–(g) are standard error of the mean. Further details in ref. (Tang et al., 2013). (For color figure see eBook figure.)

calcium influx using computational approaches: homology modeling and MD (Feng et al., 2019). The steered MD simulations on the model revealed four barriers for ion permeation: three calcium-binding sites formed by the EEEE and TTTT rings within the SF region, and a large barrier rendered by the hydrophobic internal gate. Figure 1.21 presents the results of the simulations showing water molecules and calcium ions forming H-bond with residues at the three SF sites.

## 1.3  LIGAND-GATED ION CHANNELS

LGICs are integral membrane proteins containing a pore that allows the regulated flow of selected ions across the plasma membrane. The ion flux is passive and driven by the electrochemical gradient for permeant ions. These channels are gated by the binding of a neurotransmitter to an orthosteric site(s) known to trigger a conformational change resulting in the conducting state.

LGICs are large, multi (four or five) subunit receptors and their OS create the passage of Na$^+$, K$^+$, Ca$^{+2}$, or Cl$^-$ charges. Each subunit is found to contain four hydrophobic transmembrane domains linking with

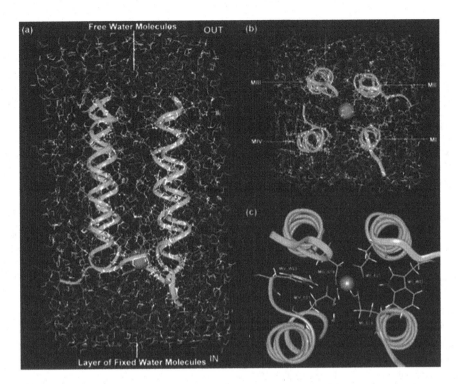

**FIGURE 1.20** The VGCC pore model with one $Ca^{2+}$ ion at the binding site. (a) A vertical vision. (b) A horizontal vision from the bottom of the channel. (c) A close vision of the interaction between the $Ca^{2+}$ ion and the carboxylate oxygen atoms of motifs I–IV, as well as the presence of the W residues of motifs II and IV. Details in Barreiro et al. (2002).

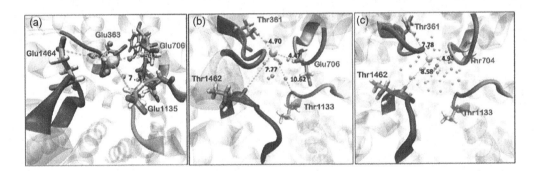

**FIGURE 1.21** Water molecules and calcium ion form H-bond with residues at the three SF sites. (a) In site-1, calcium ions form H-bonds with carboxyl of Glu706, Glu1135, and Glu1464 through water molecules. (b) In site-2, the calcium ions form H-bonds with carboxyl of Thr361, Glu706, and Thr1462 through water molecules. (c) In site-3, the calcium ions form H-bonds with carboxyl of Thr361 and Thr704 through water molecules. However, Glu363 in domain-I directly interacts with $Ca^{2+}$ through electrostatic interaction. Thr1133 and Thr1462 interact with calcium ions through multiple water molecules. Details in Feng et al. (2019).

hydrophilic groups. During activation of the channel complex, depending on the type and the direction of ion flow, the membrane potential may become depolarized or hyperpolarized. The channel activation may lead to a rapid response, allowing ions to flow down their electrochemical gradients.

Following many seminal works, such as the postulation of ligand binding and channel gating (del Castillo and Katz, 1957) and biochemical purification of an LGIC using snake venom toxins

(Changeux et al., 1970), Nobel laureates Neher and Sakmann used single-channel recordings and demonstrated that this receptor was an ion channel activated by a ligand (here acetylcholine) with discrete channel openings and closings (Neher and Sakmann, 1976). The 3D structures of the receptors and the first crystal structures of extracellular domains were later published (Armstrong et al., 1998), followed by the atomic structures of the nearly full-length receptors (Hilf and Dutzler, 2008).

### 1.3.1 LIGAND-GATED ION CHANNEL FAMILIES

LGICs constitute an important class of plasma membrane proteins that are critical for mediating cell–cell communication and cellular excitability. LGICs have been divided into the following families (first three are major members):

- "Cys-loop" LGIC which include nicotinic acetylcholine receptors, γ-aminobutyric acid (GABA) A and C receptors, 5-hydroxytryptamine-3 (5HT3) receptors, and glycine receptors.
- Glutamate receptors of the α-amino-3-hydroxy-5-methylisoxazole-4-propionic acid (AMPA), kainate, and N-methyl-D-aspartate (NMDA)-sensitive classes.
- ATP-sensitive P2X receptors (P2XRs).
- Members of the transient receptor potential (TRP) family that respond to endogenous or exogenous extracellular signals.
- Acid-sensitive ion channels. These LGICs are known to mediate neurotransmitter/ligand responses in central and peripheral nervous systems, skeletal and smooth muscle, and the endocrine system. Additionally, some of these LGICs appear to function in nonexcitable cells (e.g., glia, endothelial, and T-cells).

### 1.3.2 LIGAND-GATED ION CHANNEL STRUCTURES AND FUNCTIONS

Structural components of LGIC, their association in plasma membranes, and functions are addressed rigorously, and a large amount of information is already available. Based on the available information, we can predict concrete models and other crucial structural and functional aspects.

#### 1.3.2.1 General Models

LGICs consist of three major vertebrate superfamilies, each having unique folding architecture and mechanism. As mentioned earlier, these are Cys-loop receptors, ionotropic glutamate receptors (iGluRs), and P2XRs. See Figure 1.22 for a model representation, based on the established research findings over many decades, of these subunit assemblies.

Cys-loop receptors form pentameric ion channels which are gated by acetylcholine (ACh), GABA, glycine (Gly), and serotonin or 5-hydroxytryptamine (5-HT). This superfamily comprises excitatory, cation-selective nicotinic acetylcholine receptors (nAChRs); 5-HT3 receptors; zinc-activated channel inhibitory, anionselective

GABAA; strychnine-sensitive glycine receptors; and invertebrate glutamate-gated chloride channels (GluCl). The iGluRs are tetrameric nonselective cation channels to transport $Na^+$, $K^+$, $Ca^{2+}$ and are activated by glutamate. This superfamily comprises the excitatory NMDA, AMPA, and kainate receptors. P2XRs are trimeric ion channels gated by ATP. This superfamily of LGICs comprises the excitatory ATP-gated P2X receptor.

#### 1.3.2.2 LGIC Structures – Computer Simulations

A selection of various computational tools, including first principles method, classical MD, and enhanced sampling techniques, contribute to the picture of how LGICs function (Crnjar et al., 2019). There are numerous simulation studies using various computational techniques that have produced data to address the various aspects of LGIC structure, binding sites, and functions that are related to membrane transport of various electrolytes. We present here just one specific case study on

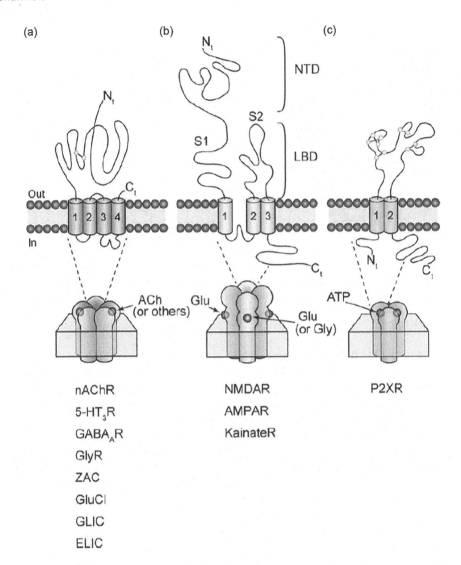

**FIGURE 1.22** Schematic representation of the membrane topology and subunit assembly of the three families of LGICs: (a) Cys-loop receptors, (b) iGluRs, and (c) P2X receptors. Conserved cysteine residues engaged in disulfide bridges are shown as yellow circles; N- and C-termini and transmembrane spans are also indicated. Out and in stand for, respectively, outside and inside the cell (Lemoine et al., 2012).

GABA-binding site, see Figure 1.23 (Comitani et al., 2014). This MD simulations allowed the analysis of the behavior of GABA at binding sites, including the hydrogen bond and cation-π interaction networks it formed, the conformers it visited, and the possible role of water molecules in mediating the interactions. The binding free energies were also estimated. For details, see Comitani et al. (2014).

### 1.3.3 RECORDING LGIC CHANNEL ACTIVITY

Here, we present a case study to address the activity of GABA receptors using EPRs across membrane hosting LGICs (Tierney, 2011). Significant progress has been made in recording LGICs in plasma membrane regarding understanding various molecular mechanisms. Tierney presents the novel use of the patch-clamp method to understand the influence of protein interactions on the

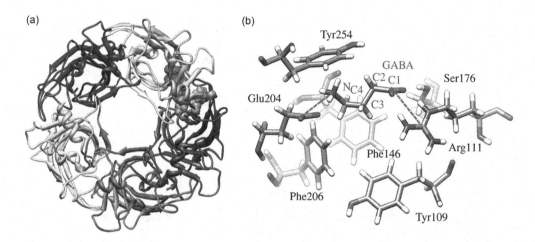

**FIGURE 1.23** An exemplary orthosteric binding site in a pentameric LGIC. (a) Top view of a homology model of the ECD of the resistance to dieldrin (RDL) receptor, with residues in the orthosteric binding site in red and orange (atoms represented as van der Waals spheres). (b) GABA in the binding site, including aromatic residues which form a "cage" around the GABA amine group). Details in Comitani et al. (2014). (For color figure see eBook figure.)

activity of individual GABAA receptors, which is a crucial aspect not addressed well earlier. Ion conduction is found to be a dynamic property of a channel, and protein interactions in a cytoplasmic domain underlie the channel's ability to alter ion permeation. Figure 1.24 demonstrates protein–protein interactions between adjacent GABAA receptors and the effects of the interactions on channel activity. A structural model describing a reorganization of the conserved cytoplasmic gondola domain and the influence of drugs on this process are also presented.

Thus, protein interactions have the potential to alter the disposition of the helices and therefore alter the gondola structure beneath each pore. A possible explanation for conductance increase observed in native GABAA receptors and clustered recombinant receptors could be that physical interactions between the γ2 subunits of adjacent receptors provide the potential to alter the structural organization beneath the pore, see Figure 1.24.

There is another important aspect of protein–protein interactions – ion channel interactions with intracellular proteins. For GABAA receptors, multiple interacting proteins are characterized, which might play a role in cytoskeletal coupling by gephyrin, radixin, and F-actin, as well as in signaling modulation by RAFT1 and collybistin (Chen and Olsen, 2007). For example, GABARAP, GODZ, and Plic1 interact with the intracellular parts of GABAA receptors, while GRIP1 and gephyrin are localized to sub-membrane and intracellular compartments, for details see Li et al. (2014).

## 1.4 MECHANICALLY GATED (MECHANOSENSITIVE) ION CHANNELS

Specialized proteins and their complexes can sense and respond to mechanical forces or stresses that organisms, from bacteria (single-celled) to animals and plants, experience in the environment. Mechanosensitive or mechanically gated ion channels (MSICs/MGICs) may be constructed out of these mechanical force-responsive proteins in cell membranes. These channels provide an excellent opportunity to study the interplay between various membrane mechanical properties and protein structure and function. Figure 1.25 presents a model sketch of these channels in membrane.

Membrane is a dynamic medium that directly affects the function and spatial distribution of the proteins embedded in it through physical interactions and/or coupling (electrostatic and mechanical)

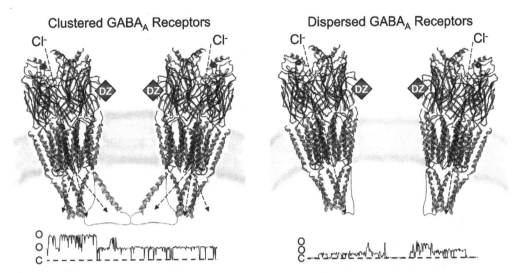

**FIGURE 1.24** Model depicting protein–protein interactions between adjacent GABAA receptors that enable receptor cross-talk, facilitating the enhanced biophysical properties of the ion channels. It is postulated that trafficking of GABAA receptors by GABARAP facilitates interactions between γ-subunit intracellular loops of adjacent receptors. Drug binding provides the potential to alter the disposition of the gondola structure beneath the pore, which, in these clustered receptors, increases ion permeation through each channel because of the reduced steric hindrance. In contrast, the unregulated expression of GABAA receptors produces dispersed receptors in the plasma membrane, which lack the inter-receptor interactions. These channels exhibit unitary conductance levels and drugs increase only the open probability of the channel(s). Single-channel traces depicted beneath each receptor model illustrate the large difference in current size between clustered and nonclustered GABAA receptors, which is likely to impose functional differences in cellular signaling and variations in physiological responses. C = closed, O = open.

(Ashrafuzzaman and Tuszynski, 2012a; 2012b). Here, we develop a theoretical formalism to address the membrane regulation of membrane protein functions and that the theory concomitantly addresses both electrostatic coupling and mechanical coupling-based regulation that the membrane integral proteins sense while being dynamic in the hydrophobic environment. Therefore, this theory may be generalized to address the mechanosensitivity of channels to respond to the mechanical properties of the hosting bilayers or the environment thereof.

### 1.4.1 SMALL AND LARGE CONDUCTANCE MGICs

The interplay between lipid or membrane properties and protein functions has been demonstrated for two channel proteins found in *E. coli* plasma membrane, mechanosensitive channel of large conductance (MscL) and mechanosensitive channel of small conductance (MscS).

MscL and MscS are found to directly respond to changes in membrane tension by opening nanoscale protein pores. Interaction with the membrane is thus predicted to be an integral aspect of their function (Kung et al., 2010). Variations in phospholipid bilayer thickness and the addition of compounds into the bilayer that induce spontaneous membrane curvature are found to directly impact the tension required to open, or gate, MscL (Perozo et al., 2002). This hints that the membrane may play a crucial role in regulating the integral channels.

MscL and MscS were first identified in their membrane-stretch-regulated electrophysiological activities in the plasma membrane of giant *E. coli* spheroplasts (Martinac et al., 1987; Sukharev et al., 1993). The MscL gene was cloned in a biochemical purification scheme (Sukharev et al., 1994). Then, the MscS encoding gene was identified (Levina, 1999).

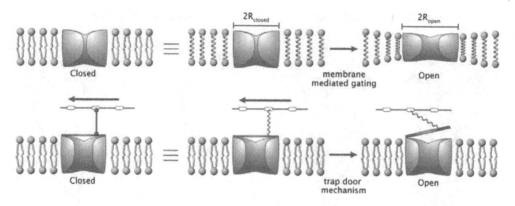

**FIGURE 1.25** Schematic representations of two models of gating for mechanosensitive channels. (a) The membrane-mediated mechanism and (b) the trapdoor mechanism. (a) The closed state of a channel embedded in a bilayer will be at equilibrium in the absence of applied tension (left). The application of tension to a membrane (right) serves to stabilize conformational states with greater cross-sectional areas of an embedded channel, with the larger areas favored by increasing tension. However, the change in protein conformation perturbs the membrane–protein interactions (schematically indicated by the compressed bilayer dimensions adjacent to the protein), which will contribute an unfavorable free energy term proportional to the interaction area between protein and membrane. As a consequence of these two competing effects, the interplay between tension and protein conformation can give rise to a rich variety of outcomes for the effect of tension on channel function, without the membrane pulling directly on the channel. (b) In the trapdoor mechanism, tension is coupled to the channel through an extramembrane component, such as the cytoskeleton or peptidoglycan (depicted by the bar with embedded ovals above the membrane), connected to a gate (trapdoor) covering the channel in the closed state (left). Movement of the extramembrane component relative to the membrane stretches the connecting spring, resulting in opening of the trapdoor (right). In this simplified mechanism, the membrane and channel are more passive participants with the applied tension doing work outside the membrane. It should be emphasized that these representations depict idealized mechanisms that are not mutually exclusive; activation barriers could be present that require direct interaction between components even when the overall transition is energetically favorable. Details in Haswell et al. (2011).

## 1.4.2  MGIC STRUCTURES AND FUNCTIONS

Crystal structures of MscL and MscS have been known for quite some time now. Figures 1.26 and 1.27 model the structure of MscL (Haswell et al., 2011). For each structure, three representations are provided:

(left) α-helices and β-sheets are shown as cylinders and arrows, respectively. The symmetry axis of each channel, assumed to be parallel to the membrane normal, is oriented vertically, with the cytoplasmic region positioned at the bottom.

(middle) Chain traces of the transmembrane region of each channel. The view (down the membrane normal) is rotated 90° about the horizontal axis. The cytoplasmic helical bundle of MtMscL has been omitted.

(right) Scale bars 3.4 and 5.0 nm are indicated. The cross-sections of the membrane-spanning region of each channel may be approximated as regular polygons (pentagon or square), giving values for the corresponding areas of MtMscL, SaMscL, and the SG-model as 20, 25, and 43 nm², respectively.

## 1.4.3  SIMULATION OF MGICs

MD simulations revealed the gating mechanisms of the bacterial mechanosensitive channel MscL (Sawada et al., 2012). We take this as an example case to explain the regulation of channel gating

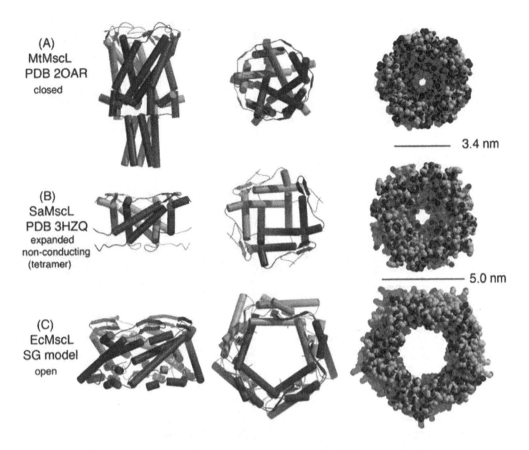

**FIGURE 1.26** Structurally characterized forms of MscL as observed in the crystal structures of (a) *M. tuberculosis* MscL (Chang et al., 1998) and (b) *S. aureus* MscL (Liu et al., 2009), and (c) the model of the open state of EcMscL developed by Sukharev and Guy ("SG-model"; Sukharev et al. (2001).

due to the influence of tensions. The bacterial MscL got its 3D protein structure in the CS revealed on the atomic scale and large amounts of electrophysiological data on its gating kinetics (Sukharev et al., 1994; Chang et al., 1998). MD simulations on the initial process of MscL opening in response to a tension increase in the lipid bilayer were addressed here, see Figure 1.28 (Sawada et al., 2012). To identify the tension-sensing site(s) in the channel protein, this group calculated interaction energy between membrane lipids and candidate amino acids facing the lipids. Phe78 was found to have a conspicuous interaction with the lipids, suggesting that Phe78 is the primary tension sensor of MscL. Increased membrane tension by membrane stretch dragged radially the inner (TM1) and outer (TM2) helices of MscL at Phe78, and the force was transmitted to the pentagon-shaped gate that is formed by the crossing of the neighboring TM1 helices in the inner leaflet of the bilayer. The radial dragging force-induced radial sliding of the crossing portions led to a gate expansion. The calculated energy for this expansion was comparable to an experimentally estimated energy difference between the closed and the first subconductance state.

During a 2 ns simulation, the transmembrane α-helices tilted and radially expanded in the membrane plane and the channel pore opened gradually. The average pore radius calculated (5.8 Å) of the most constricted part of the pore (the ostensible gate region of MscL) formed with the residues from Leu19 to Val23 in TM1 helix of each subunit at 2 ns simulation was found much smaller than the open pore size estimated by electrophysiological analyses or channel-mediated protein efflux measurements (Sukharev et al., 1999). This suggests that the pore detected in 2 ns is yet to grow bigger with time, and longer time simulation may reveal further details.

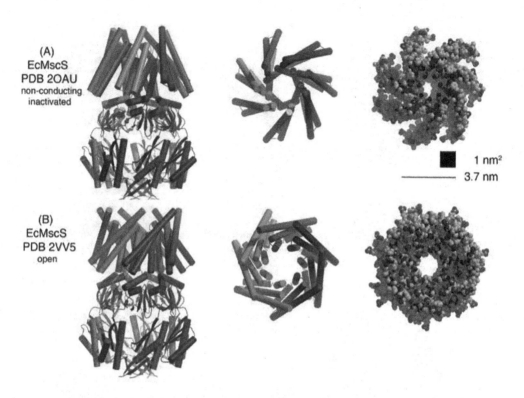

**FIGURE 1.27** Structurally characterized forms of *E. coli* MscS as observed in the crystal structures of (a) nonconducting/inactivated (Bass, 2002) and (b) open (Wang et al., 2008) states. The organization of this figure parallels that represented in Figure 1.3 for MscL. The scale bar between the space-filling models equals 3.7 nm, assuming that the cross-sections of each structure are depicted as regular heptagons with this side length, the corresponding areas are 50 nm$^2$. If lipids can intercalate between the splayed TM1, TM2 of adjacent subunits in the nonconducting/inactivated structure (Figure 1.27a), the cross-sectional area will be reduced from this value.

### 1.4.4   MGICs in Mammalian Hair Cells – Example of Pore Proteins to be Identified

The sensory mechanoelectrical transduction (MET) channel of hair cells is localized near stereo-cilia tips at the base of the tip link filament that connects a shorter stereocilium to its next taller neighbor (Pickles et al., 1984; Beurg et al., 2008). The inner ear hair cells are specialized mechanosensory cells, which convert mechanical stimuli provided by sound waves (cochlea) or head movement (vestibular system) into some electrical signals. Hair bundle deflection toward the tallest stereocilia leads to the MET channel open probability increase, while deflections in the opposite direction work toward decreasing the channel open probability (Hudspeth and Corey, 1977; Nicolson et al., 1998). In addition to the sensory MET channels at tip links, another mechanically activated channel has been identified in hair cells, that is located at their apical cell surface where stereocilia emanate from the cell body (Wu et al., 2017).

Compelling evidence has emerged in recent studies that hair cells express several molecularly distinct ion channels with a different function. The best of these studied is the sensory MET channel at the tips of stereocilia, suggesting that TMC1/2, TMHS, and TMIE are integral components of the sensory MET channel (Figure 1.29a); however, which protein(s) form the channel pore remains an open question. Almost nothing about the molecular composition and function of stretch-activated ion channels in the cell body of hair cells is known. PIEZO2 is shown to be an integral component

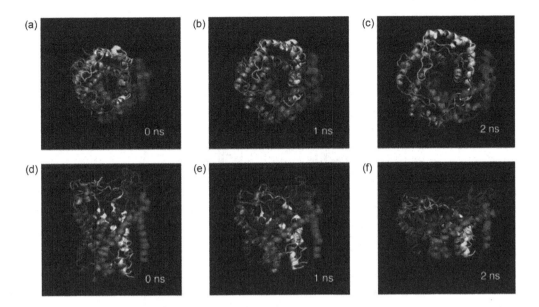

**FIGURE 1.28** Snapshots of MscL structural changes upon tension increase. Top views taken at (a) 0 ns, (b) 1 ns, and (c) 2 ns, and the corresponding side views (d, f). Eco-MscL is shown in a ribbon representation with different colors for each subunit. The lipid and water molecules are not shown here. (For color figure see eBook figure.)

of the stretch-activated MET channel in the apical surface of hair cells (Figure 1.29b), but nothing is known regarding the molecular compositions of stretch-activated MET channels located on the basolateral surface of hair cells.

## 1.5   LIGHT-GATED ION CHANNELS

Ion channels that are regulated by electromagnetic radiation are known as light-gated ion channels. Under the influence of an electrochemical gradient, ions move through transmembrane protein(s)-formed pores across a membrane.

Channelrhodopsins, a subfamily of retinylidene proteins, function as light-gated ion channels (Nagel et al., 2002). Retinylidene proteins contain seven membrane-embedded alpha-helices that form an internal pocket binding the chromophore retinal inside. Besides being in photoreceptor cells of animal eyes, they are found in other places, such as unicellular eukaryotic microbes, archaeal prokaryotes, dermal tissues of frogs, pineal glands of lizards and birds, the hypothalamus of toads, and the human brain. Functions of these proteins include light-driven ion transport and phototaxis signaling in microorganisms, retinal isomerization, and various photosignal transductions in higher animals.

Retinylidene proteins, which are integral membrane proteins, covalently bind a retinal chromophore. As per the amino acid sequence, these proteins are divided into two families with distinct functions, as follows (Spudich et al., 2000):

- Type I rhodopsins, for example, bacteriorhodopsin from the archaeon *Halobacterium salinarum*, re-found to function as light-driven ion transporters/pumps, channels, and phototaxis receptors.
- Type II rhodopsins, known for the visual pigment of mammalian rod photoreceptor cells, are found to function primarily as photosensitive receptor proteins in metazoan eyes and in specific extraocular tissues.

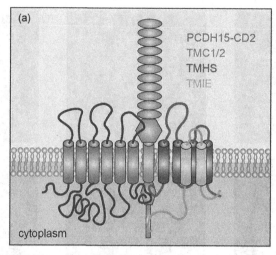

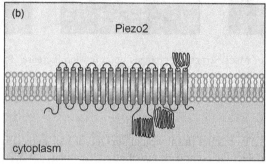

**FIGURE 1.29**  Model of the sensory transduction channel and for PIEZO2. (a) Transmembrane channel-like proteins 1 and 2 (TMC1/2), tetraspan membrane protein in hair cell stereocilia (TMHS)/LHFPL5 and transmembrane inner ear (TMIE) bind to PCDH15 and are constituents of the sensory mechanoelectrical transduction (MET) machinery. TMC1/2 and TMHS/LHFPL5 bind to PCDH15. TMIE binds to TMHS/LHFPL5 as well as to the unique C-terminal domain of one specific PCDH15 isoform in stereocilia. (b) Model of the PIEZO2 channel, which contain at least 18 transmembrane domains and potentially up to 38 transmembrane domains. Details in Qiu and Müller (2018).

### 1.5.1  CHANNEL SENSITIVITY AND PHOTOCURRENTS

Phototaxis and photophobic responses of green algae were found to be mediated by rhodopsins with microbial-type chromophores two decades ago (Nagel et al., 2002). Here, a complementary DNA sequence in the green alga *Chlamydomonas reinhardtii* was reported that encodes a microbial opsin-related protein, termed "Channelopsin-1." Expression of Channelopsin-1, or only the hydrophobic core of the protein (showing homology to the light-activated proton pump bacteriorhodopsin) in *Xenopus laevis* oocytes in the presence of all-trans retinal was found to produce a light-gated conductance with characteristics of a proton permeable channel selectively.

Channelrhodopsin permeation of cations can be triggered fast, repetitively, reproducibly, and noninvasively by light. This opens up novel ways to address crucial aspects of channel on- and off-gating. The first identified channelrhodopsins were from *C. reinhardtii* (Nagel et al., 2002; 2003), which are naturally hosted in eyespots of the unicellular alga *C. reinhardtii*.

Green (and not red) light illumination induced inward currents in Chop1 RNA–injected oocytes under a membrane potential −100 mV (figures cannot be reproduced due to copyright issues). An identical process was observed in truncated Chop1 RNAs encoding amino acids 1–346 or 1–517.

At an external pH of 7.5, the green light-induced inward current reversed at a voltage near −15 mV, with clearly visible outward photocurrents at positive membrane potentials. The photocurrent direction dependence on applied potentials suggests that the reconstituted Channelrhodopsin-1 (ChR1) mediates a light-induced passive ion conductance. A reversal potential −15 mV is close to the Nernst potential for $Cl^-$ or $H^+$, but far from Nernst potentials for $Na^+$, $K^+$, or $Ca^{2+}$ (Nagel et al., 2002).

Recently, up to 13 channelrhodopsin sequences have been identified in other green algae, which differ mainly in cation selectivity, kinetics, light wavelength, and intensity sensitivities (Govorunova et al., 2011; Lórenz-Fonfría and Heberle, 2014). Channelrhodopsins, being illuminated, transiently increase their conductance for various monovalent and divalent cations, leading to cell membrane depolarization within milliseconds, mostly to control neural activity.

The photocurrents under continuous illumination show desensitization, which is not observed in single turnover experiments (Lórenz-Fonfría and Heberle, 2014). A few milliseconds after the start of illumination, the photocurrent typically reaches a peak that decays within some tens of milliseconds to a steady level, see Figure 1.30a. Desensitization (inactivation) seemingly originates from channelrhodopsin 2 (ChR2) being temporally in a late low conductive state and/or in a mixture of conductive/nonconductive states. Desensitization is quantified by the ratio of the steady current to the peak current, which can be as low as 0.2–0.4 for ChR2. The desensitized fraction increases with illumination intensity, and as the holding potential becomes more positive, it is reduced at low extracellular pH and in some ChR2 variants. Channel desensitization is almost absent in ChR1 and in some ChR1–ChR2 chimeras.

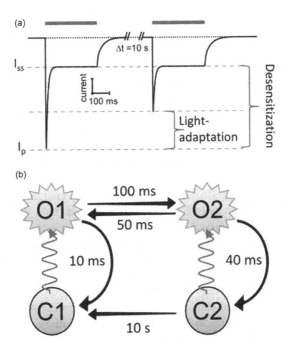

**FIGURE 1.30** Photocurrent response and photocycle of ChR2 under continuous illumination. (a) Illustrative (synthetic) photocurrent of ChR2 expressed in a host cell under negative voltage-clamp conditions and rectangular blue light excitation (blue bars). After an initial peak, $I_p$, the photocurrent decays (desensitizes) to a steady state, Iss, and relaxes to zero when the light is switched off. A second pulse after a certain time delay, Δt, generates a smaller peak current (light adaptation). (b) Four-state model used to describe the photocurrents under continuous illumination. The model includes two closed states (C1 and C2) and two open conductive states (O1 and O2). Approximate time constants are indicated for thermally driven transitions (plain arrows). Light-driven transitions (wavy arrows) depend on, among other factors, wavelength and intensity. For details, see Lórenz-Fonfría and Heberle (2014). (For color figure see eBook figure.)

## 1.5.2 Channelrhodopsin Structure

High-resolution structures of ChR2 and the C128T mutant (having a markedly increased OS life-time) have been identified by Volkov and colleagues using X-ray diffraction (Volkov et al., 2017). The general structure of ChR2 and structures showing its alignment with the structure of a chimera (C1C2) are presented (due to copyright issues we cannot present them here).

The structure reveals two cavities on both intracellular and extracellular sides. They are be connected by extended hydrogen-bonding networks that involve water molecules and side-chain residues. Central is the retinal Schiff base controlling and synchronizing three gates to separate the cavities. Additionally, DC gate comprising a water-mediated bond between C128 and D156 and interacting directly with the retinal Schiff base is found. A structural comparison with C128T reveals a direct connection of the DC gate to the central gate, suggesting that the gating mechanism is affected by subtle tuning of the Schiff base interactions.

The X-ray crystallographic structures of the wild-type ChR2 and its slow C128T mutant (solved to 2.39- and 2.7-Å resolution, respectively) indicate that the rhodopsin-derived channels are different from other ion channels. The ChR2 structure also exhibits considerable differences from that of the C1C2 chimera, as follows:

  i. ChR2 has two conformations in its ground state,
  ii. there are additional gates found in the extracellular and intracellular parts,
  iii. there is a water-mediated hydrogen bond found between C128 and D156, explaining the mechanism of DC gating.

These structural features give insights into the molecular mechanism of ChR2 and other channelrhodopsins.

In 2012, the first considerable structure was revealed by Kato and colleagues (Kato et al., 2012). The crystal structure of a ChR (a C1C2 chimera between ChR1 and ChR2 from *C. reinhardtii*) at 2.3-A° resolution was observed. The structure revealed the essential molecular architecture of ChRs almost similar to those in Volkov et al. (2017), including the retinal-binding pocket and cation conduction pathway.

Kato also calculated the electrostatic surface potential of C1C2, which revealed an electronegative pore formed by TM1, 2, 3, and 7, see Figure 1.31a. Several negatively charged residues, including Glu 129 (90), Glu 136 (97), and Glu 140 (101), as well as Glu 162 (123) and Asp 292 (253), are aligned along the pore, see Figure 1.31b. Because most negatively charged residues are derived from TM2, ion conductance and selectivity of C1C2 are mainly defined by TM2.

As four residues are highly conserved in both ChRs and bacteriorhodopsins (BRs) (Arg 82, Tyr 83, Glu 194 and Glu 204, respectively), and the corresponding residues in BR reportedly have an important role in proton pumping, the R159A mutant in C1C2 was generated. This mutant did not produce a photocurrent despite robust membrane expression, see Figure 1.31c and d.

The light-induced channel opening of channelrhodopsin has recently been simulated (Cheng et al., 2018). An atomic structural model of a chimeric ChR in a precursor state of the channel opening was determined using an accurate hybrid MD simulation technique (Takemoto et al., 2015) and a statistical theory of internal water distribution. The photoactivated structure features extensive tilt of the chromophore (Figure 1.32) accompanied by a redistribution of water molecules in its binding pocket (Figure 1.33). The atomistic model manifests a unique photoactivated ion conduction pathway, markedly different from a previously proposed one, and explains experimentally observed mutagenic effects on key channel properties (Cheng et al., 2018).

The cytoplasmic cavity in the vicinity of the binding pocket is known to be more pronounced in the X-ray crystallographic structure of ChR2 determined recently (Volkov et al., 2017). Furthermore, the tilt of the 13-cis polyene chain in the bent form is generally expected to create a cavity in its concave side, see Figure 1.34. This can connect the newly formed cytoplasmic cavity and the existing channel

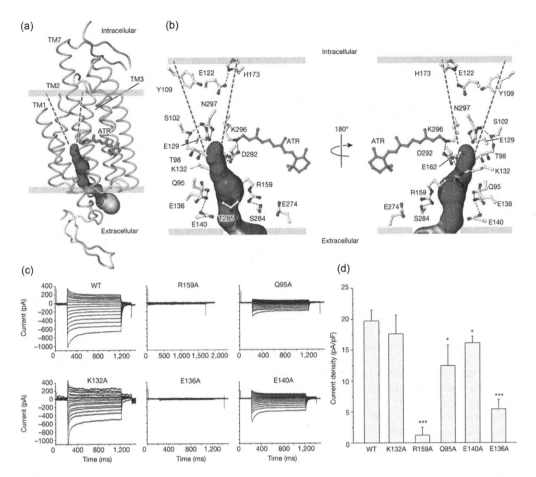

**FIGURE 1.31** Cation-conducting pathway formed by TM1, 2, 3, and 7. (a) Pore lining surface calculated by theCAVER37 program, colored by the electrostatic potential. Dashed red lines indicate putative intracellular vestibules. (b) Close-up views of the surface of the pore, with 17 polar lining residues (subtract 39 from the C1C2 residue number to obtain ChR2 numbering). Hydrogen bonds are shown as black dashed lines. (c), Photocurrents of mutants of the five residues within the pathway, measured under the same conditions as in Figure 1.3 (c, d) The peak amplitudes of the photocurrents, as in Figure 1.3 (d). *P,0.05, ***P,0.001. Error bars represent SEM. For details, see Kato et al. (2012).

in the extracellular half as the tilt is more developed. Thus, the photoinduced 13-cis polyene chain generally acts as the gate, and its tilt opens the channel that is found to detour the constriction site.

### 1.5.3 VIRAL CHANNELRHODOPSINS

Recently, viral channelrhodopsins have been reported to mediate phototaxis of algae, enhancing the host anabolic processes to support virus reproduction, and thereby serving key role in global phytoplankton dynamics (Zabelskii et al., 2020). A function-structure characterization of two homologous proteins, representatives of family 1 of viral rhodopsins, OLPVR1 and VirChR1 was presented. VirChR1 is a highly selective, $Ca^{2+}$-dependent, $Na^+/K^+$-conducting channel. In contrast to known cation channelrhodopsins, this is impermeable to $Ca^{2+}$ ions. In human neuroblastoma cells, VirChR1, upon illumination, depolarizes the cell membrane to a level sufficient to fire neurons, suggesting its unique optogenetic potential. 1.4-Å resolution structure of OLPVR1 reveals their

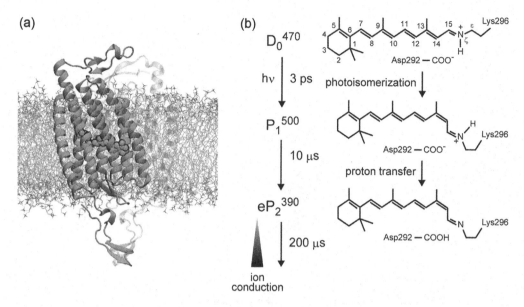

**FIGURE 1.32**   Protein structure and photoactivation scheme of ChR. (a) A simulation system. A dimeric form of ClC2 is embedded in a lipid bilayer. One of the monomers is shown in a transparent representation for clarity. A retinal chromophore is drawn in orange. (b) A scheme of the photoactivated channel opening. In the dark state (D0470), the retinal chromophore in the all-trans conformation resides in the protein. The photoisomerization immediately leads to the P1500 intermediate, in which the conformation of the chromophore is changed to the 13-cis form. A proton transfer from the Schiff base to its counterion carboxylate of Asp292 with a time constant of 10 μs and generates the early stage of P2390 state (eP2390), in which the ion conduction is not yet started. The ion conduction gradually develops after the formation of eP2390 with a time constant of 200 μs. For details, Cheng et al. (2018).

remarkable differences from the known channelrhodopsins and a unique ion-conducting pathway. A representative sketch to show the ion-conducting pathway of OLPVR1 is presented in Figure 1.35.

## 1.6   TEMPERATURE-GATED ION CHANNELS

Thermal regulation of ion channel functions is nothing new. Ion channels that are found directly gated by heat are called temperature-gated ion channels. Temperature-gated or heat-gated ion channels enable the transmembrane transfer of ions in response to a temperature stimulus. About three decades ago, these channels started getting attention (Treede et al., 1992; Cesare et al., 1999).

Heat has been reported to open a nonselective cation channel in primary sensory neurons. An ion channel gated by capsaicin was cloned from sensory neurons. This channel (vanilloid receptor subtype 1, VR1) has been reported to be gated by heat in a manner similar to the native heat-activated channel, probably responsible for the detection of painful heat. The heat channel response is potentiated by phosphorylation by protein kinase C, whereas VR1 is potentiated by externally applied protons (Cesare et al., 1999).

Shifting the heat-sensitive channel, from CS to OS, is expected to depend on the general thermodynamic equation, as follows:

$$\Delta G = \Delta H + T\Delta S$$

The change in the equilibrium between closed ↔ open states of the channel depends on the Gibbs free energy change (ΔG). This can be markedly temperature-dependent only if there is a large

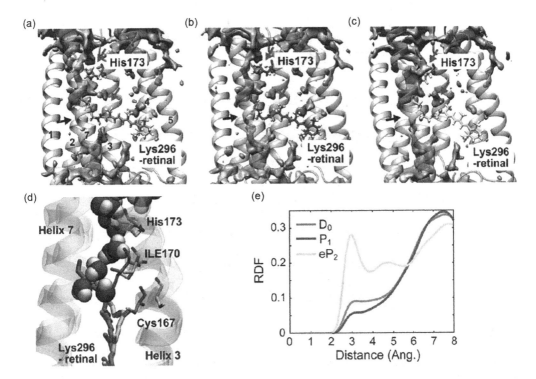

**FIGURE 1.33** Distributions of water molecules inside C1C2. (a–c) Distributions of water molecules in the D0 state (a), the P1 state (b), and the eP2 state (c) calculated by the 3D-RISM method are shown. Red arrows indicate regions where water populations newly appear upon the formation of the eP2 state. Black arrows indicate the constriction sites. (d) Snapshot structures of MD simulations of the D0 state (orange) and the eP2 state (yellow). Water molecules in the eP2 state are shown. (e) Radial distribution functions of oxygen atoms of water molecules around the oxygen atom of the main-chain carbonyl group of His173 calculated by the 3D-RISM method. To see this figure in color, go online. For details, see Cheng et al. (2018). (For color figure see eBook figure.)

entropy difference ($\Delta S$) between OS and CS. Therefore, changes (elevations) in temperature (T) must cause heat-sensitive ion channels to change from an ordered to a more disordered state, similar to the melting of ice or dissolving of a salt in water.

It is less likely that any accessory protein or signaling pathway is needed to gate these channels because heat-sensitive ion channels may be seen in cell-free membrane patches from nociceptors. Therefore, the temperature-sensitive gating unit is likely to be intrinsic to heat-sensitive proteins.

Thermosensitive transient receptor potential proteins (TRPs) such as vanilloid subfamily of TRP channels (TRPV1–4) are all heat-activated nonselective cation channels that are modestly permeable to $Ca^{2+}$. The permeation and gating mechanisms of this heat-sensitive channel have been revealed by understanding the structures (Yuan, 2019). The membrane currents and $[Ca^{2+}]_i$ transients induced by thermal and agonist TRPV1 and four stimulations in human corneal epithelial cells (HCEC) were characterized (Mergler et al., 2011). Their activation confers temperature sensitivity at the ocular surface, which may protect the cornea against stresses. Results of RT-PCR, fluorescence calcium imaging, and the planar patch-clamp technique helped demonstrate that there is functional thermo-TRPV1, TRPV2, and TRPV4 expression in HCEC. For example, the expression of thermo-TRPs in HCEC has been shown in the whole-cell configuration of the planar patch-clamp technique in Figure 1.36. The temperature was maximally raised to $\approx 45°C$, as above this level most whole-cell planar patch-clamp recordings were unstable or seals broke above 50°C. In response to a suitable temperature increase from 20°C to >40°C, a rapid increase in nonselective cation channel

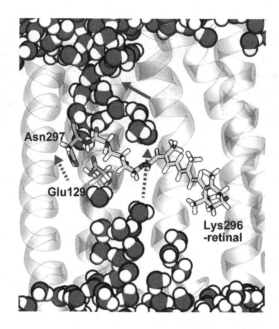

**FIGURE 1.34** Ion conduction pathways proposed in the eP2 structure. A snapshot structure of an equilibrium MD trajectory of the eP2 state is shown. A newly proposed pathway goes by the 13-cis retinal chromophore, and thus is different from a previously proposed pathway through the constriction site composed of Asn297 and Glu129, which is suggested to open upon the deprotonation of Glu129 (Cheng et al., 2018).

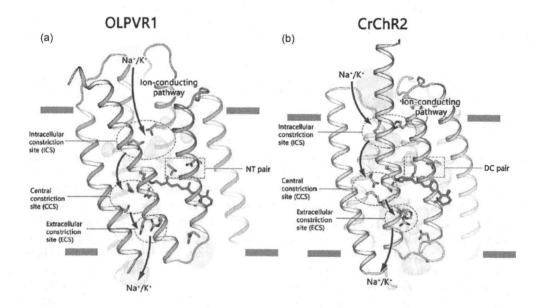

**FIGURE 1.35** Proposed ion-conducting pathway of OLPVR1. Three consecutive constriction sites and proposed cation pathway of (a) OLPVR1 and (b) CrChR2 (Nagel et al., 2002, 2003) proteins. The proposed ion pathway is shown only in one direction for clarity. Important residues are depicted as sticks. All-trans retinal is depicted with orange color. NT and DC pairs are additionally indicated. Details in Zabelskii et al. (2020). (For color figure see eBook figure.)

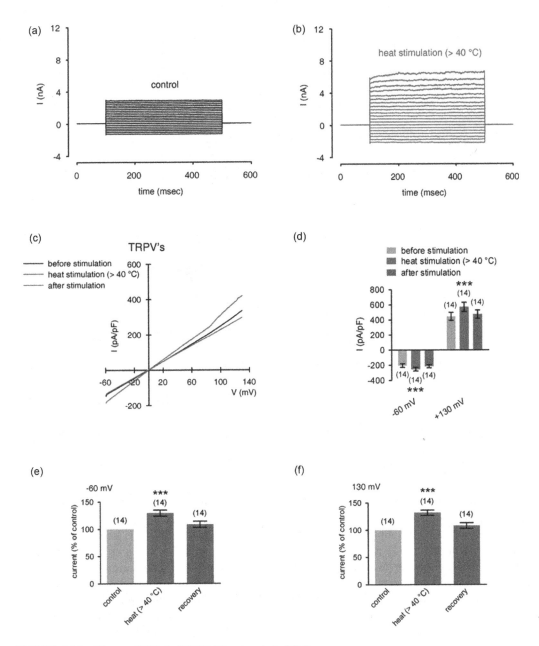

**FIGURE 1.36** Thermo-TRPs in HCEC (Mergler et al., 2011).

outward and inward currents occurred (Figure 1.36a and b). Figure 1.36c demonstrates current responses induced by a voltage ramp protocol from −60 to +130 mV (duration 500 ms). A summary of voltage ramp experiments is shown in Figure 8d–f. At −60 mV, the maximum nonselective cation inward currents increased to 130%±6% of control (set to 100%) (p<0.005; n=14; paired tested). At 130 mV, the maximum nonselective cation inward currents increased to 132%±5% of control (p<0.005; n=14; paired tested) (Mergler et al., 2011).

Chowdhury and colleagues adopted a bottom-up protein design approach to rationally engineer ion channels to activate in response to thermal stimuli. Varying amino acid polarities at sites undergoing state-dependent changes in solvation helps to systematically confer temperature sensitivity

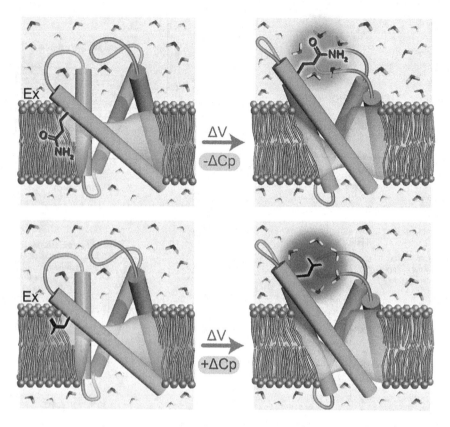

**FIGURE 1.37** Modeling the state-dependent change in solvation, a possible mechanism of temperature-dependent gating (Chowdhury et al., 2014).

to a canonical voltage-gated ion channel. The specific heat capacity change ($\Delta C_p$) during channel gating is a major determinant of thermosensitive gating, see Figure 1.37 (Chowdhury et al., 2014).

The role of $\Delta C_p$ on temperature-dependent gating of ion channel was tested by developing a model system that allows to modulate the $\Delta C_p$ associated with the gating process and test its effects on channel gating. $\Delta C_p$ of solvation of polar and nonpolar residues are opposite in sign, and during voltage-dependent activation, certain regions of the Shaker potassium channel undergo changes in water accessibility. The study also showed that the reduction of gating charges amplifies temperature sensitivity of designer channels accounting for low-voltage sensitivity in all known temperature-gated ion channels.

## 1.7   PEPTIDE-GATED ION CHANNELS

In patch-clamp experiments on the C2 neuron of helix aspersa the neuropeptide Phe-Met-Arg-Phe-NH2 (FMRFamide) was found (probably first ever as claimed) to directly gate a $Na^+$ channel (Cottrell, 1997). The channel is amiloride-sensitive. The activation of this channel is responsible for the fast excitatory action of the peptide. The cDNA sequence (FaNaCh) is predicted to have just two membrane-spanning regions and a large extracellular loop. When expressed in *X. laevis* oocytes, the channel showed a response to FMRFamide (due to copyright issues we cannot reproduce figures here). These data altogether provided the first evidence for a peptide-gated ion channel. Thus, it is clear that channels of this group may serve diverse roles in transporting epithelia and regulating $Na^+$ transport; in neurons, at least one (the HelixFaNaCh) is peptide-gated, besides finding other proton-gated channels (Kellenberger and Schild, 2002) and yet others involving mechano-sensation.

Recently, *Hydra* was found to have a large variety of peptide-gated ion channels that are activated by a restricted number of related neuropeptides (Assmann et al., 2014). The existence and expression pattern of these channels, as well as behavioral effects induced by channel blockers, suggest that *Hydra* co-opted neuropeptides for fast neuromuscular transmission. The nervous system of the cnidarian *Hydra* contains a large repertoire of neuropeptides. It is known that neuropeptides are the principal transmitters of *Hydra*. An ion channel directly gated by *Hydra*-RFamide neuropeptides has already been identified in *Hydra* –the *Hydra* Na$^+$ channel (HyNaC) 2/3/5, which is expressed at the oral side of the tentacle base. *Hydra*-RFamides are more widely expressed, being found in neurons of the head and peduncle region. Assmann and colleagues explored whether further peptide-gated HyNaCs exist, with positive results (Assmann et al., 2014). Figure 1.38 presents one representative example.

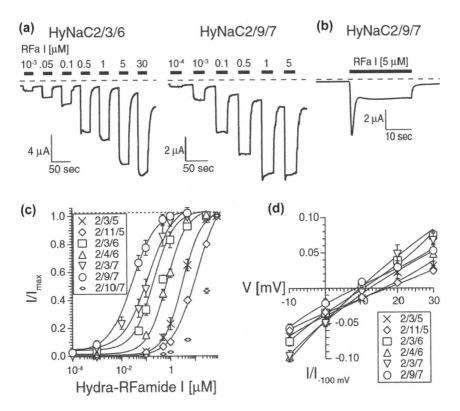

**FIGURE 1.38** HyNaCs are activated by Hydra-RFamide I. (a) Representative EP recorded current traces showing concentration-dependent activation of HyNaC2/3/6 and HyNaC2/9/7 by Hydra-RFamide I when oocytes had been injected with EGTA. Dashed lines represent the zero current level. (b) When oocytes had not been injected with EGTA, Hydra-RFamide I elicited biphasic currents; an oocyte expressing HyNaC2/9/7 is shown as an example. (c) Concentration response curves for HyNaCs and Hydra-RFamide I. Currents were normalized to the currents at the highest agonist concentration, which had amplitudes of 9.2±1.9 μA (n=9; 2/3/5), 2.6±0.4 μA (n=12; 2/11/5), 15.7±3.2 μA (n=8; 2/3/6), 10.1±1.7 μA (n=12;2/4/6), 19.6±1.8 μA (n=10; 2/3/7), 7.2±0.9 μA (n=10; 2/9/7), and 10.6±1.8 μA (n=8; 2/10/7). Lines represent fits to the Hill equation. (d) I/V curves for putative physiological HyNaCs, revealing slightly positive reversal potentials. Voltage ramps were run from −100 to +30 mV in 2 s. Background currents had been subtracted by voltage ramps in the absence of agonist. Currents were normalized to the current at −100 mV. EGTA, ethylene glycol tetra acetic acid; HyNaC, Hydra Na$^+$ channel. Details in Assmann et al. (2014).

## 1.8  SMALL PEPTIDE CHANNELS IN MODEL MEMBRANE SYSTEMS

Antimicrobial peptides (AMPs) are produced by bacteria (e.g., bacteriocin, and others), fungi (e.g., peptaibols, plectasin, and others), and cnidaria (e.g., hydramacin, aurelin). AMPs are abundant and produced by many tissues and cell types in a variety of invertebrate, plant, and animal species. Thousands of different AMPs have already been identified or predicted. Many of them are natural or their synthetic counterparts.

In recent years, significant research has been done to find new insights into focused aspects of the development of synthetic AMPs with lower toxicity and improved activity compared to their endogenous counterparts. AMPs appear as promising therapeutic options for the treatment of skin and soft tissue infections and wounds as they show a broad spectrum of antimicrobial activity, low resistance rates, and display pivotal immunomodulatory as well as wound healing promoting activities such as induction of cell migration and proliferation and angiogenesis. Some of the vital molecular actions are found in their ability to induce versatile membrane transport events (membrane-spanning channels) by modifying membrane physical properties. We shall pinpoint only a few of them to demonstrate how our knowledge is progressing in regard to their involvement in the formation of ion channels with AMP specificity in nature. We do not aim to summarize information regarding AMP actions in thousands of publications available to date since the early discovery of their channel forming potency many decades ago.

### 1.8.1  PEPTIDE-INDUCED ION CHANNELS IN MEMBRANES

Peptides are observed to line across the bilayer interior hydrophobic region and create a stable structure capable of conducting ion currents across the membrane at the influence of a transmembrane potential. Here, we present examples of two AMPs, gramicidin A (gA) and alamethicin (Alm), which form β-helix channels and barrel stave pores inside lipid bilayers. Figure 1.1 (presented earlier) and Figure 1.39 present EP current trances and the models of the channels, respectively. For details see Ashrafuzzaman et al. (2008; 2012). These channels are peptide-lined in nature.

We have significant evidence on the construction of peptide-induced lipid-lined channels, called toroidal channels. An example is presented here. MD simulations of melittin peptides interacting with a zwitterionic DPPC bilayer were performed (Sengupta et al., 2008). As peptides were initially placed randomly in the solution, the spontaneous formation of transmembrane water pores was observed repeatedly. In the pore state, the lipids bend into the bilayer to continuously line the water pore in a toroidal shape. Only one or two peptides insert into the pore. The remaining peptides line the mouth of the pore, helping its curvature to stabilize. No regular packing of the peptides is observed, rather the peptide/lipid pore complex is intrinsically disordered. A snapshot of the disordered toroidal pore is given in Figure 1.40, together with a cartoon image comparing the disordered toroidal pore to the traditional model. The arrangement of the melittin peptides and lipids within the pore is similar to that seen in the case of magainin-H2 (Leontiadou et al., 2006). A striking characteristic of the disordered toroidal pore is that the orientations of the peptides are not well defined.

### 1.8.2  DRUG-INDUCED ION PORES LOOK LIKE PEPTIDE-INDUCED TOROIDAL PORES

We discovered a new class of ion channels induced by chemotherapy drugs (CDs) in lipid bilayers (Ashrafuzzaman et al., 2011; 2012b). These were originally branded as lipid-lined ion channels forced constructed due to drug interactions with lipids.

EP recordings across lipid membranes in aqueous phases containing CDs were used to investigate the drug effects on membrane conductance. Figure 1.41 demonstrates the drug effects on membrane bilayer through induction of lipid-lined toroidal-type pores/channels.

Here lipids are forced to align across the pore opening to facilitate conduction of ions through the pore. The pore opening is a continuous process experiencing back-and-forth changes. This means

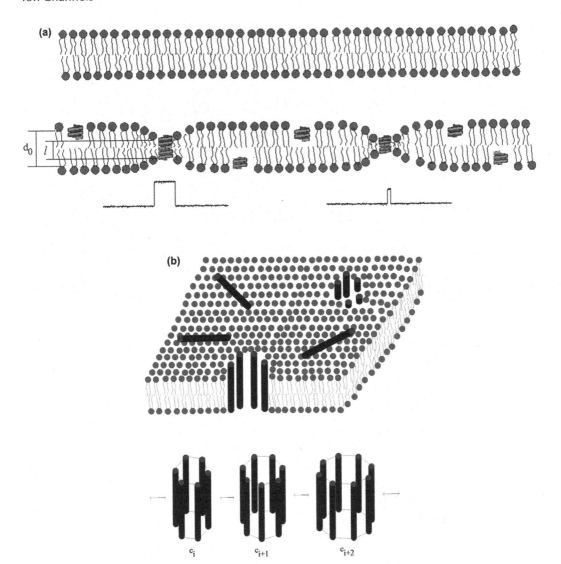

**FIGURE 1.39** (a) gA channels (lower panel) deform lipid bilayer's resting thickness. $d_0$ and $l$ represent hydrophobic bilayer thickness and gA channel length, respectively. (b) In Alm channels, cylindrical rods represent monomers on a "barrel stave" pore. Transitions between different Alm conductance pores by addition/release of monomer(s) from/to the surrounding space are shown in the lower panel.

the cross-sectional diameter of the pore opening region is not constant with time. This is the first discovery of a time-dependent pore opening change, unlike OS and CS for other channels. We have modeled this pore structure in Figure 1.42. For details readers should consider going through Ashrafuzzaman et al. (2011; 2012b).

## 1.9  LIPID-GATED ION CHANNELS

Membrane signaling lipids exert lipid-specific regulation of membrane integral ion channels (addressed earlier in this chapter). Additionally, membrane lipids are recently found as bona fide agonists of LGICs. These freely diffusing signals are found to reside in the plasma membrane, occasionally bind to the transmembrane domain of proteins, and cause a membrane conformational

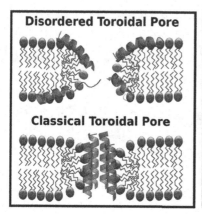 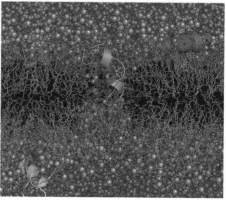

**FIGURE 1.40**  (Left) Toroidal pore induced by mellittin. A cartoon image comparing the disordered toroidal pore state to the traditional view. A striking characteristic of the disordered toroidal pore is the lack of a well-defined peptide orientation as opposed to ordered, transmembrane helices in the classic toroidal pore model. (Right) A snapshot of the disordered toroidal pore from our simulations (simulation 21–200 ns). For details see Sengupta et al. (2008).

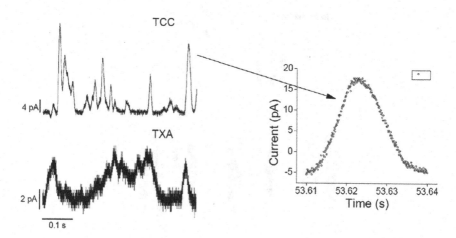

**FIGURE 1.41**  The upper panel shows triangular conductance events induced by CDs thiocolchicoside (TCC) and taxol (TXL), both at 90 μM; pH=5.7, V=100 mV. Both traces were filtered at 20 kHz but the lower one shows higher noise due to its presentation (current axis) at an amplified scale. In a high-resolution plot (shown in the right side of the arrow) of a single event only showing individual points (in Origin 8.5 plot), we observe all points (open circle) with increasing and decreasing, respectively, corresponding values of conductance at both left and right lateral sides of the chemotherapy drug-induced triangular conductance events. Details in Ashrafuzzaman et al. (2011, 2012).

change leading to allosteric gating of an ion channel. These are known as lipid-gated ion channels (LpGICs) (Hansen, 2015).

Synthetic lipid membranes are found to display channel-like ion conduction events without the presence or requirements of proteins. These constitute another class of synthetic lipid-constructed LpGICs (Blicher and Heimburg, 2013).

A lack of binding constants for lipids and ion channels challenged the thought on lipids as ligands. Biological systems employ a catalog of diverse signaling lipids that are ultimately controlled by lipid enzymes and raft localization. Recent investigations have developed various models of lipid signaling to ion channels, suggesting that the formation of anionic lipids caused a change

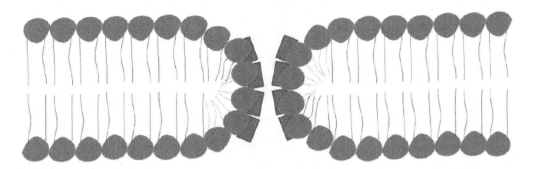

**FIGURE 1.42** CD-induced toroidal pore model. I proposed this model considering that the bilayer thickness vanishes at the pore due to the energetics caused by interactions of CD molecules (red blocks) with lipids. With this toroidal-type lipid-lined structure, they have a chance to spontaneously increase or decrease, which is reflected in the current event, as seen in EP records presented in earlier Figure 1.41. (For color figure see eBook figure.)

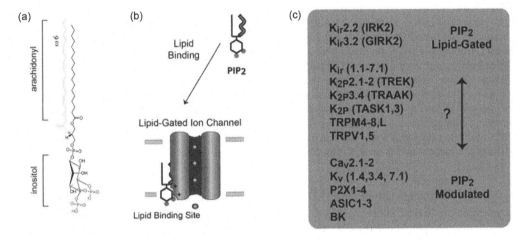

**FIGURE 1.43** $PIP_2$ lipid regulation of ion channels. (a) The chemical structure of plasma membrane $PIP_2$ is shown with an arachidonyl acyl chain (green) and inositol phosphates at the 4′ and 5′ position (red). (b) A cartoon representation of a $PIP_2$ lipid-gated ion channel. $PIP_2$ is shown bound to a lipid-binding site in the transmembrane domain of an ion channel. (c) List of ion channels with lipid-gating properties. $K_{ir}2.2$ and 3.2 are the most clearly "lipid-gated." A second group appears to be dual regulated or "$PIP_2$ modulated." $PIP_2$ modulates channel gating, but gating also requires either voltage or a second ligand. A third group of channels behave similar to $K_{ir}$ but await definitive proof of lipid gating versus $PIP_2$ modulation (?). The list of channels is exemplary and not comprehensive.

in the plasma membrane surface charge. Two decades ago, Hilgemann and colleagues showed that a signaling lipid could directly activate an ion channel (Huang et al., 1998). They found that phosphatidylinositol 4,5-bisphosphate ($PIP_2$), a minor constituent of the plasma membrane, was required and sufficient for the activation of a potassium channel. Recent protein structure and electrophysiological data on understanding lipid regulation have defined inward rectifying potassium channels ($K_{ir}$) as a new class of $PIP_2$ lipid-gated ion channels, see Figure 1.43 (Hansen, 2015).

An X-ray crystal structure complex of $K_{ir}2.2$ with $PIP_2$ was revealed in 2011, where a $PIP_2$ binding site in the transmembrane domain of the channel was detected, see Figure 1.44 (Hansen, 2015). This also supports the role of $PIP_2$ predicted in the construction of LpGICs.

The ligand-like characteristics of $PIP_2$ binding to the entire family of inward rectifiers warrant these channels to be classified as ligand-gated. The unique properties of lipids giving rise to a lipid subclass are suggested here as these events "lipid-gated," so are lipid-gated ion channels.

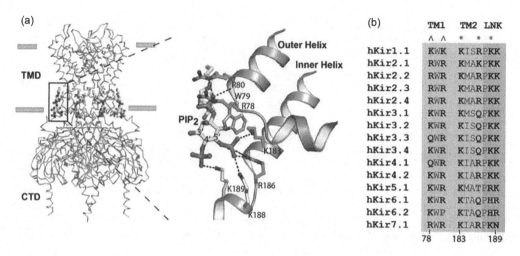

**FIGURE 1.44** Conserved $PIP_2$ binding site in $K_{ir}2.2$. $PIP_2$ binds the transmembrane domain (TMD) of $K_{ir}$ and causes a conformational change that allosterically gates the channel. (a) The $PIP_2$ binding site is specific for inositol 5′ phosphate. (b) A sequence alignment of all $K_{ir}$ family members reveals a highly structured $PIP_2$ binding site comprising basic residues. Amino acid residues that directly contact $PIP_2$ are shown in bold type. Only two residues (brown type) at the conserved site lack a positive charge. Residues originating from the TMD and a linker (LNK) are shaded green and gray, respectively. ^ indicates residues that strongly coordinate the lipid backbone phosphate, and * indicates the residues that strongly (red) and weakly (gray) bind the $PIP_2$ 5′ phosphate. $PIP_2$ atoms are colored yellow for carbon, orange for phosphate, and red for oxygen. Amino acid side chains with carbons colored green are located on transmembrane outer helix 1 (TM1) or inner helix 2 (TM2). Lysines colored gray are located on the start of a linker helix (LNK) or "tether helix" connecting the TMD and the cytoplasmic domain (CTD). Residue numbering is according to $K_{ir}2.2$. (For color figure see eBook figure.)

Almost a decade ago, EP recordings of channel traces and current histograms on lipid membranes were made and evidence of pore formation was found (Blicher and Heimburg, 2013). The voltage dependence of these LpGICs is comparable to that of protein channels. Lifetime distributions of OS and CS events indicate interesting features that the channel open distribution does not follow exponential statistics, but rather power-law behavior for the long open duration. Figure 1.45 presents the data showing evidence of the formation of LpGICs in synthetic lipid membranes. One of the most important EP recorded characteristics is the voltage dependence of the channel mechanism. The free energy of the open pore of these LpGICs has been detected to be a quadratic dependence on voltage. This quadratic voltage dependence is characteristic of the charging of a capacitor with electrical breakthroughs at critical voltages (Heimburg, 2010; Blicher and Heimburg, 2013).

In a latest study, lipid-gated monovalent ion fluxes were found to regulate endocytic traffic and support immune surveillance (Freeman et al., 2019). Here, osmotically driven increases in the surface-to-volume ratio of endomembranes were found to promote traffic between compartments and to ensure tissue homeostasis. A high-resolution video imaging was used to analyze the fate of macropinosomes formed by macrophages. $Na^+$ internalized, exited endocytic vacuoles via two pore channels, accompanied by parallel efflux of $Cl^-$, and osmotically coupled water.

When $Na^+$ efflux from the endocytic pathway was prevented *in vivo*, the ability of interstitial resident tissue macrophages (RTM) to survey their environment was impaired. Blocking β1 and β2 integrins similarly inhibited surveillance. When microlesions are made by targeted laser ablation to adjacent fibroblasts, RTM normally contains the damage, preventing the recruitment and activation of neutrophils. When PIKfyve or TPCs were inhibited, the cells failed to resorb vacuoles and were unable to respond to the damage. Consequently, neutrophil swarming ensued. Thus, vacuole resolution, mediated by the lipid-gated $Na^+$ efflux, underpins membrane traffic necessary to maintain

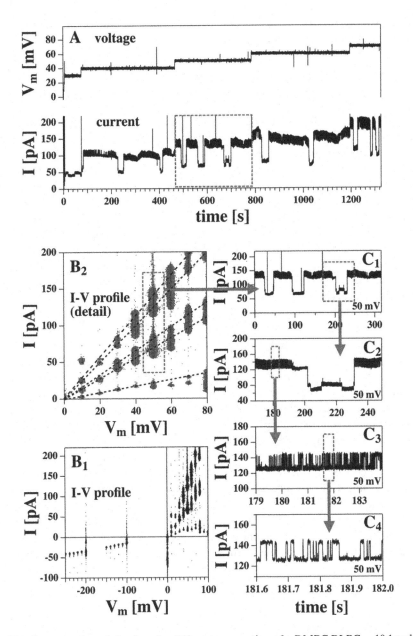

**FIGURE 1.45**  Current voltage behavior of a different preparation of a DMPC:DLPC = 10:1 mol:mol membrane at 30°C showing many channel events. A. Section of the raw voltage and current traces (20 min out of 1 h recording). The section in the red box is amplified in panel $C_1$. $B_1$: I–V profile of the complete data set. At positive voltages one can recognize ~5 individual current levels. $B_2$: Amplified section of panel $B_1$. The black dashed lines are guides to the eye to mark the five current levels. The red box marks the four current levels that can be seen in panel $C_1$. $C_1$: Current I at $V_m$ = 50 mV over 5 min. Panels $C_2$–$C_4$: Consecutive enlargement of small sections of the profile in panel $C_1$. In panel $C_4$, two individual currents levels can easily be recognized. Membrane constructing synthetic lipids include 1,2-dimyristoyl-sn-glycero-3-phosphocholine (DMPC) and 1, 2-dilaureoyl-sn-glycero-3-phosphocholine (DLPC). For details, see Blicher and Heimburg (2013). (For color figure see eBook figure.)

cellular responsiveness. Unfortunately, due to copyright issues, we cannot represent figures demonstrating any of these results (Freeman et al., 2019).

## 1.10  FUTURE PERSPECTIVES LED BY CURRENT UNDERSTANDING OF ION CHANNELS

Ion channels have been at the forefront of research for many decades. Using various biological, biochemical, biophysical, physiological, and engineering techniques, we have understood many aspects related to channel structures and functions. We have also understood channel conditions emerging as a result of genetic mutations occurring in channel subunits. We find the channels to be versatile molecular tools helping us in understanding channel-based cellular disease conditions, as well as target them to discover chemicals with channel inhibiting or regulating potency, as well as discover drugs.

This chapter will hopefully be a background to help understand and process huge information accumulated so far using mostly classical mechanical treatment of ion channels in hazy biological systems. The information may be considered enough or helpful to understand overall biophysical states and physiological roles of channels in various membranes of different types of cells. Numerous underlying molecular mechanisms active in channel construction and function are understood, but some crucial ones are not yet understood. Using the classical treatment of these complex molecular structures, we have addressed the hidden energetics pinpointing vital physical (mostly Newtonian and Statistical Mechanical) principles in subsequent chapters. Still all of the tuned, often hidden mechanisms cannot be addressed. Structural bioinformatics and modeling and Quantum mechanics are two fields which have emerged quite recently as applicable in addressing crucial biological processes. I plan to go beyond using classical mechanical principles to address ion channels and pinpoint some of their structural and functional aspects to be explained using bioinformatics and quantum mechanical principles. Additionally, while addressing the naturally occurring channels in biological systems, I also aim to address the technical aspects in engineering the channel structures using artificial means. Channel building blocks with novel physicochemical properties can be synthesized using chemical techniques, which may mimic their biological peers. These novel molecules may be found considerably important in especially drug discovery science dealing with channelopathies. We shall be able to pinpoint to a lot of unresolved nanoscale functioning events and processes active in the field of ion channel biophysics. The readers will find the subsequent chapters highly helpful to not only enrich their understanding of ion channels but also occasionally find opportunities to go beyond and adopt quality thought-provoking and brain-stimulating questions that will emerge in the future.

## REFERENCES

Aldrich, R., & Stevens, C. (1987). Voltage-dependent gating of single sodium channels from mammalian neuroblastoma cells. *The Journal of Neuroscience, 7*(2), 418–431. doi:10.1523/jneurosci.07-02-00418.1987

Armstrong, N., Sun, Y., Chen, G., & Gouaux, E. (1998). Structure of a glutamate-receptor ligand-binding core in complex with kainate. *Nature, 395*(6705), 913–917. doi:10.1038/27692

Ashrafuzzaman, M., Andersen, O., & Mcelhaney, R. (2008). The antimicrobial peptide gramicidin S permeabilizes phospholipid bilayer membranes without forming discrete ion channels. *Biochimica Et Biophysica Acta (BBA) - Biomembranes, 1778*(12), 2814–2822. doi:10.1016/j.bbamem.2008.08.017

Ashrafuzzaman, M., Duszyk, M., & Tuszynski, J. A. (2011). Chemotherapy drugs thiocolchicoside and taxol permeabilize lipid bilayer membranes by forming ion pores. *Journal of Physics: Conference Series, 329*, 012029. doi:10.1088/1742-6596/329/1/012029

Ashrafuzzaman, M., Lampson, M. A., Greathouse, D. V., Koeppe, R. E., & Andersen, O. S. (2006). Manipulating lipid bilayer material properties using biologically active amphipathic molecules. *Journal of Physics: Condensed Matter, 18*(28). doi:10.1088/0953-8984/18/28/s08

Ashrafuzzaman, M., Tseng, C., Duszyk, M., & Tuszynski, J. A. (2012). Chemotherapy drugs form ion pores in membranes due to physical interactions with lipids. *Chemical Biology & Drug Design, 80*(6), 992–1002. doi:10.1111/cbdd.12060

Ashrafuzzaman, M., & Tuszynski, J. A. (2012a). Regulation of channel function due to coupling with a lipid bilayer. *Journal of Computational and Theoretical Nanoscience, 9*(4), 564–570. doi:10.1166/jctn.2012.2062

Ashrafuzzaman, M., & Tuszynski, J. A. (2012b). *Membrane Biophysics.* Berlin, Heidelberg: Springer-Verlag. doi: 10.1007/978-3-642-16105-6

Assmann, M., Kuhn, A., Dürrnagel, S., Holstein, T. W., & Gründer, S. (2014). The comprehensive analysis of DEG/ENaC subunits in Hydra reveals a large variety of peptide-gated channels, potentially involved in neuromuscular transmission. *BMC Biology, 12*(1), 84. doi:10.1186/s12915-014-0084-2

Barreiro, G., Guimarães, C. R., & Alencastro, R. B. (2002). A molecular dynamics study of an L-type calcium channel model. *Protein Engineering, Design and Selection, 15*(2), 109–122. doi:10.1093/protein/15.2.109

Bass, R. B. (2002). Crystal structure of Escherichia coli MscS, a voltage-modulated and mechanosensitive channel. *Science, 298*(5598), 1582–1587. doi:10.1126/science.1077945

Bentele, K., & Falcke, M. (2007). Quasi-steady approximation for ion channel currents. *Biophysical Journal, 93*(8), 2597–2608. doi:10.1529/biophysj.107.104299

Bernstein, J. (1902). Untersuchungen zur Thermodynamik der bioelektrischen Ströme. *Pflüger Archiv Für Die Gesammte Physiologie Des Menschen Und Der Thiere, 92*(10–12), 521–562. doi:10.1007/bf01790181

Beurg, M., Nam, J., Crawford, A., & Fettiplace, R. (2008). The actions of calcium on hair bundle mechanics in mammalian cochlear hair cells. *Biophysical Journal, 94*(7), 2639–2653. doi:10.1529/biophysj.107.123257

Bezanilla, F. (2000). The voltage sensor in voltage-dependent ion channels. *Physiological Reviews, 80*(2), 555–592. doi:10.1152/physrev.2000.80.2.555

Bezanilla, F., & Armstrong, C. M. (1977). Inactivation of the sodium channel. I. Sodium current experiments. *The Journal of General Physiology, 70*(5), 549–566. doi:10.1085/jgp.70.5.549

Blicher, A., & Heimburg, T. (2013). Voltage-gated lipid ion channels. *PLoS ONE, 8*(6). doi:10.1371/journal.pone.0065707

Catterall, W. A. (2011). Voltage-gated calcium channels. *Cold Spring Harbor Perspectives in Biology, 3*(8). doi:10.1101/cshperspect.a003947

Cesare, P., Moriondo, A., Vellani, V., & Mcnaughton, P. A. (1999). Ion channels gated by heat. *Proceedings of the National Academy of Sciences, 96*(14), 7658–7663. doi:10.1073/pnas.96.14.7658

Chang, G., Spencer, R. H., Lee, A. T., Barclay, M. T., & Rees, D. C. (1998). Structure of the MscL homolog from mycobacterium tuberculosis: A gated mechanosensitive ion channel. *Science, 282*(5397), 2220–2226. doi:10.1126/science.282.5397.2220

Changeux, J., Kasai, M., & Lee, C. (1970). Use of a snake venom toxin to characterize the cholinergic receptor protein. *Proceedings of the National Academy of Sciences, 67*(3), 1241–1247. doi:10.1073/pnas.67.3.1241

Chen, Z., & Olsen, R. W. (2007). GABA A receptor associated proteins: A key factor regulating GABA A receptor function. *Journal of Neurochemistry, 100*(2), 279–294. doi:10.1111/j.1471-4159.2006.04206.x

Cheng, C., Kamiya, M., Takemoto, M., Ishitani, R., Nureki, O., Yoshida, N., & Hayashi, S. (2018). An atomistic model of a precursor state of light-induced channel opening of channelrhodopsin. *Biophysical Journal, 115*(7), 1281–1291. doi:10.1016/j.bpj.2018.08.024

Chowdhury, S., Jarecki, B., & Chanda, B. (2014). A molecular framework for temperature-dependent gating of ion channels. *Cell, 158*(5), 1148–1158. doi:10.1016/j.cell.2014.07.026

Comitani, F., Cohen, N., Ashby, J., Botten, D., Lummis, S. C., & Molteni, C. (2014). Insights into the binding of GABA to the insect RDL receptor from atomistic simulations: A comparison of models. *Journal of Computer-Aided Molecular Design, 28*(1), 35–48. doi:10.1007/s10822-013-9704-0

Cottrell, G. A. (1997). The first peptide-gated ion channel. *Journal of Experimental Biology, 200*, 2377–2386.

Crnjar, A., Comitani, F., Melis, C., & Molteni, C. (2019). Mutagenesis computer experiments in pentameric ligand-gated ion channels: The role of simulation tools with different resolution. *Interface Focus, 9*(3), 20180067. doi:10.1098/rsfs.2018.0067

Curtis, B. M., & Catterall, W. A. (1985). Phosphorylation of the calcium antagonist receptor of the voltage-sensitive calcium channel by cAMP-dependent protein kinase. *Proceedings of the National Academy of Sciences, 82*(8), 2528–2532. doi:10.1073/pnas.82.8.2528

Del Castillo, J., & Katz, B. (1957). Interaction at end-plate receptors between different choline derivatives. *Proceedings of the Royal Society of London. Series B - Biological Sciences, 146*(924), 369–381. doi:10.1098/rspb.1957.0018

Dolphin, A. C. (2018). Voltage-gated calcium channels: Their discovery, function and importance as drug targets. *Brain and Neuroscience Advances, 2.* doi:10.1177/2398212818794805

Doyle, D. A., Cabral, J. M., Pfuetzner, R. A., Kuo, A., Gulbis, J. M., Cohen, S. L., . . . MacKinnon, R. (1998). The structure of the potassium channel: Molecular basis of K+ conduction and selectivity. *Science, 280*(5360), 69–77. doi:10.1126/science.280.5360.69

Elsässer, T., & Bakker, H. J. (2002). *Ultrafast Hydrogen Bonding Dynamics and Proton Transfer Processes in the Condensed Phase.* Dordrecht: Springer. doi:10.1007/978-94-017-0059-7

Feng, T., Kalyaanamoorthy, S., Ganesan, A., & Barakat, K. (2019). Atomistic modeling and molecular dynamics analysis of human CaV1.2 channel using external electric field and ion pulling simulations. *Biochimica Et Biophysica Acta (BBA) - General Subjects, 1863*(6), 1116–1126. doi:10.1016/j.bbagen.2019.04.006

Fenwick, E. M., Marty, A., & Neher, E. (1982). Sodium and calcium channels in bovine chromaffin cells. *The Journal of Physiology, 331*(1), 599–635. doi:10.1113/jphysiol.1982.sp014394

Freeman, S. A., Uderhardt, S., Saric, A., Collins, R. F., Buckley, C. M., Mylvaganam, S., . . . Grinstein, S. (2019). Lipid-gated monovalent ion fluxes regulate endocytic traffic and support immune surveillance. *Science, 367*(6475), 301–305. doi:10.1126/science.aaw9544

Goldman, D. E. (1943). Potential, impedance, and rectification in membranes. *The Journal of General Physiology, 27*(1), 37–60. doi:10.1085/jgp.27.1.37

Govorunova, E. G., Spudich, E. N., Lane, C. E., Sineshchekov, O. A., & Spudich, J. L. (2011). New channelrhodopsin with a red-shifted spectrum and rapid kinetics from Mesostigma viride. *MBio, 2*(3). doi:10.1128/mbio.00115-11

Hansen, S. B. (2015). Lipid agonism: The PIP2 paradigm of ligand-gated ion channels. *Biochimica Et Biophysica Acta (BBA) - Molecular and Cell Biology of Lipids, 1851*(5), 620–628. doi:10.1016/j.bbalip.2015.01.011

Haswell, E., Phillips, R., & Rees, D. (2011). Mechanosensitive channels: What can they do and how do they do it? *Structure, 19*(10), 1356–1369. doi:10.1016/j.str.2011.09.005

Heimburg, T. (2010). Lipid ion channels. *Biophysical Chemistry, 150*(1–3), 2–22. doi:10.1016/j.bpc.2010.02.018

Hell, J. W., Yokoyama, C. T., Wong, S. T., Warner, C., Snutch, T. P., & Catterall, W. A. (1993). Differential phosphorylation of two size forms of the neuronal class C L-type calcium channel $\alpha 1$ subunit. *Journal of Biological Chemistry, 268*, 19451–19457.

Hilf, R. J., & Dutzler, R. (2008). X-ray structure of a prokaryotic pentameric ligand-gated ion channel. *Nature, 452*(7185), 375–379. doi:10.1038/nature06717

Hille, B. (2001). *Ion channels of Excitable Membranes.* Sunderland, MA: Sinauer.

Hodgkin, A. L., & Huxley, A. F. (1945). Resting and action potentials in single nerve fibres. *The Journal of Physiology, 104*(2), 176–195. doi:10.1113/jphysiol.1945.sp004114

Hodgkin, A. L., & Keynes, R. D. (1955). The potassium permeability of a giant nerve fibre. *The Journal of Physiology, 128*(1), 61–88. doi:10.1113/jphysiol.1955.sp005291

Horn, R., & Vandenberg, C. A. (1984). Statistical properties of single sodium channels. *The Journal of General Physiology, 84*(4), 505–534. doi:10.1085/jgp.84.4.505

Huang, C., Feng, S., & Hilgemann, D. W. (1998). Direct activation of inward rectifier potassium channels by PIP2 and its stabilization by $G\beta\gamma$. *Nature, 391*(6669), 803–806. doi:10.1038/35882

Hudspeth, A. J., & Corey, D. P. (1977). Sensitivity, polarity, and conductance change in the response of vertebrate hair cells to controlled mechanical stimuli. *Proceedings of the National Academy of Sciences, 74*(6), 2407–2411. doi:10.1073/pnas.74.6.2407

Jongh, K. S., Merrick, D. K., & Catterall, W. A. (1989). Subunits of purified calcium channels: A 212-kDa form of alpha 1 and partial amino acid sequence of a phosphorylation site of an independent beta subunit. *Proceedings of the National Academy of Sciences, 86*(21), 8585–8589. doi:10.1073/pnas.86.21.8585

Kasimova, M. A., Tarek, M., Shaytan, A. K., Shaitan, K. V., & Delemotte, L. (2014). Voltage-gated ion channel modulation by lipids: Insights from molecular dynamics simulations. *Biochimica Et Biophysica Acta (BBA) - Biomembranes, 1838*(5), 1322–1331. doi:10.1016/j.bbamem.2014.01.024

Kato, H. E., Zhang, F., Yizhar, O., Ramakrishnan, C., Nishizawa, T., Hirata, K., . . . Nureki, O. (2012). Crystal structure of the channelrhodopsin light-gated cation channel. *Nature, 482*(7385), 369–374. doi:10.1038/nature10870

Kellenberger, S., & Schild, L. (2002). Epithelial sodium channel/degenerin family of ion channels: A variety of functions for a shared structure. *Physiological Reviews, 82*(3), 735–767. doi:10.1152/physrev.00007.2002

Kim, D. M., & Nimigean, C. M. (2016). Voltage-gated potassium channels: A structural examination of selectivity and gating. *Cold Spring Harbor Perspectives in Biology, 8*(5). doi:10.1101/cshperspect.a029231

Kung, C., Martinac, B., & Sukharev, S. (2010). Mechanosensitive channels in microbes. *Annual Review of Microbiology, 64*(1), 313–329. doi:10.1146/annurev.micro.112408.134106

Lemoine, D., Jiang, R., Taly, A., Chataigneau, T., Specht, A., & Grutter, T. (2012). Ligand-gated ion channels: New insights into neurological disorders and ligand recognition. *Chemical Reviews, 112*(12), 6285–6318. doi:10.1021/cr3000829

Lenaeus, M. J., El-Din, T. M., Ing, C., Ramanadane, K., Pomès, R., Zheng, N., & Catterall, W. A. (2017). Structures of closed and open states of a voltage-gated sodium channel. *Proceedings of the National Academy of Sciences, 114*(15). doi:10.1073/pnas.1700761114

Leontiadou, H., Mark, A. E., & Marrink, S. J. (2006). Antimicrobial peptides in action. *Journal of the American Chemical Society, 128*(37), 12156–12161. doi:10.1021/ja062927q

Levina, N. (1999). Protection of Escherichia coli cells against extreme turgor by activation of MscS and MscL mechanosensitive channels: Identification of genes required for MscS activity. *The EMBO Journal, 18* (7), 1730–1737. doi:10.1093/emboj/18.7.1730

Li, S., Wong, A. H., & Liu, F. (2014). Ligand-gated ion channel interacting proteins and their role in neuroprotection. *Frontiers in Cellular Neuroscience, 8.* doi:10.3389/fncel.2014.00125

Li, Y. (2019). Almost periodic functions on the quantum time scale and applications. *Discrete Dynamics in Nature and Society, 2019*, 1–16. doi:10.1155/2019/4529159

Lieu, R., & Hillman, L. W. (2003). The phase coherence of light from extragalactic sources: Direct evidence against first-order planck-scale fluctuations in time and space. *The Astrophysical Journal, 585*(2). doi:10.1086/374350

Liu, Z., Gandhi, C., & Rees, D. (2009). Structure of a tetrameric MscL in an expanded intermediate state. doi:10.2210/pdb3hzq/pdb

Long, S. B., Tao, X., Campbell, E. B., MacKinnon, R. (2007). Atomic structure of a voltage-dependent $K^+$ channel in a lipid membrane-like environment. *Nature 450*, 376–382.

Lórenz-Fonfría, V. A., & Heberle, J. (2014). Channelrhodopsin unchained: Structure and mechanism of a light-gated cation channel. *Biochimica Et Biophysica Acta (BBA) - Bioenergetics, 1837*(5), 626–642. doi:10.1016/j.bbabio.2013.10.014

Martinac, B., Buechner, M., Delcour, A. H., Adler, J., & Kung, C. (1987). Pressure-sensitive ion channel in Escherichia coli. *Proceedings of the National Academy of Sciences, 84*(8), 2297–2301. doi:10.1073/pnas.84.8.2297

Mergler, S., Garreis, F., Sahlmüller, M., Reinach, P. S., Paulsen, F., & Pleyer, U. (2011). Thermosensitive transient receptor potential channels in human corneal epithelial cells. *Journal of Cellular Physiology, 226*(7), 1828–1842. doi:10.1002/jcp.22514

Meves, H. (1979). Inactivation of the sodium permeability in squid giant nerve fibres. *Progress in Biophysics and Molecular Biology, 33*, 207–230. doi:10.1016/0079-6107(79)90029-4

Nagel, G., Ollig, D., Fuhrmann, M., Kateriya, S., Musti, A.M., Bamberg, E., & Hegemann, P. (2002). Channelrhodopsin-1: A light-gated proton channel in green algae. *Science, 296*(5577), 2395–2398. doi:10.1126/science.1072068

Nagel, G., Szellas, T., Huhn, W., Kateriya, S., Adeishvili, N., Berthold, P., . . . Bamberg, E. (2003). Channelrhodopsin-2, a directly light-gated cation-selective membrane channel. *Proceedings of the National Academy of Sciences, 100*(24), 13940–13945. doi:10.1073/pnas.1936192100

Neher, E., & Sakmann, B. (1976). Single-channel currents recorded from membrane of denervated frog muscle fibres. *Nature, 260*(5554), 799–802. doi:10.1038/260799a0

Nicolson, T., Rüsch, A., Friedrich, R. W., Granato, M., Ruppersberg, J. P., & Nüsslein-Volhard, C. (1998). Genetic analysis of vertebrate sensory hair cell mechanosensation: The zebrafish circler mutants. *Neuron, 20*(2), 271–283. doi:10.1016/s0896-6273(00)80455-9

Opatowsky, Y., Chen, C., Campbell, K. P., & Hirsch, J. A. (2004). Structural analysis of the voltage-dependent calcium channel β subunit functional core and its complex with the α1 interaction domain. *Neuron, 42* (3), 387–399. doi:10.1016/s0896-6273(04)00250-8

Payandeh, J., Scheuer, T., Zheng, N., & Catterall, W. A. (2011). The crystal structure of a voltage-gated sodium channel. *Nature, 475*(7356), 353–358. doi:10.1038/nature10238

Perozo, E., Kloda, A., Cortes, D. M., & Martinac, B. (2002). Physical principles underlying the transduction of bilayer deformation forces during mechanosensitive channel gating. *Nature Structural Biology, 9*(9), 696–703. doi:10.1038/nsb827

Petegem, F. V., Clark, K. A., Chatelain, F. C., & Minor, D. L. (2004). Structure of a complex between a voltage-gated calcium channel β-subunit and an α-subunit domain. *Nature, 429*(6992), 671–675. doi:10.1038/nature02588

Pickles, J., Comis, S., & Osborne, M. (1984). Cross-links between stereocilia in the guinea pig organ of Corti, and their possible relation to sensory transduction. *Hearing Research, 15*(2), 103–112. doi:10.1016/0378-5955(84)90041-8

Qiu, X., & Müller, U. (2018). Mechanically gated ion channels in mammalian hair cells. *Frontiers in Cellular Neuroscience, 12.* doi:10.3389/fncel.2018.00100

Reuter, H. (1979). Properties of two inward membrane currents in the heart. *Annual Review of Physiology, 41*(1), 413–424. doi:10.1146/annurev.ph.41.030179.002213

Sawada, Y., Murase, M., & Sokabe, M. (2012). The gating mechanism of the bacterial mechanosensitive channel MscL revealed by molecular dynamics simulations. *Channels, 6*(4), 317–331. doi:10.4161/chan.21895

Sengupta, D., Leontiadou, H., Mark, A. E., & Marrink, S. (2008). Toroidal pores formed by antimicrobial peptides show significant disorder. *Biochimica Et Biophysica Acta (BBA) - Biomembranes, 1778*(10), 2308–2317. doi:10.1016/j.bbamem.2008.06.007

Spudich, J. L., Yang, C., Jung, K., & Spudich, E. N. (2000). Retinylidene proteins: Structures and functions from archaea to humans. *Annual Review of Cell and Developmental Biology, 16*(1), 365–392. doi:10.1146/annurev.cellbio.16.1.365

Sula, A., Booker, J., Ng, L. C., Naylor, C. E., Decaen, P. G., & Wallace, B. A. (2017). The complete structure of an activated open sodium channel. *Nature Communications, 8*(1). doi:10.1038/ncomms14205

Sukharev, S., Durell, S. R., & Guy, H. R. (2001). Structural models of the MscL gating mechanism. *Biophysical Journal, 81*(2), 917–936. doi:10.1016/s0006-3495(01)75751-7

Sukharev, S., Martinac, B., Arshavsky, V., & Kung, C. (1993). Two types of mechanosensitive channels in the Escherichia coli cell envelope: Solubilization and functional reconstitution. *Biophysical Journal, 65*(1), 177–183. doi:10.1016/s0006-3495(93)81044-0

Sukharev, S. I., Blount, P., Martinac, B., Blattner, F. R., & Kung, C. (1994). A large-conductance mechanosensitive channel in E. coli encoded by mscL alone. *Nature, 368*(6468), 265–268. doi:10.1038/368265a0

Sukharev, S. I., Sigurdson, W. J., Kung, C., & Sachs, F. (1999). Energetic and spatial parameters for gating of the bacterial large conductance mechanosensitive channel, MscL. *Journal of General Physiology, 113*(4), 525–540. doi:10.1085/jgp.113.4.525

Takemoto, M., Kato, H. E., Koyama, M., Ito, J., Kamiya, M., Hayashi, S., . . . Nureki, O. (2015). Molecular dynamics of channelrhodopsin at the early stages of channel opening. *Plos One, 10*(6). doi:10.1371/journal.pone.0131094

Tang, L. (2019). Structure basis for Diltiazem block of a voltage-gated calcium channel. doi:10.2210/pdb6keb/pdb

Tang, L., El-Din, T. M., Payandeh, J., Martinez, G. Q., Heard, T. M., Scheuer, T., . . . Catterall, W. A. (2013). Structural basis for $Ca^{2+}$ selectivity of a voltage-gated calcium channel. *Nature, 505*(7481), 56–61. doi:10.1038/nature12775

Tang, L., El-Din, T. M., Swanson, T. M., Pryde, D. C., Scheuer, T., Zheng, N., & Catterall, W. A. (2016). Structural basis for inhibition of a voltage-gated $Ca^{2+}$ channel by $Ca^{2+}$ antagonist drugs. *Nature, 537*(7618), 117–121. doi:10.1038/nature19102

Tierney, M. L. (2011). Insights into the biophysical properties of GABAA ion channels: Modulation of ion permeation by drugs and protein interactions. *Biochimica Et Biophysica Acta (BBA) - Biomembranes, 1808*(3), 667–673. doi:10.1016/j.bbamem.2010.11.022

Treede, R., Meyer, R. A., Raja, S. N., & Campbell, J. N. (1992). Peripheral and central mechanisms of cutaneous hyperalgesia. *Progress in Neurobiology, 38*(4), 397–421. doi:10.1016/0301-0082(92)90027-c

Volkov, O., Kovalev, K., Polovinkin, V., Borshchevskiy, V., Bamann, C., Astashkin, R., . . . Gordeliy, V. (2017). Structural insights into ion conduction by channelrhodopsin 2. *Science, 358*(6366). doi:10.1126/science.aan8862

Wang, W., Black, S. S., Edwards, M. D., Miller, S., Morrison, E. L., Bartlett, W., . . . Booth, I. R. (2008). The structure of an open form of an E. coli mechanosensitive channel at 3.45 A resolution. *Science, 321*(5893), 1179–1183. doi:10.1126/science.1159262

Wu, B. M., Wang, X. H., Zhao, B., Bian, E. B., Yan, H., Cheng, H., Lv, X. W., Xiong, Z. G., & Li, J. (2013). Electrophysiology properties of voltage-gated potassium channels in rat peritoneal macrophages. *International Journal of Clinical and Experimental Medicine, 6*(3), 166–173.

Wu, J., Yan, Z., Li, Z., Zhou, Q., & Yan, N. (2016). Structure of the mammalian voltage-gated calcium channel Cav1.1 complex at near atomic resolution. doi:10.2210/pdb5gjv/pdb

Wu, Z., Grillet, N., Zhao, B., Cunningham, C., Harkins-Perry, S., Coste, B., . . . Müller, U. (2017). Mechanosensory hair cells express two molecularly distinct mechanotransduction channels. *Nature Neuroscience, 20*(1), 24–33. doi:10.1038/nn.4449

Yu, F. H., & Catterall, W. A. (2003). *Genome Biology, 4*(3), 207. doi:10.1186/gb-2003-4-3-207

Yuan, P. (2019). Structural biology of thermoTRPV channels. *Cell Calcium, 84*, 102106. doi:10.1016/j.ceca.2019.102106

Zabelskii, D., Alekseev, A., Kovalev, K., Oliviera, A., Balandin, T., Soloviov, D., . . . Gordeliy, V. (2020). Viral channelrhodopsins: Calcium-dependent $Na^+/K^+$ selective light-gated channels. doi:10.1101/2020.02.14.949966

# 2 Mitochondrial Membrane Channels – Physical Structures and Gating Mechanisms

Mitochondria are generally known as energy production centers of biological cells. These organelles generate most of the chemical energy, which is stored in small molecules as adenosine triphosphates (ATPs). This energy is used for biochemical reactions in cells. In addition to producing ATPs for energy supply, mitochondria participate in pinpointed biomolecular processes. These so-called "powerhouses" of cells are surrounded by double membranes, both of which maintain independent structural and functional integrity. They host ion channels in both mitochondrial outer membrane (MOM) and mitochondrial inner membrane (MIM) that help in maintaining a multitude of processes determining the biological life (Mannella and Kinnally, 1997; O'Rourke, 2007; Ashrafuzzaman, 2018). The normal functioning of channels and transporters helps mitochondria to maintain the physiological balance of electrical conditions including potentials, ion concentrations, and protein densities across the mitochondrial membranes. Physiological actions of mitochondrial membrane channels are often found to play deterministic roles in many crucial cellular functions including cell signaling pathways. Mutations in channel proteins are often found to disrupt (temporarily or, occasionally, permanently) the channel function. Consequently, the associated cellular signaling that is linked directly or indirectly to mitochondrial transport of specific and nonspecific materials and information may get partially compromised. The statics and dynamics of mitochondrial channels including their structural integrity (and/or mutation-based disintegration) and transport phenomena will be addressed in detail.

## 2.1 INTRODUCTION TO MITOCHONDRIAL CHANNELS

Mitochondrial membrane-hosted ion channels are listed in Table 2.1 (reproduced from O'Rourke, 2007). The table contents have been modestly modified due to specific needs. However, ongoing progresses are not done yet. Continued investigations are making new discoveries on various aspects of channels in different systems, for example, the recently discovered "brain mitochondrial sodium-sensitive potassium channel" (Fahanik-Babaei et al., 2020).

Transport, mostly from mitochondrial inner regions to cytosol through mitochondrial membrane ion channels, helps maintain physiological conditions related to cell health. These channels are often referred to as the gatekeepers of life and death of cells (O'Rourke et al., 2005). The channel malfunction is commonly associated with aging, which may lead to the rise in disease conditions (Strickland et al., 2019).

Specific channels in either mitochondrial outer or inner membrane are known to perform specific types of cellular activities. Table 2.1 has summarized some of these aspects briefly. As an example, I wish to briefly explain one such case, the role of $Ca^{2+}$-activated potassium ($K^+$) channels ($K_{Ca}$). $K_{Ca}$ channel subtypes (e.g., BK, IK, and SK) have been detected at the MIM of several cell types. Figure 2.1 explains these features schematically (Krabbendam et al., 2018).

The mitochondrial $K_{Ca}$ channels primarily regulate the production of mitochondrial reactive oxygen species (ROS), participate in maintaining the mitochondrial membrane potential, and preserving the mitochondrial calcium homeostasis. Therefore, the channels are thought to contribute to cellular protection against oxidative stress through mitochondrial mechanisms of preconditioning.

DOI: 10.1201/9781003010654-2

**TABLE 2.1**

**Mitochondrial Ion Channels.**

| Location | Type | | Conductance (~150 mM Salt) | Modulators/Inhibitors | Putative Role(s) | Selected Refs. |
|---|---|---|---|---|---|---|
| Outer membrane | VDAC (porin) | | 0.5–4 nS | Bax/Bak/Bcl-xL/Bcl-2, TOM20, $Ca^{2+}$, pH, $\Delta V$, NADH, VDAC modulator | Metabolite transport, cytochrome c release/apoptosis, PTP complex | Colombini, (2004) |
| | TOM40 (PSC) | | 0.5–1 nS | Signal peptides | Protein transport | Grigoriev et al. (2004) |
| | MAC (BH proteins) | | 2.5 nS | Bax/Bak | Cytochrome c release/apoptosis | Dejean et al. (2006) |
| | Miscellaneous | | 10–307 pS | $\Delta V$ (for >100 pS) | – | Moran et al. (1992) |
| Inner membrane | $Ca^{2+}$ uniporter | | 6 pS | Divalents, nucleotides, RuRed, ryanodine | $Ca^{2+}$ uptake | Kirichok et al. (2004) |
| | PTP | MCC | 0.03–1.5 nS | $Ca^{2+}$, $\Delta V$, signal peptides, CsA | Protein transport | Zorow et al. (1992) |
| | | MMC | 0.3–1.3 nS | CsA, pH, $Ca^{2+}$, thiols, Bax, ANT inhibitors | Necrosis, apoptosis | Szabò and Zoratti (1992) |
| | UCP | | 75 pS | Fatty acids | Thermogenesis | Huang and Klingenberg (1996) |
| | $K_{Ca}$ | | 295 pS | $Ca^{2+}$, $\Delta V$, ChTx, IbTx | Volume regulation | Siemen et al. (1999); Xu et al. (2002) |
| | $K_{ATP}$ | | 9.7 pS | ATP, GTP, palmitoyl CoA, $Mg^{2+}$, $Ca^{2+}$ | Volume regulation, protection against apoptosis/ischemic injury | Inoue et al. (1991); Dahlem et al. (2004) |
| | $K_V 1.3$ | | 17 pS | Margatoxin | Cell death | Szabò et al. (2005) |
| | IMACs | | 45, 450 pS | ATP | Volume regulation (in yeast) | Ballarin, and Sorgato (1995) |
| | | | 15 pS(LCC)107 pS (centum pS) | $Mg^{2+}$, pH, $P_i$, thiols, DIDS, cationic amphiphiles | Volume regulation | Sorgato et al. (1987); Ježek and Borecký (1996) |

Krabbendam et al. (2018) summarized the current knowledge on mitochondrial $K_{Ca}$ channels and their role in mitochondrial function concerning cell death and survival pathways. Here, the role of these mitochondrial $K_{Ca}$ channels in pharmacological preconditioning and protective effects on ischemic insults to the brain and heart are discussed.

Mitochondrial function is regulated due to the physiological actions of various associated ion channels (O'Rourke, 2007). The major role of mitochondria is to create energy for cells which is a complex process, depending highly on the ions that drive the ATPase machinery and phosphorylate ATP. The flow of electrons through the electron transport chain causes cationic hydrogen ($H^+$) ions to be pumped from the mitochondrial matrix into the intermembrane space (IMS) (Mitchell, 1961). In addition to being involved in energy production, the movement

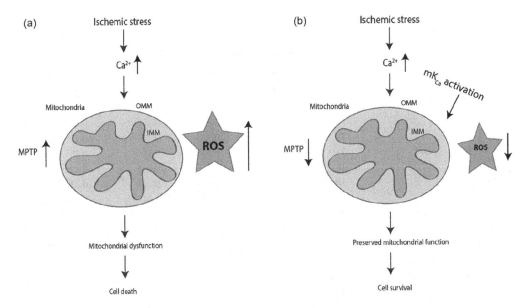

**FIGURE 2.1** Schematic overview of cell death and survival pathways. Ischemic stress insults on cell function and, in particular, on mitochondrial function are schematically shown when mitochondrial KCa channels are closed (a) or opened/activated (b). Ischemic stress facilitates elevated levels of mito [Ca²⁺], increased reactive oxygen species (ROS) production, and higher probability of mitochondrial permeability transition pore (MPTP) opening and alterations, thereby leading to mitochondrial dysfunction and cell death. However, mitoKCa activation preserves cell viability by lowering the generation of mitochondrial ROS and opening the probability of MPTP. OMM and IMM in figure stand for MOM and MIM, respectively.

of these ions across the mitochondrial membrane serves as essential phenomena to establish membrane potential, to maintain proton (H⁺) flux, and to transport other ions such as potassium (K⁺), sodium (Na⁺), and calcium (Ca²⁺) (O'Rourke, 2007). The voltage-dependent anion channel (VDAC) has been well studied to address the primary route of metabolites and ion exchanges across MOMs (Colombini, 2004).

## 2.2 MITOCHONDRIAL OUTER MEMBRANE CHANNELS

The transport process of materials and information across the MOM is poorly understood. MOM channels are structurally diverse. Their general functional modes ensuring cross-border transport and physiological roles related to normal cell health and disease conditions are versatile. We will not be able to detail all of them but certainly will touch most of the crucial aspects.

### 2.2.1 MOM CHANNELS: CLASSIFICATIONS AND GENERAL DESCRIPTION

MOM hosts a variety of channels whose transport mechanisms appear to be more diverse than originally thought (Becker and Wagner, 2018). Four protein-conducting channels are known to promote the transport of precursor proteins. VDAC is the most abundant protein in the MOM and serves as the major pathway for the metabolite/ion transport between the cytosol and IMS. VDAC transports small hydrophilic molecules. Moreover, three channels with yet unknown substrate specificity exist in the MOM.

The MOM hosts a limited number of channel-forming proteins with broad substrate specificity. β-barrel structure transmembrane proteins form a central hydrophilic pore that helps in the passage

**TABLE 2.2**

**Secondary Structure of Renatured Tom40 Estimated from CD Spectra**

| Protein | Method | α-Helix | β-Sheet | Turn | Random Coil |
|---|---|---|---|---|---|
| *N.c.* Tom40 in DM | I | 0.18 | 0.37 | 0.19 | 0.25 |
| | II | 0.186 | 0.300 | 0.203 | 0.280 |
| | III | 0.192 | 0.314 | 0.206 | 0.289 |
| | IV | 0.164 | 0.337 | 0.213 | 0.276 |
| *S.c.* Tom40 in DM | I | 0.15 | 0.39 | 0.17 | 0.28 |
| | II | 0.093 | 0.378 | 0.265 | 0.250 |
| | III | 0.083 | 0.395 | 0.263 | 0.258 |
| | IV | 0.067 | 0.390 | 0.247 | 0.298 |
| *N.c.* Tom40 in liposomes | I | 0.12 | 0.42 | 0.13 | 0.33 |
| | II | 0.049 | 0.308 | 0.185 | 0.420 |
| | III | 0.036 | 0.314 | 0.196 | 0.453 |
| | IV | 0.003 | 0.457 | 0.224 | 0.313 |
| *S.c.* Tom40 in liposomes | I | 0.16 | 0.41 | 0.15 | 0.28 |
| | II | 0.027 | 0.413 | 0.222 | 0.364 |
| | III | 0.041 | 0.325 | 0.201 | 0.434 |
| | IV | 0.013 | 0.429 | 0.227 | 0.137 |

of preproteins, hydrophilic small metabolites, and various ions. Protein Tom40 (TOM complex) forms channels for conducting and importing the majority of mitochondrial proteins (Hill et al., 1998; Ahting et al., 1999; Suzuki et al., 2004; Becker et al., 2005). Most of the mitochondrial proteins are translocated from cytosolic compartments and imported into mitochondria. Proteins destined for the outer or inner membrane, the IMS, or the matrix are recognized and translocated by the TOM machinery, containing protein import channel Tom40. This channel translocates proteins destined for MOM, MIM, or IMS. Tom40, suggested to have evolved from a porin-type protein, which has a β-barrel shape. Table 2.2 presents the information on the secondary structure of renatured Tom40, estimated from circular dichroism (CD) spectra in different systems (Becker et al., 2005). The data clearly demonstrate that the β-sheet structure dominates over other types in all investigated systems.

Secondary structures were calculated from the CD spectra by a neural network approach (I) (Böhm et al., 1992), and three different methods according to Sreerama and Woody (2000; Sreerama, 2003; 2004), termed SELCON3 (II), CONTIN/LL (III), CDSSTR (IV).

For a homology model of Tom40, see Figure 2.2 (Gessmann et al., 2011). Two conserved polar slides in the pore interior have been found in this study (Gessmann et al., 2011). One is possibly involved in a pore-inserted helix positioning. The other might serve in mitochondrial presequence peptide binding as it is present only in Tom40, but not in VDAC proteins. The outer surface of Tom40 barrel reveals two conserved amino acid clusters. They are thought to be involved in binding other components of the TOM complex, or bridging components of the TIM machinery of the MIM.

Presequence TOM channel sensitivity from purified MOMs, TOM complex, and Tom40 has been demonstrated in Figure 2.3 (Becker et al., 2005). This group compared the expressed and renatured Tom40 from two species *Saccharomyces cerevisiae* and *Neurospora crassa*, and found a high content of β-structure in CD measurements, in agreement with refined predictions of the secondary structure. The electrophysiological characterization of the renatured Tom40 reveals the same characteristics as the purified TOM complex or MOM vesicles.

MOM protein Sam50 has an essential role in the maintenance of the cristae structure. A cristae is a fold in the MIM. Cristae provides a large surface area for chemical reactions. Sam50 is a part of the sorting and assembly machinery (SAM) necessary for the assembly of β-barrel proteins in

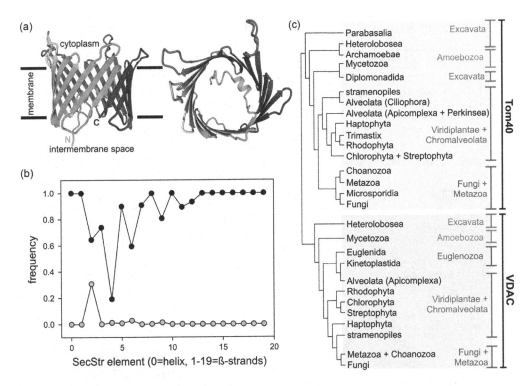

**FIGURE 2.2** Homology model of Tom40. (a) The homology model of Tom40 from *Neurospora crassa* is colored from the N- to C-terminus in a gradient from cyan, green, yellow, red, and magenta, to blue. The membrane-inserted region of NcTom40 is approximately indicated on the left panel, and on the right panel the model is rotated by 90° around the x axis, resulting in a view through the pore from the cytoplasm to the IMS. (b) Multiple sequence alignment strategy was applied to the full, nonredundant set (black lines) of Tom40 and VDAC sequences. The frequency of occurrence of equal alignments of NcTom40 regions to the helix and β-strand core regions of mVDAC-1 is shown for the most frequently (black circles) and for the second most frequently occurring aligned regions (gray circles) of NcTom40. (c) A schematic representation of the maximum likelihood (ML) phylogeny of Tom40 and VDAC. The branch lengths in this schematic tree do not correspond to the frequency of amino acid substitutions, but only indicate the topology of the ML tree. (For color figure see eBook figure.)

the MOM (Paschen et al., 2003; Bredemeier et al., 2006). The SAM components are found to exist in a large protein complex together with the MIM proteins mitofilin and CHCHD3, termed as the mitochondrial IMS bridging (MIB) complex (Ott et al., 2012). Interactions between MOM and MIM components of the MIB complex are crucial for the preservation of cristae.

The VDACs ensure the passage of small hydrophilic ions and metabolites. The VDAC channel transport is driven by the concentration gradient of the materials across MOM.

During apoptosis, a pore is formed in the MOM of mammalian mitochondria to release cytochrome c from the IMS. This process leads to the activation of cytosolic caspases and eventually cell death. The proapoptotic factor Bax and Bak assemble into oligomers that contribute to MOM permeabilization (Antignani and Youle, 2006).

Studies in model organism bakerś yeast *S. cerevisiae* has recently revealed that the number of channel-forming proteins in the MOM and their functional diversity is considerably larger than thought so far. Two new channels have been identified that are involved in protein transport. The mitochondrial division and morphology protein 10 (Mdm10) are found to form a β-barrel channel mediating the biogenesis of the TOM complex (Ellenrieder et al., 2016). Mim1 of the mitochondrial import machinery (MIM) promotes the biogenesis of α-helical MOM proteins. This constitutes the first MOM channel with a predicted α-helical structure (Doan et al., 2020). The

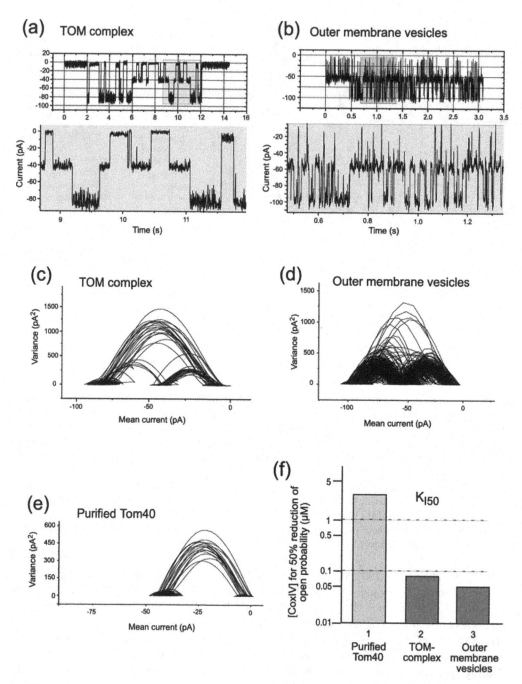

**FIGURE 2.3** Presequence sensitivity of the TOM channel from purified mitochondrial outer membranes, TOM complex, and Tom40. (a) Current recording of purified *S. cerevisiae* TOM complex (VhZ140 mV). (b) Current recording of purified *S. cerevisiae* outer membrane vesicles (Vh = 100 mV). (c) Mean variance plot of purified *S. cerevisiae* TOM complex, demonstrating the connective gating of a two-pore channel. (d) Mean variance plot of purified *S. cerevisiae* outer membrane vesicles, demonstrating the connective gating of a two-pore channel. (e) Mean variance plot of purified renatured *S. cerevisiae* Tom40. (f) Reduction of the mean open probability by addition of a mitochondrial presequence peptide. The concentration of the CoxIV presequence peptide is shown at which the mean Popen of the Tom40 channel is reduced to 50% (KI50) (averages of at least three different experiments).

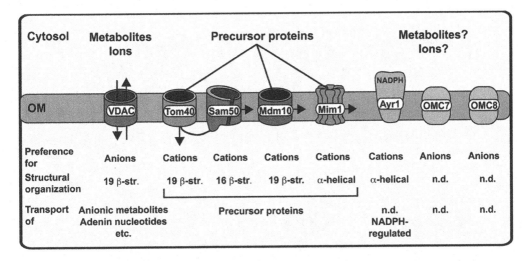

**FIGURE 2.4** The diversity of MOM channels. Depicted are the function and properties of the eight channel-forming proteins that have been found so far in outer membrane vesicles of yeast mitochondria. Tom40, Sam50, Mdm10, and Mim1 are involved in protein import, whereas VDAC is the main channel for small metabolites and ions. Ayr1, OMC7, and OMC8 might function in transport of metabolites and ions; however, the transported substrate remains unknown. Tom40, Sam50, and VDAC are highly conserved and present in the MOM of plant and mammalian mitochondria.

acyl-dihydroxyacetone phosphate reductase (Ayr1) forms an NADPH-regulated channel. Two additional anion-selective channels, OMC7 and OMC8, have been identified in MOM vesicles (Krüger et al., 2017).

Eight channels explained above having large hydrophilic pores with defined electrostatic pore properties are known to exist in the MOM (see Figure 2.4) (Becker and Wagner, 2018). These channels with distinct specific pore properties are believed to maintain selective and regulated transport of small hydrophilic molecules across MOM.

## 2.2.2 PROTEIN TRANSPORT OF MOM

MOM contains two major translocators, TOM40 (TOM) and TOB/SAM complexes for translocating proteins across and/or insertion into the MOM. The TOM40 complex works like an entry gate for most mitochondrial proteins. Figure 2.5 presents the models for the TOM40 complex translocation of precursor proteins (Endo and Yamano, 2010). Figure 2.5a shows a presequence-less, Tom70-dependent precursor protein. Many of the presequence-less precursor proteins including carrier proteins require Hsp70/Hsp90 for maintenance of their import competence. Hsp70/Hsp90 docks onto the Tom70, and the precursor proteins are found to be transferred to Tom70, which may also prevent precursor aggregation. The precursor proteins are then found to be transferred to the Tom40 channel, likely via Tom22 and Tom5. Following the translocation through the TOM40 channel in a loop-type conformation, the precursor proteins are found to bind to small Tim proteins in the IMS. Figure 2.5b presents a presequence-containing, Tom70-dependent precursor protein. Some of the presequence-containing precursor proteins require cytosolic chaperones and dock onto the Tom70. As the Tom70 prevents the aggregation of the precursor proteins, Tom20 and Tom22 are found to recognize the targeting signal in the presequence. Then the precursor proteins get transferred to the TOM40 channel, likely via the Tom5. The inner wall of the TOM40 channel may function as a chaperone for the precursor proteins' unfolded mature part, while the N-terminal presequence are found to bind to the trans site of TOM40 complex. Figure 2.5c presents a presequence-containing, Tom70-independent precursor protein. Many of the presequence-containing proteins are directly

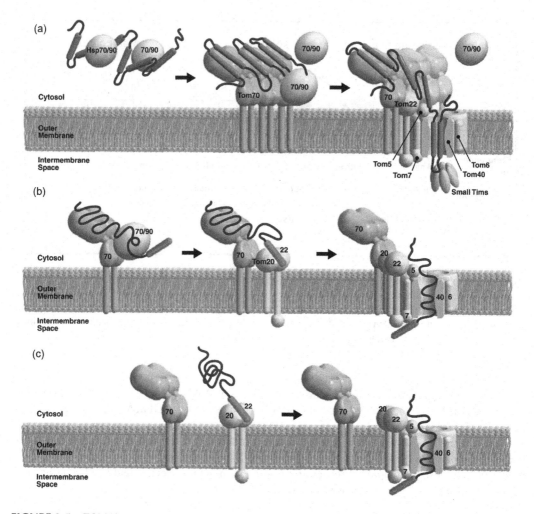

**FIGURE 2.5**  TOM40 complex translocation of precursor proteins.

recognized by Tom20 and Tom22 through their interactions with the presequence. Subsequent steps are found to be similar to Figure 2.5b.

For biophysical characterization, Tom40 channel ion conductance was measured in the presence of the mitochondrial presequence peptide $pF_1\beta$ (Mahendran et al., 2012). This study revealed the peptide binding kinetics. The rates for association $k_{on}$ and dissociation $k_{off}$ were found to strongly depend on the applied transmembrane potential, and both of these kinetic constants were found to increase with increasing applied voltage. Figure 2.6 presents ion current recordings through a single Tom40 channel in the absence and presence of peptide $pF_1\beta$ (Mahendran et al., 2012). For details of the experimental techniques and condition, please refer to the original article. The appearance of channel conductance events clearly demonstrates that the conductance events are appearing as a result of peptide binding to the Tom40 channel. The average residence time of mitochondrial presequence peptide $pF_1\beta$ within the Tom40 channel is found to decrease with the increase in applied transmembrane voltage. The equivalent increase in the peptide dissociation rates with applied transmembrane voltage demonstrates the translocation of the peptide. This method may help address the characterization of the protein translocation pathway through the Tom40 channel.

The TOB/SAM complex works as an insertion machinery for β-barrel membrane proteins (MPs). The translocators cooperate with chaperones in the cytosol and IMS and handle the loosely folded or

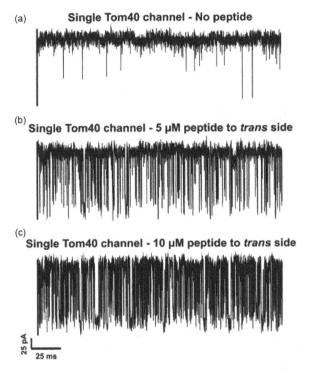

(a) **Single Tom40 channel - No peptide**

(b) **Single Tom40 channel - 5 µM peptide to *trans* side**

(c) **Single Tom40 channel - 10 µM peptide to *trans* side**

25 pA

25 ms

**FIGURE 2.6** Electrophysiology record of ion currents through a single Tom40 channel. (a) 0 µM, (b) 5 µM, and (c) 10 µM pF1β peptide added to the trans side of the lipid membrane. Applied voltage = 50 mV. Experimental conditions: 1 M KCl, 20 mM MES, pH 6, at room temperature.

unfolded precursor polypeptides, exhibiting chaperone-like functions on their own. Many α-helical MPs take "non-standard" routes to be inserted into the MOM.

Precursors of the β-barrel proteins are transferred further to another MOM complex that mediates their topogenesis (TOB complex). Tob55 is an essential component of the TOB complex. It is known to constitute the core element of the pore that conducts proteins. The other two TOB complex components are Tob38, building a functional TOB core complex with Tob55, and Mas37, which is a peripheral member of the complex. A biogenesis investigation of the TOB complex was done by Habib et al. (2004). Reduced insertion of the Tob55 precursor, in the absence of Tom20 and Tom70, argues for the initial recognition of the precursor of Tob55 by import receptors. Next, it is transferred through the import channel formed by Tom40.

Historically, mitochondria evolved through endosymbiosis of Gram-negative progenitor with a cell to generate eukaryotes. The MOM and Gram-negative bacteria are found to contain pore proteins with a β-barrel topology. After getting synthesized in the cytosol, β-barrel precursor proteins are transported first into the MIS. Folding and membrane integration of these β-barrel proteins depend on the mitochondrial SAM located in the MOM, which is related to the β-barrel assembly machinery in bacteria. The SAM complex is known to recognize β-barrel proteins by a β-signal in the C-terminal β-strand, required to initiate β-barrel protein insertion into the MOM.

Similar to other mitochondrial precursors, β-barrel proteins are recognized first by TOM receptors (Keil et al., 1993) before they are imported across MOM with the help of the TOM complex (Wiedemann et al., 2003). Subsequently, β-barrel precursors are exported with the help of SAM/TOB, required for β-barrel protein's membrane insertion into MOM. Near the C-terminal end, mitochondrial β-barrel proteins harbor a β-signal, required for SAM complex recognition of the β-barrel precursor (Kutik et al., 2008). SAM consists of three core subunits: Sam50 (Tob55,

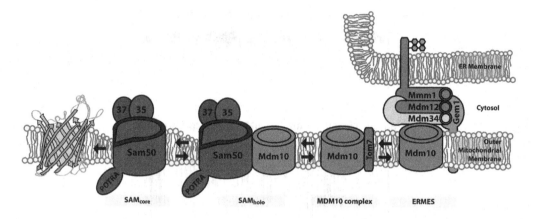

**FIGURE 2.7** The SAM of the MOM and the endoplasmic reticulum–mitochondria encounter structure (ERMES) are linked by Mdm10. For detailed description, see Höhr et al. (2015).

Omp85), Sam37 (Mas37, Tom37), and Sam35 (Tob38, Tom38) in a 1:1:1 stoichiometry, as well as the auxiliary subunit Mdm10 (mitochondrial distribution and morphology protein). Figure 2.7 presents the SAM of the MOM and the endoplasmic reticulum–mitochondria encounter structure (ERMES) linked by Mdm10 (Höhr et al., 2015). The SAM complex is essential for the biogenesis of MOM β-barrel proteins and several other subcomplexes. The $SAM_{core}$ complex consists of β-barrel protein Sam50, the receptor subunits Sam35 and Sam37. Upon binding of the β-barrel MP, Mdm10, the $SAM_{holo}$ complex is formed. The Mdm10 dissociation from the $SAM_{holo}$ complex is regulated by its binding to Tom7, a small TOM complex subunit. Mdm10 is also required for the formation of the ERMES complex. The ERMES complex tethers the endoplasmic reticulum (ER) membrane to MOM. Further ERMES subunits are Mmm1, embedded in the ER membrane, as well as Mdm12, Mdm34, and Gem1.

The import of proteins with multiple α-helical membrane spans is promoted by the mitochondrial import machinery (MIM complex). Precursors of proteins having a single α-helical membrane anchor are usually imported via several distinct routes. Figure 2.8 presents a model for the import of a precursor of α-helically embedded MOM proteins (Ellenrieder et al., 2015).

## 2.3 MITOCHONDRIAL INNER MEMBRANE CHANNELS

The MIM hosts various ion-conducting channels. These channels actively participate in maintaining controlled permeabilization of small ions. Some of these channels have been found to be the mitochondrial counterpart of them present in other cellular membranes as well. Two reviews of O'Rourke and collaborators (2005; O'Rourke, 2007) listed all the known MOM and MIM channels, as presented in Table 2.1. The current state of knowledge on major MIM channels, their properties, identity, and known or proposed functions have been briefed in many studies (Zoratti et al., 2009; Szabò and Zoratti, 2014).

### 2.3.1 MIM CHANNELS-CURRENT STATE

More than three decades ago by patch-clamping the MIM, a slightly anion-selective channel was identified (Sorgato et al., 1987). These early-stage recordings are presented in Figure 2.9 (Sorgato et al., 1987). This important conducting event was identified as voltage-dependent channel having a mean conductance of 107 pS in the presence of a symmetrical physiological condition, 150 mM KCl.

Since the first patch-clamp of the MIM, several developments have so far been made by discovering several MIM conductance events. As mentioned earlier, the details of MIM channels are already

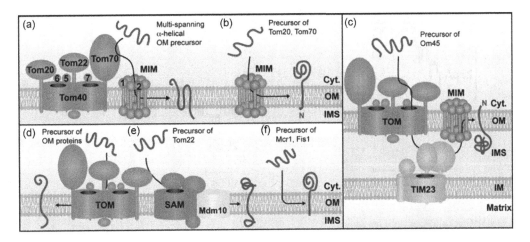

**FIGURE 2.8** Models for the import of precursor of α-helically embedded MOM proteins. (a) Precursors of multispanning proteins are recognized by Tom70 and transferred to the MIM complex, which is crucial for their assembly into the outer membrane. (b) The MIM complex mediates the biogenesis of precursors of the signal-anchored Tom20 and Tom70. (c) The precursor of Om45 is imported via the TOM channel with the help of the TIM23 complex into the IMS. The MIM complex promotes Om45 assembly into the MOM. (d) TOM subunits promote the biogenesis of several single-spanning outer MPs. (e) The precursor of Tom22 is recognized by TOM receptors and inserted into the outer membrane by the SAM-Mdm10 complex. (f) The membrane integration of the Fis1 and Mcr1 precursors occur independently of proteins.

explained in various articles (O'Rourke, 2005; Zoratti et al., 2009; Szabò et al., 2011; Szabò and Zoratti, 2014). Figure 2.10 presents a summary (Szabò and Zoratti, 2014).

### 2.3.2 POTASSIUM CHANNELS OF THE MIM

We wish to avoid providing the details on all MIM channels, as shown Figure 2.10, as there are many review articles detailing the channels' pinpointed aspects, including their physiological roles and roles concerning a multitude of diseases. However, as an example, we wish to briefly present a few crucial characteristics on the physiology of potassium channels in MIM (Szabò et al., 2011).

Szewczyk and colleagues reviewed in detail on the following MIM potassium channels: the channels that regulate ATP, the large conductance $Ca^{2+}$-activated channels, the voltage-gated Kv1.3 channels, and the twin-pore domain TASK-3 channels (Szewczyk et al., 2009; Szewczyk, 2016). Figure 2.11 illustrates these channels in MIM (Szewczyk et al., 2009). These channels primarily participate in changing the mitochondrial matrix volume, respiration of mitochondria, and membrane potential and synthesis of ROS by mitochondria. The fundamental biophysical properties of MIM potassium channels (Figure 2.11), for example, mitoK$_{ATP}$ (Inoue et al., 1991), mitoBKCa (Siemen et al., 1999), mitoKv1.3 (Szabò et al., 2005), and TASK-3 channels (Rusznák et al., 2007), resemble the properties of some potassium channels found in the plasma membrane of various cells.

### 2.3.3 CALCIUM SIGNALING OF THE MIM

Mitochondrial uptake of $Ca^{2+}$ serves as a crucial regulator of important cellular functions, such as cytoplasmic $Ca^{2+}$ signaling, energy metabolism, and apoptosis. Mitochondrial uniporter uptake of $Ca^{2+}$ is regulated by $Ca^{2+}$ in a temporally complex manner (Putney and Thomas, 2006). The potassium channels may constitute a link between cellular and mitochondrial ATP or calcium signaling and reactions, a molecular process that depends on mitochondrial membrane potentials. For detail analysis, see Szewczyk et al. (2009).

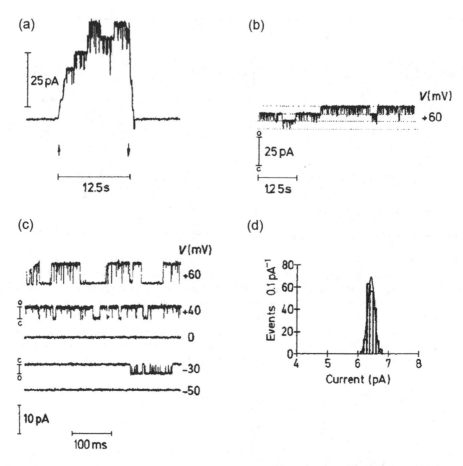

**FIGURE 2.9**  Single-channel record of MIM channels. (a) Single-channel current in response to a voltage pulse. (b) Record over an extended time scale. (c) Records against various voltage pulses. (d) Amplitude histogram of the channel record. The peak location on current axis corresponds to the current of a single-channel conduct, central peak value ~6.4 pA.

Mitochondria are mobile inside cells. $Ca^{2+}$ inhibits the mitochondrial mobility, thus raising a mechanism that helps retain mitochondria at specific $Ca^{2+}$ signaling sites (Yi et al., 2004). More than three decades ago, biochemical analysis of mitochondrial $Ca^{2+}$ content and the studies on the regulation of mitochondrial $Ca^{2+}$-sensitive enzymes found that mitochondria are more likely to be targets for $Ca^{2+}$ signaling, rather than sources (Shears and Kirk, 1984; Denton and Mccormack, 1985). The first direct in situ mitochondrial $Ca^{2+}$ measurements clarified that the receptor-activated $Ca^{2+}$ signals cause rapid and large $Ca^{2+}$ signals in the mitochondrial matrix (Rizzuto et al., 1992a; 1992b). Mitochondrial $Ca^{2+}$ uptake may serve to compartmentalize $Ca^{2+}$ signaling in appropriate cellular domains. Mitochondrial uptake of $Ca^{2+}$ may initiate a key step in apoptosis through activating the mitochondrial permeability transition pore that permits the escape of cytochrome c and many other proapoptotic factors to cytoplasm. The $Ca^{2+}$-uptake process occurs through a channel "uniporter." The MIM has a $Ca^{2+}/2H^+$ exchanger and/or a $Ca^{2+}/3Na^+$ exchanger, analogous to that found in plasma membrane of the cell. These efflux pathways may become saturated due to the high matrix $Ca^{2+}$ loads, such that the sustained and rapid $Ca^{2+}$ influx may lead to the mitochondrial $Ca^{2+}$ overload. Regulation of mitochondrial $Ca^{2+}$ uptake is demonstrated in Figure 2.12 (Putney and Thomas, 2006).

In a recent review, crucial discoveries on the regulation of interorganellar $Ca^{2+}$ homeostasis and its role in pathophysiology, including regulation and the role of the mitochondrial calcium

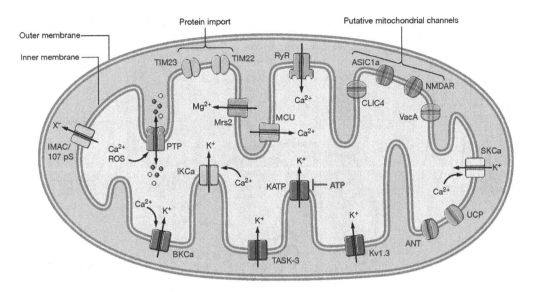

**FIGURE 2.10** Proteins mediating ion fluxes in the MIM. Bona fide ion channels are shown with colored filled forms. Proteins giving rise to channel activities under certain circumstances (ANT, UCP, Tim22, Tim23) are also shown. Putative channels whose channel function has not been proven in MIM are also listed. The main factors activating (red arrow) or inhibiting (violet arrows) channel activities are shown. (For color figure see eBook figure.)

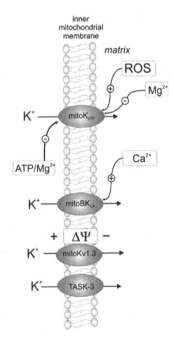

**FIGURE 2.11** Potassium channels present in MIM. For general description on all mentioned here, see the text and the various references quoted.

uniporter complex have been addressed (Raffaello et al., 2016). For a deeper understanding on this topic, other articles may also be consulted (Tsai and Tsai, 2020; Marchi et al., 2019; Mishra et al., 2017; García-Sancho, 2014). Further on a molecular level, a recent claim has just been revealed (Tsai and Tsai, 2020). The mitochondrial calcium uniporter contains the pore-forming MCU protein, containing a conserved DIME sequence thought to form a calcium selectivity filter and

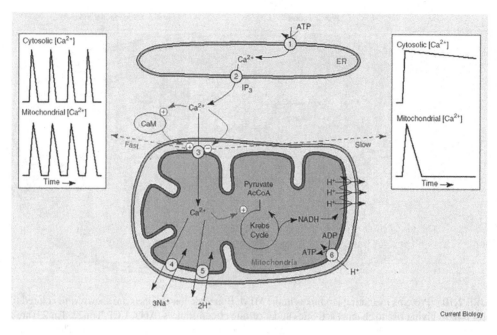

**FIGURE 2.12**  Regulation of mitochondrial $Ca^{2+}$ uptake. Close apposition between mitochondria and endoplasmic reticulum (ER) $Ca^{2+}$ release sites facilitates $Ca^{2+}$ uptake into the mitochondria. A fast activation of the uniporter, mediated by calmodulin (CaM), enhances mitochondrial $Ca^{2+}$ accumulation. A slower $Ca^{2+}$-dependent inactivation of the uniporter serves to limit mitochondrial $Ca^{2+}$ uptake during a prolonged increase in cytosolic $Ca^{2+}$. The fast activation allows the mitochondrial matrix $[Ca^{2+}]$ to closely follow changes in the cytosolic $[Ca^{2+}]$, as occurs during cytosolic $Ca^{2+}$ oscillations (left inset). The slower inactivation process limits mitochondrial $Ca^{2+}$ uptake when the elevation of cytosolic $Ca^{2+}$ is more sustained (right inset). (1) SERCA $Ca^{2+}$ pump, (2) IP3 receptor $Ca^{2+}$-release channel, (3) $Ca^{2+}$ uniporter, (4) mitochondrial $Ca^{2+}/Na^+$ exchanger, (5) mitochondrial $Ca^{2+}/H^+$ exchanger, (6) ATP synthetase. The mitochondrial respiratory chain is shown as three cycles of proton pumping from the matrix.

regulatory subunits EMRE, MICU1, and MICU2. Tsai et al. reveals that MICU1 is found to interact with MCU through the Asp residue in the DIME motif. A mutagenesis screening of MICU1 identifies two highly conserved Arg residues that might contact the DIME-Asp. Disrupting the MCU and MCIU1 interactions renders the uniporter hyperactive to constitutively load calcium into mitochondria. Altogether, these results demonstrate that MICU1 opens/closes the uniporter by blocking/unblocking the MCU pore.

### 2.3.4  PROTEIN TRANSPORT OF THE MIM

Trafficking of the mitochondrial proteins requires coordinated operation of specific protein translocator complexes in mitochondrial membrane. The TIM23 complex is known to translocate and insert proteins into the MIM. An important analysis of the IMS domains of Tim23 and Tim50, which are essential subunits of the TIM23 complex, was conducted to address their roles in protein translocations (Tamura et al., 2009). The interactions of Tim23 and Tim50 in IMS facilitates the transfer of precursor proteins from the TOM40 complex, a general protein translocator in the MOM, to the TIM23 complex. Tim23–Tim50 interactions is also found to facilitate a late step in protein translocation across MIM by promoting the motor functions of mitochondrial Hsp70 in the matrix. Therefore, the Tim23–Tim50 pair is considered to coordinate the actions of the TOM40 and TIM23 complexes together with the motor proteins to ensure the mitochondrial protein import. A model of

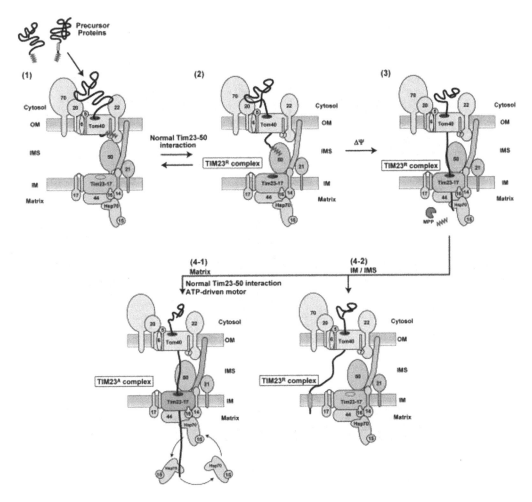

**FIGURE 2.13** Protein translocation across and insertion into the MIM via the TIM23 complex may be modeled as presented here. The interactions among some of the components are dynamic while a few may be static (more investigations on this matter needed to reveal further conclusions).

protein translocation across and insertion into MIM via the TIM23 complex has been presented in Figure 2.13 (Tamura et al., 2009).

A recent study showed that the two domain structures of Tim44 plays an important role in protein translocation process (Banerjee et al., 2015). The N-terminal domain of Tim44 is found to interact with the components of the import motor and its C-terminal domain is found to interact with the translocation channel in MIM and is in contact with translocating proteins. For experimental details and data supporting the conclusion that the translocation channel and the import motor of the TIM23 complex communicate through rearrangements of the two domains of Tim44 that are stimulated by translocating proteins, readers should refer to Banerjee et al. (2015). The underlying message is that the two domains of Tim44 have distinctive interaction partners within the TIM23 complex. Thus, Tim44 holds the TIM23 complex together. A model on how the TIM23 complex function is presented in Figure 2.14 (Banerjee et al., 2015). It describes how the translocation of precursor proteins through the channel in the MIM is coupled to their capture by the ATP-dependent import motor at the channel's matrix face. As mentioned earlier, Tim44 is found to play a central role here. Two domains of Tim44 are predicted to be connected by the central segment that contains membrane-recruitment helices, similar to the two cherries on the stalks, see the insert of

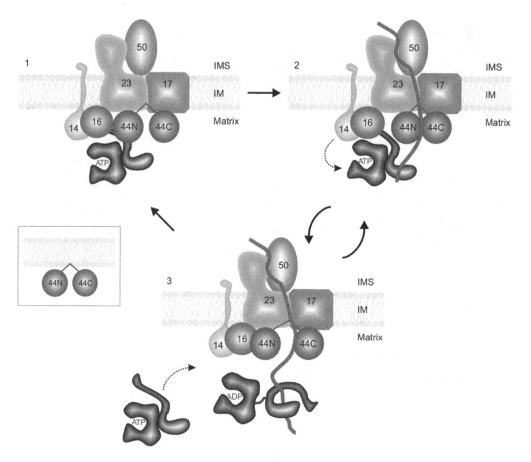

**FIGURE 2.14**    A model of the function of the TIM23 complex.

Figure 2.14. The central segment of Tim44 recruits the protein to the cardiolipin-containing membranes. Through the direct protein–protein interactions, Tim44's C-terminal domain binds to Tim17 and the N-terminal domain to mtHsp70 and to Tim14-Tim16 subcomplex, see Figure 2.14. Tim44 creates a central platform that connects the translocation channel in the MIM with the import motor at the face of the matrix. Additional interactions are likely to stabilize the complex, in particular, that between the N-terminal domain of Tim44 and Tim23 (Ting et al., 2014), as well as the other one between Tim17 and the IMS-exposed segment of Tim14 (Chacinska et al., 2005). The translocation channel, in the resting state, is closed to maintain the permeability barrier of MIM. During protein translocation (Figure 2.14 (2)), the translocation channel of MIM needs to open to allow the passage of the proteins. The channel opening will likely change the Tim17 conformation that could be conveyed to the C-terminal domain of Tim44. It is quite tempting to speculate that the change in the conformation is transduced to the N-terminal domain of Tim44 through the central, membrane-bound Tim44 segment, leading to the relative rearrangements of two Tim 44 domains. This change allows Tim14-Tim16 complex toward stimulating the ATPase activity of mtHsp70, leading to the stable binding of translocating protein to mtHsp70. mtHsp70, after binding with the polypeptides, moves into the matrix, opening a site of binding on Tim44 for another molecule of mtHsp70, see Figure 2.14(3). For further details, see Banerjee et al. (2015).

Tim50 is the central receptor of the TIM23 complex that recognizes the precursor proteins in the IMS. Additionally, Tim50 interacts with the IMS domain of the channel-forming subunit, Tim23, an essential interaction for protein import across the MIM. To gain deeper insight into the molecular function of Tim50, random mutagenesis was used to determine residues important for its function (Dayan et al., 2019). The temperature-sensitive isolated mutants were defective in import of TIM23-dependent precursor proteins. The residues mutated map to two distinct patches on the surface of Tim50. Mutations in both patches were found to impair the Tim23 interaction of Tim50. Two regions of Tim50 are predicted to play a role in its interaction with Tim23, and thereby affecting the import function of the complex. Figure 2.15 presents the molecular insight into regions of Tim50 involved in interaction with Tim23 (Dayan et al., 2019). Figure 2.15 (a), residue A221 (orange) identified in the screen is close to residues R214 and K217 (both in cyan) that were previously implicated in Tim50 interaction with Tim23 (Qian et al., 2011). The electrostatic interaction between K217 and D222, see figure inset, is likely changed by A221D mutation (Dayan et al., 2019). Amino group in the lysine side chain, shown in blue, and two oxygen atoms in the carboxyl group of the aspartate side chain, shown in red. (b) Residues D278, N283, D337, and R339 (all orange) mutated in ts mutants identified in the present screen map closely to residues L279, L282, and L286 (all cyan), implicated previously in Tim23–Tim50 interaction (Tamura et al., 2009). The model in Figure 2.15 (b) is rotated by 180° relative to the model in Figure 2.15 (a). Figure 2.15 (c) shows the mapping of residue D293, identified in the screen on the structure of Tim50. The electrostatic interaction (depicted in inset) between K277 and D293 is likely to be affected by D293N mutation (Dayan et al., 2019). Amino group in the lysine side chain (shown in blue) and two oxygen atoms in the carboxyl group of the aspartate side chain (shown in red).

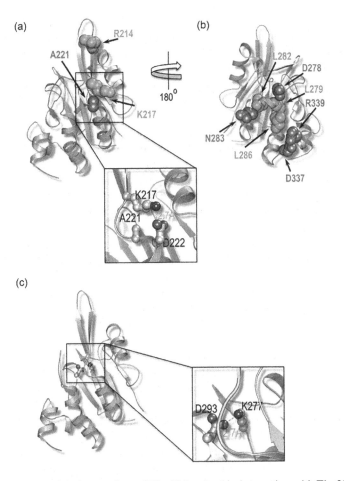

**FIGURE 2.15** Molecular insight into regions of Tim50 involved in interaction with Tim23.

The interactions of Tim50 with the IMS domain of the channel-forming subunit, Tim23 and other related molecular mechanisms are further investigated regarding the influence might get drawn due to certain lipid compositions in MIM. Phospholipid cardiolipin is found to mediate membrane and channel interactions of the mitochondrial TIM23 protein import complex receptor Tim50 (Malhotra et al., 2017). Cardiolipin promotes the naturally strong interaction of Tim50 receptor with lipid bilayer. It is shown here that the soluble receptor domain of Tim50 interacts with membranes in general, and, in particular, with specific sites on the Tim23 channel, and that the interactions and related mechanisms are directly modulated by insertion of cardiolipin in membrane.

The first small-angle X-ray scattering-based structure of the soluble Tim50 receptor was utilized in molecular dynamics (MD) simulations, combined with a range of biophysical measurements that confirmed the role of cardiolipin in driving the Tim50 receptor association with the lipid bilayers (Malhotra et al., 2017). The concomitant structural changes highlight the key roles of the structural elements in mediating the interaction. The results confirm that the cardiolipin is required to mediate the TIM23 complex-hosted specific receptor-channel associations. Thus, the results are found to support a new working model favoring the dynamic structural changes that occur within the TIM23 complex during transport processes.

Figure 2.16 presents a model depicting the Tim50$^{IMS}$-bilayer and Tim50–Tim23 interactions (Malhotra et al., 2017). Tim50$^{IMS}$ represents the large globular receptor domain of Tim50 in the IMS that is found to be anchored to the IM by a single transmembrane segment. Figure 2.16 (a) shows the interaction between Tim50$^{IMS}$ and cardiolipin-containing bilayers is presented. Tim50$^{IMS}$, in solution, is initially attracted (due to charge-based interactions) to the negatively charged bilayer by long-range Coulombic force. To understand this kind of Coulomb interactions, which often are found not just Coulomb interactions but screened Coulomb interactions in biological systems, one may read the first ever published article (Ashrafuzzaman and Tuszynski, 2012a), and for further details in our book, see Ashrafuzzaman and Tuszynski (2012b). The electric field is expected to orient a basic face of the protein, including the β-hairpin, toward the membrane bilayer. Upon electrostatic docking, hydrophobic interactions between the nonpolar residues at the binding interface and the nonpolar bilayer core stabilize the interaction in a manner that is stimulated by the $H_{II}$ propensity of cardiolipin. Coupled to the membrane association are changes in Tim50$^{IMS}$ conformation relative to its water-solvated state. The membrane association of the Tim50$^{IMS}$ gets promoted if these Coulombic and hydrophobic interactions are favorably high enough to offset the energetic penalty of normal desolvation. The left panel presents the Tim50$^{IMS}$ homology model of the study; subsequent panels show hypothetical the outlines of Tim50$^{IMS}$ structure (in cyan), either in the membrane-bound state or in the presence of TIM23 complex. Figure 2..16 shows the cardiolipin- and substrate-dependent dynamics of Tim50 in the context of TIM23 complex. The presence of cardiolipin (left) favors the membrane-associated process of Tim50$^{IMS}$, facilitating interaction with the channel domain of Tim23. In the presence of the substrate, Tim50$^{IMS}$ is found to retain its interaction with both Tim23$^{IMS}$ and Tim23 channel regions. The absence of cardiolipin (right) is found to favor the membrane-dissociated process of Tim50$^{IMS}$, blocking its interaction with the region of the Tim23 channel. The presence of substrate is found to further promote the membrane-dissociated state of Tim50$^{IMS}$. The stippled region of Tim50 shown in Figure 2.16 depicts the β-hairpin.

The analyses of the interaction between Tim50$^{IMS}$ and membranes of different lipid compositions provide insights into the molecular basis of the protein-bilayer association, an important molecular mechanism in MIM. Similar to other membrane-interactive amphitropic proteins, the binding of Tim50$^{IMS}$ entails an electrostatic as well as a hydrophobic component (Mulgrew-Nesbitt et al., 2006). The electrostatic interaction between Tim50$^{IMS}$ and membrane was revealed by the fact that all of the anionic lipids resulted in some degree of Tim50$^{IMS}$-bilayer interaction: PG and MLCL effected relatively weak attraction, but cardiolipin and dCL attracted the protein relatively stronger. The weaker attraction of PG can be rationalized because it is a monoanionic lipid, compared to

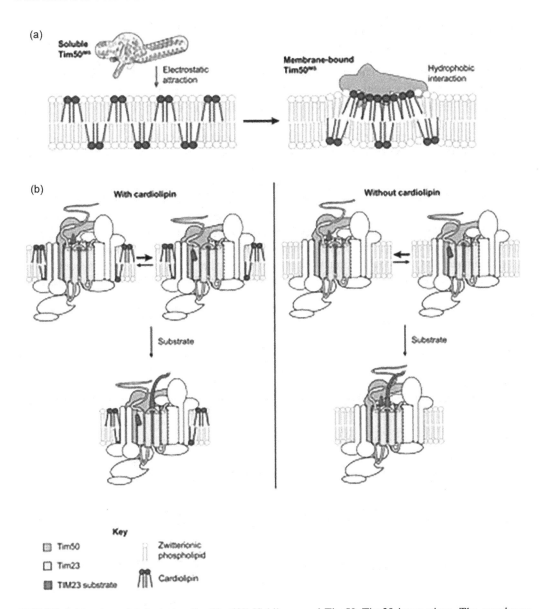

**FIGURE 2.16** A model depicting the Tim50IMS bilayer and Tim50–Tim23 interactions. The membrane presence of cardiolipin may regulate the interactions.

cardiolipin, which is branded as dianionic (Olofsson and Sparr, 2013; Sathappa and Alder, 2016; Kooijman et al., 2017; 87) at physiological pH.

The MIM has long been considered as poorly permeable to all sorts of cations and anions. The strict control of the MIM permeability mechanisms is, however, crucial for efficient synthesis of ATP. More than three decades of research along with a huge amount of produced data has helped us in understanding now that various ion channels – along with antiporters and uniporters – are present in MIM. However, these channels exist at rather low abundance. These channels are not only important for maintaining energy supply chain for cells, but also produce decisive factors in determining various crucial aspects of the cell health condition, even deciding whether a cell lives or dies.

## 2.4   MITOCHONDRIAL CHANNELS TRANSPORT SPECIFIC PROTEINS LINKED TO APOPTOSIS

MOM transport is directly and indirectly linked to the opening/closing of various channels formed due to the presence of a multitude of proteins and lipids involved in cell death mechanisms. Mitochondrial membrane permeabilization (MMP) may get triggered during apoptosis due to convergence of many signals. MMP is a rate-limiting step, usually active during cell death process. The MMP triggering signals are mainly a collection of endogenous proteins, translocating from an intracellular compartment, for example, nucleus, cytosol, lysosomes, etc., to MOM. Various lipids (natural or modified), for example, oxidized cardiolipin, ceramide, etc., and ions, for example, $Ca^{2+}$, contribute into promoting and regulating MMP. The physiological environmental condition determining intracellular milieu, for example, pH, ROS, ATP levels, etc. can directly or indirectly contribute to determining the strength of MMP. MMP may lead and control or regulate the release of a multitude of IMS factors into cytosol, caspase-dependent proteins, for example, cytochrome c, Smac/DIABLO, pro-caspases, and caspase-independent proteins, for example, apoptosis-inducing factor, endonuclease G (EndoG) (a protein primarily participates in the caspase-independent apoptosis via DNA degradation during translocating from the mitochondrion to the nucleus under the oxidative stress), etc.

### 2.4.1   APOPTOTIC PROTEIN TRANSPORT: AN OVERVIEW

The release of IMS proteins results in the loss of integrity of the MOM caused by the proapoptotic members of the Bcl-2 family. Figure 2.17 presents the classification and structures of Bcl-2 family proteins (Martinou and Youle, 2011). As shown in Figure 2.17a, the Bcl-2 family is divided into three

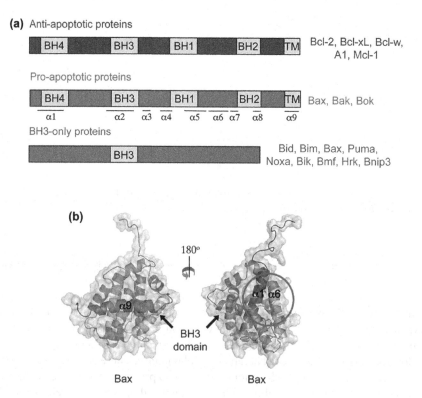

**FIGURE 2.17**   Bcl-2 family protein classification and structures.

groups considering their Bcl-2 homology domains (BH). Proapoptotic and antiapoptotic proteins contain four BH domains, but BH3-only proteins. As their name indicates, they contain only the BH3 domain. As shown in Figure 2.17b, Bax structure showing the alpha-helix 9 (transmembrane domain) embedded in the hydrophobic groove (left structure). A 180° rotation (right-sided structure) shows the alpha-helices 1 and 6. The red circle represents the binding site for Bim BH3 domain.

BH3 proteins are known to integrate and transmit death signals, emanating from defective cellular processes to other members of the Bcl-2 family. These proteins, through the BH3 domain, either interact with antiapoptotic proteins and work toward inhibiting their function and/or to interact with multidomain proteins, for example, Bax or Bak, and thus stimulate their activity. The former are often referred to as "sensitizers," whereas the later as "activators" (Giam et al., 2008). The multidomain proapoptotic proteins Bax and Bak, and perhaps Bok in some tissues, are responsible for MOM permeabilization. These oligomers form pores using mechanisms that are yet unknown. They are also the master effectors of apoptosis as cells lacking Bax and Bak fail to undergo MOM permeabilization and apoptosis due to the response of many cell death stimuli (Wei, 2001). Another article reviews in detail how Bax and Bak change conformation and oligomerize, as well as how oligomers might form a pore (Westphal et al., 2011).

Various models have been proposed to demonstrate the Bax and Bak-induced MOM permeabilization. These models predict that Bax, alone or combined with other proteins, forms large channels to allow passage of cytochrome c and various other proteins. Alternatively, Bax or Bak could act as modulator for opening existing channels, such as the permeability transition pore, to induce MOM permeabilization. However, opening of the permeability transition pore has been seriously questioned by a few of the genetic studies (Tait and Green, 2010; Westphal et al., 2011). The MOM permeability induction of Bax and Bak seems to reside, in part, in the nature of their structures' central alpha-helices 5 and 6. As the alpha-helix 5 of Bcl-xL is replaced by the equivalent in Bax, it becomes sufficient to turn Bcl-xL to a "killer" protein (George et al., 2007).

Figure 2.18 shows how Bax and Bcl-xL roles are mutually inclusive (Martinou and Youle, 2011). In normally healthy cells, Bax (blue) and Bcl-xL (red) are translocated from the cytosol to MOM,

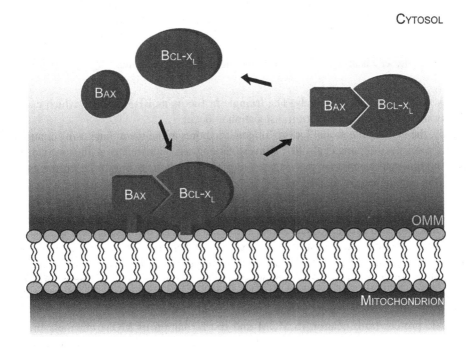

**FIGURE 2.18**  Bax moves back and forth from the cytosol to the MOM/OMM.

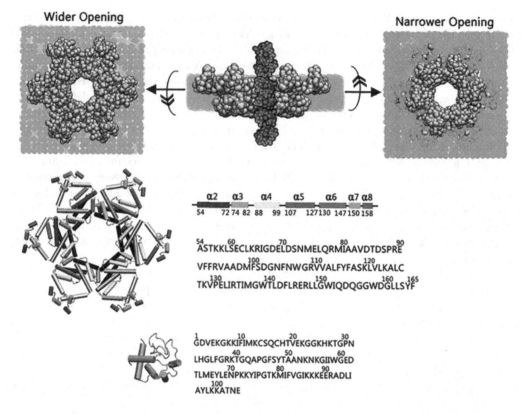

**FIGURE 2.19** Snapshots of coarse-grained Bax oligomeric pores (yellow and pink) with the cytochrome c (cyan) at the membranes (green) (top). Secondary structures (Cartoon), and the amino acid sequences of Bax and cytochrome c (bottom). For details see Zhang et al. (2017).

The free energy landscape here depicts a low barrier for permeation of cytochrome c into the Bax C-terminal mouth. The pathway proceeds all the way to the inner cavity and exits via the N-terminal mouth. Release process is guided by organized charged/hydrophilic surfaces. The hydrophilicity and negative charge of the pore surface is found to gradually increase along the release pathway from the pore entry all the way to the exit opening. For details, see Zhang et al. (2017). (For color figure see eBook figure.)

where they are found attached loosely. The triangle in Bax is its BH3 domain, which needs to be exposed for the Bcl-xL-mediated retrotranslocation to occur.

Recently, the first atomic model for Bax oligomeric pores at the membrane using computational approaches was solved (Zhang et al., 2017). Investigation was conducted using the mechanism at the microsecond time and nanometer space scale using MD simulations. Figure 2.19 presents a model depicting Bax pore (Zhang et al., 2017).

Bax participates in mitochondrial fission during apoptosis. Recently, a model has been proposed in which late-stage Bax oligomers was shown to play a functional part of the mitochondrial fragmentation machinery in apoptotic cells (Maes et al., 2019). Here, the requirement of a Bax protein for the fission process was demonstrated in Bax/Bak-deficient HCT116 cells expressing a P168A mutant of Bax. The mutant performed fusion to restore the mitochondrial network, but was not demonstrably recruited to the MOM as apoptosis was being induced. During these conditions, the mitochondrial fragmentation was blocked. Figure 2.20 presents a possible model for Bax participation in mitochondrial fission during apoptosis (Maes et al., 2019).

Apoptotic protein translocation to MOM and the formation of MOM channels involved in transporting apoptotic proteins get regulated due to localized molecular agents, certain type of

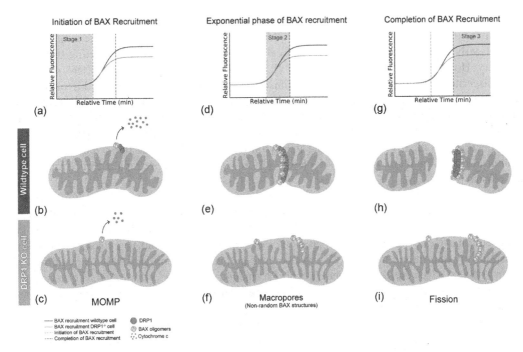

**FIGURE 2.20** Proposed model for Bax participation in mitochondrial fission during apoptosis is presented here. (a–c) Initiation of Bax recruitment (Stage 1) comprises Bax integration into the MOM, MOM permeabilization, and cytochrome c release. At this stage, (a) the time to initiate Bax recruitment is indistinguishable between wild-type and DRP1–/– cells. In wild-type cells, DRP1 has already been recruited to the MOM. (b, c) Cytochrome c is released in both cell types; however, DRP1–/– cells show a reduction in the percentage of cells releasing cytochrome c. This difference may be related to a proposed function of DRP1 in facilitating transfer of cardiolipin, which is associated with cytochrome c, from the inner membrane to the MOM (Smirnova et al., 2001). The release of other molecules, such as Smac, is not inhibited in DRP1–/– cells (Simonyan et al., 2017). Bax dimers and/or small oligomers are present at this stage. We define Bax structures at this stage as micropores. (d–f) The exponential phase of Bax recruitment (Stage 2) is defined as the period of Bax accumulation at the MOM. (d) The rate of Bax recruitment does not differ between wild-type and DRP1–/– cells. (e) At the site of DRP1 constriction of the mitochondria in wild-type cells, Bax accumulation takes on higher-order, nonrandom structures; however, (f) these structures lack organized localization at the scission site in the absence of DRP1. We define these Bax structures as macropore oligomers, which may be required for the release of larger molecules and/or evulsions of the mitochondrial inner membrane. (g–i) Completion of Bax recruitment (Stage 3) is identified as the point where the (g) Bax recruitment curve plateaus. (h) Importantly, this turning point is accompanied by mitochondrial fragmentation in wild-type cells. DRP1-mediated recruitment of Bax to the mitochondrial scission sites allows for nonrandom Bax structures or rings that are organized to completely encircle the mitochondrion. This likely weakens membrane integrity through presence of many adjacent macropores, creating a perforated membrane, and allowing constriction pressure created by DRP1 and its assemblies to complete the membrane separation. We define these Bax structures as fission oligomers. This model, where Bax perforates the membrane, is consistent with a model originally proposed about a decade ago by Martinou and Youle (Martinou and Youle, 2011). (i) Conversely, Bax structures on DRP1-deficient mitochondria are unable to organize around DRP1-identified scission sites, and therefore, cannot provide a perforated region to facilitate mitochondrial fragmentation. Loss or impairment of DRP1 function also results in fewer Bax puncta that contain approximately 25% less Bax protein. Therefore, DRP1-mediated fission oligomers may evolve both by coalescence of macropore oligomers and late-stage recruitment of inactive latent Bax monomers. For example, P168A mutant Bax cannot normally recruit to the MOM in apoptotic cells, unless in the presence of wild-type Bax protein. Kinetic studies indicate that this recruitment process occurs only when wild-type Bax nears the plateau phase of its recruitment when we predict that fission oligomers are being formed.

complexes or specific mechanisms. For example, the retromer is found to regulate apoptosis by facilitating the transport of Bcl-xL to the MOM (Farmer et al., 2019). Bcl-xL, the antiapoptotic Bcl-2 family protein, plays a critical role in cell survival mechanisms by protecting the MOM integrity. However, the mechanism that Bcl-xL utilizes to get recruited to the MOM is yet not fully discerned. The retromer complex is a conserved endosomal scaffold involved in membrane trafficking. VPS35 and VPS26, two core components of the retromer, have recently been identified as novel regulators of Bcl-xL (Farmer et al., 2019). Interactions and colocalization among Bcl-xL, VPS35, VPS26, and MICAL-L1 (a protein involved in recycling endosome biogenesis) have been observed. It is also observed that upon depletion of VPS35, the levels of nonmitochondrial Bcl-xL increase. Retromer-depleted cells also displayed more rapid Bax activation and apoptosis.

Figure 2.21 depicts the role of retromer in regulating the translocation of Bcl-xL to the MOM and impact on staurosporine-induced apoptosis (Farmer et al., 2019). Figure 2.21 shows the treatment of staurosporine-induced Bax translocation to the mitochondrial membrane under physiological conditions. Here, a competition is predicted between the Bcl-xL and Bax roles. As Bcl-xL is constitutively transported to the MOM, Bax pore formation is concomitantly inhibited and slowed down by Bcl-xL. However, as sufficient Bax pore formation occurs, cytochrome c (Cyt c) is released and apoptosis occurs. Upon VPS35 knockdown (see Figure 2.21, bottom), there is impaired generation of retromer complex and decreased constitutive transport of protein Bcl-xL to the MOM. Accordingly, upon the treatment of staurosporine, there is less inhibition of Bax protein by Bcl-xL, which leads to more rapid formation of Bax pores and increased rate of apoptosis.

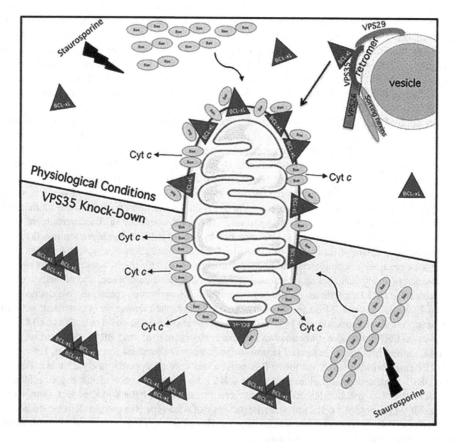

**FIGURE 2.21** Model depicting the role of retromer in regulating the translocation of Bcl-xL to the MOM and impact on staurosporine-induced apoptosis.

### 2.4.2 Lipid Regulation of Mitochondrial Channels Linked to Apoptotic Protein Transport

Mitochondrial membrane lipid compositios, in general, and presence of certain lipids, specifically, have regulatory effects on mitochondrial channels. The versatile roles of various lipids in diversified molecular processes of mitochondria, especially among mitochondrial channel structure and function determination, are crucial. For demonstration, we will address a few of them here.

#### 2.4.2.1  Lipidic Protein Channels in Mitochondrial Membrane

As an example, let us consider the nature of the pores formed by Bak and Bax and inspect the lipid arrangements in membrane hosting these pores. The activation of Bak and Bax by the BH3-only proteins is usually followed by their oligomerization in MOM to release cytochrome c and induce cell death. Bak and Bax are distinctive because Bak mostly gets inserted into MOM in healthy cells, but Bax mostly known as cytosolic, and is found to translocate to mitochondria following specific apoptotic stimuli. Translocation of Bax is triggered through binding of BH3-only proteins, a process that releases α9 from hydrophobic groove. As the released α9 inserts as a transmembrane domain into MOM, Bax resembles the same topology as the nonactivated Bak (Iyer et al., 2015).

Some investigations favor the formation of a proteinaceous channel due to these proteins (Dejean et al., 2005). However, other results are found to be consistent in favoring the lipidic pore (García-Sáez et al., 2007). An alpha-helix 5 of Bax, sufficient to perforate a membrane, was also found forming lipidic pores in synthetic membranes (Qian et al., 2008).

Subsequent developments and additional studies altogether support the in-plane model of a Bak dimer in membrane, see Figure 2.22 (Westphal et al., 2014; Uren et al., 2017a; 2017b). Bak and Bax homodimers are found to show several features that are found among antimicrobial peptides, such as human LL-37 and magainin 2. These peptides are long proposed (and now almost generally accepted) to form toroidal pores which are lipidic in nature. The lipid rearrangement in such lipidic pores are modeled and presented in Figure 2.23 (Uren et al., 2017a; 2017b).

#### 2.4.2.2  Ceramide Channels of Mitochondrial Membrane

Elevated ceramide levels in cell are known to inhibit phosphoinositide-3-kinase (PI3K) and Akt/-PBK signaling, which result in dephosphorylation, and subsequent activation of proapoptotic Bcl-2-family protein Bad (Zhu et al., 2011). Short-chain ceramides are found to be able to bind and stimulate protein phosphatase 2A (PP2A), known to dephosphorylate and is found to inactivate the antiapoptotic protein BCL 2 (Kowluru and Metz, 1997; Mukhopadhyay et al., 2008). Therefore, ceramide has considerable links to apoptotic processes, which are also linked to mitochondrial membrane transport properties including channels therein.

Ceramide forms channels in the MOM (Siskind et al., 2002; 2006). Ceramide channels are found to increase the MOM permeability to small proteins (Siskind et al., 2002). Ceramides was reported to induce the release of IMS proteins from mitochondria, the results considered to link to their ability to form large membrane channels (Siskind et al., 2006). This study also revealed that ceramide channel formation is specific to mitochondrial membranes and no channel formation was found to occur in plasma membranes of erythrocytes even at high ceramide concentrations, which was as high as 20 times required for channel formation in MOM. The MOM permeability can be regulated by controlling the size of the ceramide channels via the ceramide level in the membrane. The ceramide channel properties formed in a planar phospholipid membranes demonstrate that no proteins are required for channel formation, and that the channels are barrel-like structures. Figure 2.24 presents ceramide channels visualized by transmission electron microscopy (TEM) (Samanta et al., 2011).

Figure 2.25 is a ceramide pore model (Colombini, 2010). The size of the pore ranges between 5 and 40 nm pore diameters. The most frequent pores with 10 nm diameter were observed. These pores are large enough to comfortably allow all soluble proteins being translocated between the cytosol and the MIS (Colombini, 2016).

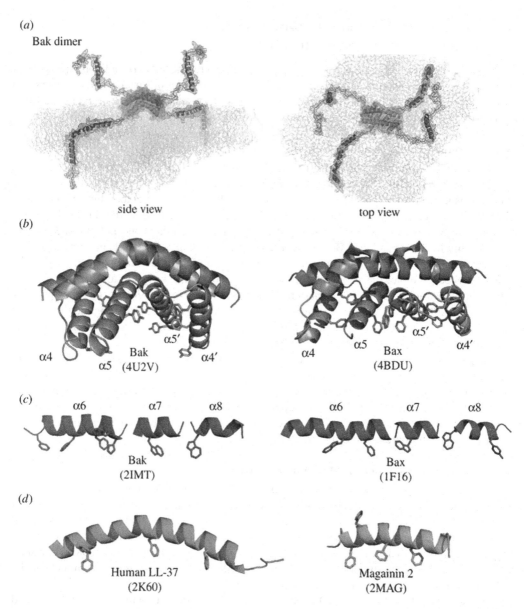

**FIGURE 2.22** Membrane topology of the Bak dimer. (a) The in-plane model of the Bak dimer. The N-terminal regions become solvent-exposed, while the remainder of the Bak dimer resembles a flexible extended amphipathic peptide that lies in-plane with the membrane, anchored at either end by transmembrane domains. Note that α1 may unfold after it dissociates, decreasing the hydrophobicity of the BH4 structural motif (VFrsYV) therein. Images were assembled in PyMol using the structures of Bak (2IMT) and the Bak dimer (4U2V), and are represented as cartoon and mesh. (b) Aromatic residues are concentrated on the bent surface of the Bak and Bax α2–α5 core dimers. (c) Aromatic residues can position on one edge of the flexible α6–α8 latch. (d) Examples of antimicrobial peptides thought to form lipidic pores, with aromatic residues indicated (Uren et al., 2017a; 2017b).

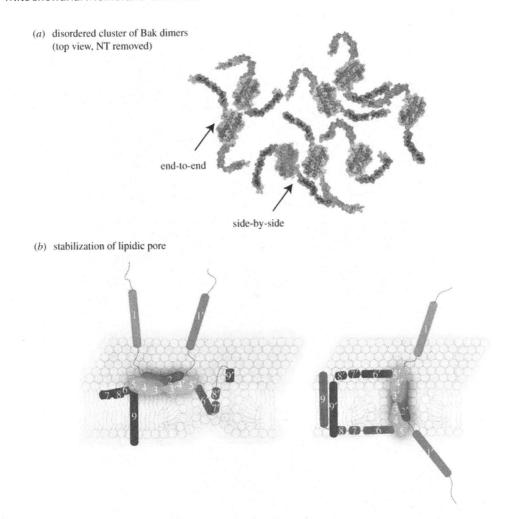

(*a*) disordered cluster of Bak dimers
(top view, NT removed)

end-to-end

side-by-side

(*b*) stabilization of lipidic pore

**FIGURE 2.23** Possible mechanisms involved in lipidic pore formation and stabilization by homodimers of Bak or Bax. (a) Schematic of Bak dimers forming a disordered cluster on the mitochondrial outer membrane, encouraged by flexibility of the α6–α8 latch. Note that end-to-end or side-by-side contact between the core regions is possible. (b) Parts of the dimer may line a lipidic pore. The flexible amphipathic latch may slide into a nascent pore to partially line and stabilize the pore. The amphipathic core dimer (α2–α5) may also line the pore generating antiparallel α9-helices. For further details, see Uren et al. (2017a; 2017b).

Ceramide channels are formed at mole fractions of the ceramides found in the MOM early in the apoptotic process, before or during protein release from mitochondria. Therefore, ceramide channels make good candidates for the protein release pathway, also supported by the fact that channel formation is inhibited by antiapoptotic proteins, but favored by Bax. Bcl-xL inhibits ceramide channel formation through binding to the apolar ceramide tails using its hydrophobic grove. Interaction of Bax with the ceramide polar regions results in MOM permeabilization through synergy with ceramide. This clearly suggests that Bcl-2 family proteins regulate the ceramide channels ability to permeabilize the MOM to proteins (Colombini et al., 2011).

Bcl-xL was engineered using point mutations that should specifically affect the interaction of ceramide with Bcl-xL (Chang et al., 2015). This helps probe the mechanism of regulation of ceramide channel and understand the physiological role of ceramide channels in apoptosis. The mutants and fluorescent ceramide molecules helped identify the hydrophobic groove on Bcl-xL, which was found to be the critical ceramide-binding site, and the regulator of the formation of

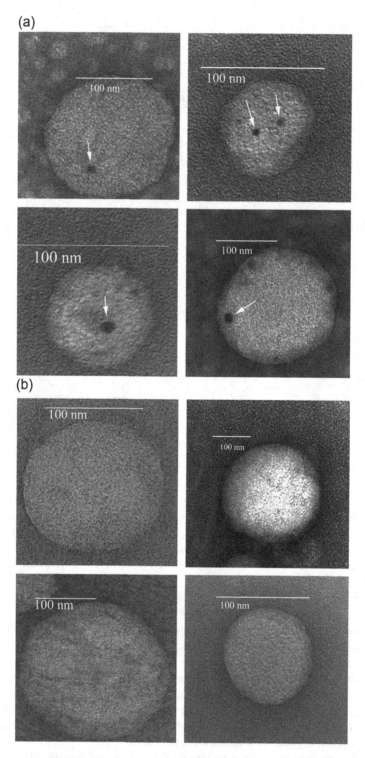

**FIGURE 2.24**  TEM scan confirming large ceramide channels in liposomes. (a) The ceramide-treated lipo-somes (some of them) were found to contain black circular regions (A). (b) The ceramide-untreated liposomes were found to contain no black circular regions, as seen in a. The black regions in, as seen in a (indicated by arrows) are interpreted as two-dimensional (2D) projections of ceramide channels filled with negative stain (uranyl acetate).

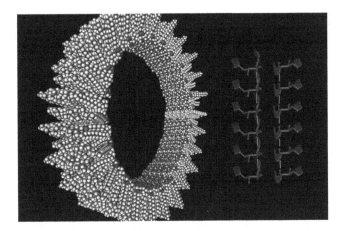

**FIGURE 2.25** Ceramide channel model of a 48-column ceramide channel. One column (in yellow) is shown to distinguish it from the entire structure. On the right, two columns are illustrated in stick mode to illustrate the hydrogen-bonding between the carbonyl oxygen of one ceramide molecule and amide hydrogen of the adjacent ceramide in the column. For details see Colombini (2010; 2016). (For color figure see eBook figure.)

ceramide channels. The hydrophobic pocket on Bcl-xL is known to bind ceramide, but the point mutations in Bcl-xL apolar pocket affects its ceramide channel inhibition potency.

Recently, a study reported that ceramides bind VDAC2 and trigger mitochondrial apoptosis (Dadsena et al., 2019). A photoactivatable ceramide probe was used to identify the voltage-dependent anion channels VDAC1 and VDAC2, and both were found to be mitochondrial ceramide-binding proteins. Coarse-grain MD simulations reveal that both VDAC1 and VDAC2 channels harbor a ceramide-binding site on one side of the barrel wall, see Figure 2.26 (Dadsena et al., 2019). This site includes a glutamate (membrane-buried) that mediates direct contact with the head group of ceramide. Substitution or any sort of chemical modification of this residue abolishes the photolabeling of both channels with the ceramide probe. Unlike the removal of VDAC1, loss of VDAC2 or replacing its membrane-facing glutamate with glutamine renders, human colon cancer cells were observed to largely become resistant to ceramide-induced apoptosis, see Figure 2.27 (Dadsena et al., 2019). This suggests that VDAC2 plays a role as the direct effector of ceramide-mediated cell death. Thus, it provides a molecular framework for ceramides on how they exert their antineoplastic activity.

### 2.4.2.3  Role of cardiolipin in Mitochondrial Membrane Transport

Cardiolipin is a unique phospholipid that is almost exclusively biosynthesized and located in MIM. Among all of the mitochondrial phospholipids, cardiolipin constitutes almost 15%–20%, and is found an essential constituent due to its vital roles in many mitochondrial processes, such as respiration and energy conversion. This phospholipid is associated with membranes generally designed to generate an electrochemical gradient used to produce ATP, for example, bacterial plasma membranes and MIM. Changes in composition of cardiolipin species is now evident in many diseases, concerning especially the signaling pathways. So, the role of cardiolipin pathology is enormous.

Cardiolipin is exposed to MOM upon mitochondrial stress. Cardiolipin domains serve as binding site(s) in cell signaling. In mitophagy (selective autophagy-based degradation of mitochondria), cardiolipin interacts with Beclin 1 and helps recruit the autophagic machinery through interaction with LC3 (a protein widely used as marker of autophagosomes). Cardiolipin is required in apoptotic signaling pathways for using it as a binding platform that helps recruit apoptotic factors, such as tBid, Bax, and caspase-8. Cardiolipin is required for activating the inflammasome, a process playing a role in inflammatory signaling. As an example, we present Figure 2.28 which demonstrates the

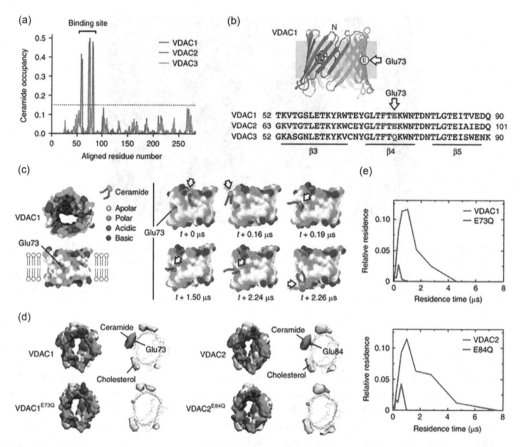

**FIGURE 2.26** MD simulations uncover a putative ceramide-binding site on VDAC1 and 2. (a) Ceramide head group contact occupancy of mouse VDAC1, VDAC2, and VDAC3 in an OMM model containing 5 mol% ceramide, with 1.0 corresponding to a ceramide contact during the entire simulation. A threshold of 15% occupancy, based on the occupancies of nonbinding site residues, is indicated by a dotted blue line. VDAC1 and VDAC2 have a clear ceramide-binding site, comprising residues 58–62, 71–75, and 81–85; this site is lacking in VDAC3. (b) Sequence alignment revealing the position of a bilayer-facing Glu residue in VDAC1 (E73) and VDAC2 (E84), which is replaced by Gln in VDAC3 (Q73). (c) Stills from an MD simulation showing the approach and binding of a ceramide molecule to VDAC1 in close proximity of the bilayer-facing Glu residue in its deprotonated state. Protein surface colors mark polar (green), apolar (white), cationic (blue), or anionic (red) residues. (d) Space-filling and wireframe models of VDAC1, VDAC1E73Q, VDAC2, and VDAC2E84Q with deprotonated E73/E84. Indicated are the volumes for which there is ceramide occupancy greater than 10% (orange) or cholesterol occupancy greater than 20% (yellow). (e) Distribution of the durations of ceramide contacts with VDAC1, VDAC1E73Q, VDAC2, and VDAC2E84Q at the preferred binding site as in (d). The y axis indicates the fraction of the total system time spent in binding events of the duration indicated by x. Summing all the y values of points yields the fraction of total simulation time when ceramide was bound. For details, see Dadsena et al. (2019). (For color figure see eBook figure.)

roles of cardiolipin in mitochondrial protein translocation and morphology, which are vital criteria in constructing channels of the mitochondrial membranes (Dudek, 2017).

$Ca^{2+}$ ions appear as the most important inducers of the mitochondrial permeability transition pore (MPTP). The exogenous added oxidized cardiolipins sensitize mitochondria to $Ca^{2+}$-induced opening of MPTP (Petrosillo et al., 2006). Oxidation of endogenous cardiolipin by tert-butyl hydroperoxide, similarly resulted in MPTP opening (Petrosillo et al., 2009). Both $Ca^{2+}$ and oxidized cardiolipin are predicted to interact with MPTP components, such as probably AAC. Cardiolipins

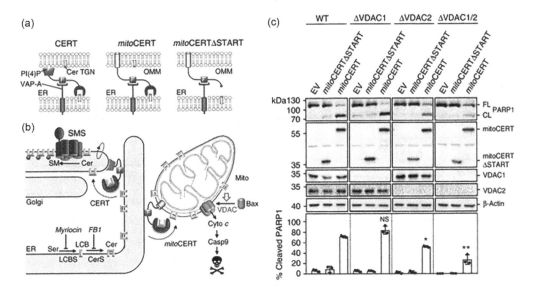

**FIGURE 2.27** VDAC2 removal disrupts ceramide-induced apoptosis. (a) Schematic outline of ceramide transfer protein CERT, mitoCERT, and mitoCERTΔSTART. MitoCERT was created by swapping the Golgi-targeting pleckstrin homology domain of CERT against the OMM anchor of AKAP1. Removal of the ceramide transfer or START domain yielded mitoCERTΔSTART. All three proteins bind the ER-resident protein VAP-A via their FFAT motif (F). Cer ceramide, PI(4)P phosphatidylinositol-4-phosphate, TGN trans-Golgi network. (b) Ceramides (Cer) are synthesized through N-acylation of long chain bases by ceramide synthases on the cytosolic surface of the ER and require CERT-mediated transfer to the Golgi for metabolic conversion into sphingomyelin (SM) by a Golgi-resident SM synthase (SMS). Expression of mitoCERT causes a diversion of this biosynthetic ceramide flow to mitochondria, triggering Bax-dependent apoptosis (Alphonse et al., 2004). (c) Wild-type (WT), VDAC1-KO (ΔVDAC1), VDAC2-KO (ΔVDAC1), and VDAC1/2 double KO (ΔVDAC1/2) human colon cancer HCT116 cells were transfected with empty vector, Flag-tagged mitoCERT, or Flag-tagged mitoCERTΔSTART. At 24h post transfection, cells were processed for immunoblotting with antibodies against PARP1, the Flag-epitope, VDAC1, VDAC2, and β-actin. The percentage of PARP1 cleavage was quantified. Data are means±SD; n = 3; *p < 0.05 and **p < 0.01 by two-tailed paired t-test. For details, see Dadsena et al., 2019.

are not only tightly associated with AAC (Beyer and Klingenberg, 1985), but also the interaction between two AAC monomers is found to be mediated by cardiolipins, perhaps stabilizing the dimeric structure (Nury et al., 2006). The oxidized cardiolipin- and $Ca^{2+}$-induced MPTP opening/-regulation is found associated with the cytochrome c release from mitochondria (Petrosillo et al., 2006). Figure 2.29 demonstrates the role of oxidized cardiolipin in cytochrome c release mechanism from mitochondria (Petrosillo et al., 2006; Paradies et al., 2014).

## 2.5 CONCLUDING REMARKS

We have addressed the various aspects of MOM and MIM-hosted channels in this chapter. These channels actively engage in maintaining mitochondrial functions including ensuring important molecular processes that involve the transport of various materials including electrolytes and proteins. The physiological roles of these channels have been briefly explained. Mitochondrial channels are linked to cellular diseases and are often considered drug targets. We have dedicated another chapter to address this issue in detail. There are a lot of important articles where considerable amount of information can be found on this topic (Hanahan and Weinberg, 2011; Leanza et al., 2014; Bachmann et al., 2019; Strickland et al., 2019). For additional details, see Chapter 8 that is especially dedicated to explaining diseases, drug targets, and drug discoveries concerning channels of mitochondria and other parts of biological cells.

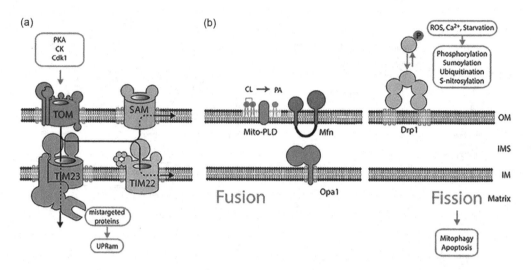

**FIGURE 2.28** CL plays an important role in mitochondrial protein translocation and morphology. (a) Protein translocases in the MOM (depicted as OM) and MIM (depicted as IM) are dependent on CL. Protein translocation is regulated by phosphorylation of components of the TOM complex. Mistargeted proteins activate the UPRam signaling pathway, creating responses that are activated by mistargeted proteins. (b) MOM fusion is mediated by MFN1 and MFN2, and fusion of the MIM requires OPA1. The phospholipase MitoPLD converts CL into phosphatidic acid (PA) facilitating MFN-mediated fusion. Fission is mediated by DRP1 oligomers recruitment to the MOM, where the oligomers interact with CL. For details, see Dudek (2017).

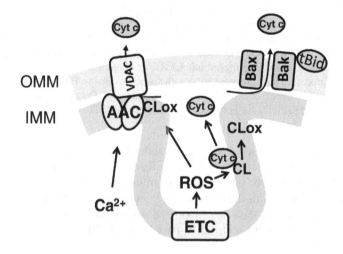

**FIGURE 2.29** The role of oxidized cardiolipin in cytochrome c release mechanism from mitochondria. ROS production causes cardiolipin oxidation which, in turn, promotes cytochrome c detachment from the IMM (should be read as MIM). Cytochrome c (Cyt c) is then released into cytosol through OMM (should be read as MOM),via MPTP or Bax/Bak oligomerization (for details see the text). ETC, electron transport chain; AAC, ADP/ATP carrier; CLox, oxidized cardiolipin.

## REFERENCES

Ahting, U., Thun, C., Hegerl, R., Typke, D., Nargang, F. E., Neupert, W., & Nussberger, S. (1999). The tom core complex. *The Journal of Cell Biology, 147*(5), 959–968. doi:10.1083/jcb.147.5.959

Alphonse, G., Bionda, C., Aloy, M., Ardail, D., Rousson, R., & Rodriguez-Lafrasse, C. (2004). Overcoming resistance to γ-rays in squamous carcinoma cells by poly-drug elevation of ceramide levels. *Oncogene, 23*(15), 2703–2715. doi:10.1038/sj.onc.1207357

Antignani, A., & Youle, R. J. (2006). How do Bax and Bak lead to permeabilization of the outer mitochondrial membrane? *Current Opinion in Cell Biology, 18*(6), 685–689. doi:10.1016/j.ceb.2006.10.004

Ashrafuzzaman, M. (2018). *Nanoscale Biophysics of the Cell.* Springer International Publishing AG, part of Springer Nature. doi:10.1007/978-3-319-77465-7

Ashrafuzzaman, M., & Tuszynski, J. (2012a). Regulation of channel function due to coupling with a lipid bilayer. *Journal of Computational and Theoretical Nanoscience, 9*(4), 564–570. doi:10.1166/jctn.2012.2062

Ashrafuzzaman, M., & Tuszynski, J. A. (2012b). *Membrane Biophysics.* Berlin, Heidelberg: Springer-Verlag. doi:10.1007/978-3-642-16105-6

Bachmann, M., Pontarin, G., & Szabo, I. (2019). The contribution of mitochondrial ion channels to cancer development and progression. *Cellular Physiology and Biochemistry, 53*(S1), 63–78. doi:10.33594/000000198

Ballarin, C., & Sorgato, M. C. (1995). An electrophysiological study of yeast mitochondria. *Journal of Biological Chemistry, 270*(33), 19262–19268. doi:10.1074/jbc.270.33.19262

Banerjee, R., Gladkova, C., Mapa, K., Witte, G., & Mokranjac, D. (2015). Protein translocation channel of mitochondrial inner membrane and matrix-exposed import motor communicate via two-domain coupling protein. *ELife, 4.* doi:10.7554/elife.11897

Becker, L., Bannwarth, M., Meisinger, C., Hill, K., Model, K., Krimmer, T., . . . Wagner, R. (2005). Preprotein translocase of the outer mitochondrial membrane: Reconstituted Tom40 forms a characteristic TOM pore. *Journal of Molecular Biology, 353*(5), 1011–1020. doi:10.1016/j.jmb.2005.09.019

Becker, T., & Wagner, R. (2018). Mitochondrial outer membrane channels: Emerging diversity in transport processes. *BioEssays, 40*(7), 1800013. doi:10.1002/bies.201800013

Beyer, K., & Klingenberg, M. (1985). ADP/ATP carrier protein from beef heart mitochondria has high amounts of tightly bound cardiolipin, as revealed by phosphorus-31 nuclear magnetic resonance. *Biochemistry, 24*(15), 3821–3826. doi:10.1021/bi00336a001

Böhm, G., Muhr, R., & Jaenicke, R. (1992). Quantitative analysis of protein far UV circular dichroism spectra by neural networks. *Protein Engineering, Design and Selection, 5*(3), 191–195. doi:10.1093/protein/5.3.191

Bredemeier, R., Schlegel, T., Ertel, F., Vojta, A., Borissenko, L., Bohnsack, M. T., . . . Schleiff, E. (2006). Functional and phylogenetic properties of the pore-forming β-barrel transporters of the Omp85 family. *Journal of Biological Chemistry, 282*(3), 1882–1890. doi:10.1074/jbc.m609598200

Chacinska, A., Lind, M., Frazier, A. E., Dudek, J., Meisinger, C., Geissler, A., . . . Rehling, P. (2005). Mitochondrial presequence translocase: Switching between TOM tethering and motor recruitment involves Tim21 and Tim17. *Cell, 120*(6), 817–829. doi:10.1016/j.cell.2005.01.011

Chang, K., Anishkin, A., Patwardhan, G. A., Beverly, L. J., Siskind, L. J., & Colombini, M. (2015). Ceramide channels: Destabilization by Bcl-xL and role in apoptosis. *Biochimica Et Biophysica Acta (BBA) - Biomembranes, 1848*(10), 2374–2384. doi:10.1016/j.bbamem.2015.07.013

Colombini, M. (2004). VDAC: The channel at the interface between mitochondria and the cytosol. *Molecular and Cellular Biochemistry, 256*(1/2), 107–115. doi:10.1023/b:mcbi.0000009862.17396.8d

Colombini, M. (2010). Ceramide channels and their role in mitochondria-mediated apoptosis. *Biochimica Et Biophysica Acta (BBA) - Bioenergetics, 1797*(6–7), 1239–1244. doi:10.1016/j.bbabio.2010.01.021

Colombini, M. (2016). Ceramide channels and mitochondrial outer membrane permeability. *Journal of Bioenergetics and Biomembranes, 49*(1), 57–64. doi:10.1007/s10863-016-9646-z

Colombini, M., Perera, M. N., Ganesan, V., & Siskind, L. J. (2011). Bcl-2 family proteins regulate the ability of ceramide channels to permeabilize the mitochondrial outer membrane to proteins. *Biophysical Journal, 100*(3). doi:10.1016/j.bpj.2010.12.301

Dadsena, S., Bockelmann, S., Mina, J. G., Hassan, D. G., Korneev, S., Razzera, G., . . . Holthuis, J. C. (2019). Ceramides bind VDAC2 to trigger mitochondrial apoptosis. *Nature Communications, 10*(1). doi:10.1038/s41467-019-09654-4

Dahlem, Y. A., Horn, T. F., Buntinas, L., Gonoi, T., Wolf, G., & Siemen, D. (2004). The human mitochondrial KATP channel is modulated by calcium and nitric oxide: A patch-clamp approach. *Biochimica Et Biophysica Acta (BBA) - Bioenergetics, 1656*(1), 46–56. doi:10.1016/j.bbabio.2004.01.003

Dayan, D., Bandel, M., Günsel, U., Nussbaum, I., Prag, G., Mokranjac, D., . . . Azem, A. (2019). A mutagenesis analysis of Tim50, the major receptor of the TIM23 complex, identifies regions that affect its interaction with Tim23. *Scientific Reports, 9*(1). doi:10.1038/s41598-018-38353-1

Denton, R. M., & Mccormack, J. G. (1985). Ca²⁺ transport by mammalian mitochondria and its role in hormone action. *American Journal of Physiology-Endocrinology and Metabolism, 249*(6). doi:10.1152/ajpendo.1985.249.6.e543

Dejean, L. M., Martinez-Caballero, S., Guo, L., Hughes, C., Teijido, O., Ducret, T., . . . Kinnally, K. W. (2005). Oligomeric Bax is a component of the putative cytochromecrelease channel MAC, mitochondrial apoptosis-induced channel. *Molecular Biology of the Cell, 16*(5), 2424–2432. doi:10.1091/mbc.e04-12-1111

Dejean, L. M., Martinez-Caballero, S., & Kinnally, K. W. (2006). Is MAC the knife that cuts cytochrome c from mitochondria during apoptosis? *Cell Death & Differentiation, 13*(8), 1387–1395. doi:10.1038/sj.cdd.4401949

Doan, K. N., Grevel, A., Mårtensson, C. U., Ellenrieder, L., Thornton, N., Wenz, L., . . . Becker, T. (2020). The mitochondrial import complex MIM functions as main translocase for α-helical outer membrane proteins. *Cell Reports, 31*(4), 107567. doi:10.1016/j.celrep.2020.107567

Dudek, J. (2017). Role of cardiolipin in mitochondrial signaling pathways. *Frontiers in Cell and Developmental Biology, 5.* doi:10.3389/fcell.2017.00090

Ellenrieder, L., Mårtensson, C. U., & Becker, T. (2015). Biogenesis of mitochondrial outer membrane proteins, problems and diseases. *Biological Chemistry, 396*(11), 1199–1213. doi:10.1515/hsz-2015-0170

Ellenrieder, L., Opaliński, Ł., Becker, L., Krüger, V., Mirus, O., Straub, S. P. . . . Becker, T. (2016). Separating mitochondrial protein assembly and endoplasmic reticulum tethering by selective coupling of Mdm10. *Nature Communications, 7*(1). doi:10.1038/ncomms13021

Endo, T., & Yamano, K. (2010). Transport of proteins across or into the mitochondrial outer membrane. *Biochimica Et Biophysica Acta (BBA) - Molecular Cell Research, 1803*(6), 706–714. doi:10.1016/j.bbamcr.2009.11.007

Fahanik-Babaei, J., Rezaee, B., Nazari, M., Torabi, N., Saghiri, R., Sauve, R., & Eliassi, A. (2020). A new brain mitochondrial sodium-sensitive potassium channel: Effect of sodium ions on respiratory chain activity. *Journal of Cell Science, 133*(10). doi:10.1242/jcs.242446

Farmer, T., O'Neill, K. L., Naslavsky, N., Luo, X., & Caplan, S. (2019). Retromer facilitates the localization of Bcl-xL to the mitochondrial outer membrane. *Molecular Biology of the Cell, 30*(10), 1138–1146. doi:10.1091/mbc.e19-01-0044

García-Sáez, A. J., Chiantia, S., Salgado, J., & Schwille, P. (2007). Pore formation by a Bax-Derived peptide: Effect on the line tension of the membrane probed by AFM. *Biophysical Journal, 93*(1), 103–112. doi:10.1529/biophysj.106.100370

García-Sancho, J. (2014). The coupling of plasma membrane calcium entry to calcium uptake by endoplasmic reticulum and mitochondria. *The Journal of Physiology, 592*(2), 261–268. doi:10.1113/jphysiol.2013.255661

George, N. M., Evans, J. J., & Luo, X. (2007). A three-helix homo-oligomerization domain containing $BH_3$ and $BH_1$ is responsible for the apoptotic activity of Bax. *Genes & Development, 21*(15), 1937–1948. doi:10.1101/gad.1553607

Gessmann, D., Flinner, N., Pfannstiel, J., Schlösinger, A., Schleiff, E., Nussberger, S., & Mirus, O. (2011). Structural elements of the mitochondrial preprotein-conducting channel Tom40 dissolved by bioinformatics and mass spectrometry. *Biochimica Et Biophysica Acta (BBA) - Bioenergetics, 1807*(12), 1647–1657. doi:10.1016/j.bbabio.2011.08.006

Giam, M., Huang, D. C., & Bouillet, P. (2008). BH3-only proteins and their roles in programmed cell death. *Oncogene, 27*(S1). doi:10.1038/onc.2009.50

Grigoriev, S. M., Muro, C., Dejean, L. M., Campo, M. L., Martinez-Caballero, S., & Kinnally, K. W. (2004). Electrophysiological approaches to the study of protein translocation in mitochondria. *International Review of Cytology,* 227–274. doi:10.1016/s0074-7696(04)38005-8

Habib, S. J., Waizenegger, T., Lech, M., Neupert, W., & Rapaport, D. (2004). Assembly of the TOB complex of mitochondria. *Journal of Biological Chemistry, 280*(8), 6434–6440. doi:10.1074/jbc.m411510200

Hanahan, D., & Weinberg, R. (2011). Hallmarks of cancer: The next generation. *Cell, 144*(5), 646–674. doi:10.1016/j.cell.2011.02.013

Hill, K., Model, K., Ryan, M. T., Dietmeier, K., Martin, F., Wagner, R., & Pfanner, N. (1998). Tom40 forms the hydrophilic channel of the mitochondrial import pore for preproteins. *Nature, 395*(6701), 516–521. doi:10.1038/26780

Höhr, A. I., Straub, S. P., Warscheid, B., Becker, T., & Wiedemann, N. (2015). Assembly of β-barrel proteins in the mitochondrial outer membrane. *Biochimica Et Biophysica Acta (BBA) - Molecular Cell Research, 1853*(1), 74–88. doi:10.1016/j.bbamcr.2014.10.006

Huang, S., & Klingenberg, M. (1996). Chloride channel properties of the uncoupling protein from brown adipose tissue mitochondria: A patch-clamp study. *Biochemistry, 35*(51), 16806–16814. doi:10.1021/bi960989v

Inoue, I., Nagase, H., Kishi, K., & Higuti, T. (1991). ATP-sensitive K+ channel in the mitochondrial inner membrane. *Nature, 352*(6332), 244–247. doi:10.1038/352244a0

Iyer, S., Bell, F., Westphal, D., Anwari, K., Gulbis, J., Smith, B. J., . . . Kluck, R. M. (2015). Bak apoptotic pores involve a flexible C-terminal region and juxtaposition of the C-terminal transmembrane domains. *Cell Death & Differentiation, 22*(10), 1665–1675. doi:10.1038/cdd.2015.15

Ježek, P., & Borecký, J. (1996). Inner membrane anion channel and dicarboxylate carrier in brown adipose tissue mitochondria. *The International Journal of Biochemistry & Cell Biology, 28*(6), 659–666. doi:10.1016/1357-2725(96)00008-8

Keil, P., Weinzierl, A., Kiebler, M., Dietmeier, K., Söllner, T., & Pfanner, N. (1993). Biogenesis of the mitochondrial receptor complex. Two receptors are required for binding of MOM38 to the outer membrane surface. *Journal of Biological Chemistry, 268*, 19177–19180.

Kirichok, Y., Krapivinsky, G., & Clapham, D. E. (2004). The mitochondrial calcium uniporter is a highly selective ion channel. *Nature, 427*(6972), 360–364. doi:10.1038/nature02246

Kooijman, E., Swim, L., Graber, Z., Tyurina, Y., Bayır, H., & Kagan, V. (2017). Magic angle spinning 31P NMR spectroscopy reveals two essentially identical ionization states for the cardiolipin phosphates in phospholipid liposomes. *Biochimica Et Biophysica Acta (BBA) - Biomembranes, 1859*(1), 61–68. doi:10.1016/j.bbamem.2016.10.013

Kowluru, A., & Metz, S. A. (1997). Ceramide-activated protein phosphatase-2A activity in insulin-secreting cells. *FEBS Letters, 418*(1–2), 179–182. doi:10.1016/s0014-5793(97)01379-3

Krabbendam, I. E., Honrath, B., Culmsee, C., & Dolga, A. M. (2018). Mitochondrial $Ca^{2+}$-activated $K^+$ channels and their role in cell life and death pathways. *Cell Calcium, 69*, 101–111. doi:10.1016/j.ceca.2017.07.005

Krüger, V., Becker, T., Becker, L., Montilla-Martinez, M., Ellenrieder, L., Vögtle, F., . . . Meisinger, C. (2017). Identification of new channels by systematic analysis of the mitochondrial outer membrane. *Journal of Cell Biology, 216*(11), 3485–3495. doi:10.1083/jcb.201706043

Kutik, S., Stojanovski, D., Becker, L., Becker, T., Meinecke, M., Krüger, V., . . . Wiedemann, N. (2008). Dissecting membrane insertion of mitochondrial β-barrel proteins. *Cell, 132*(6), 1011–1024. doi:10.1016/j.cell.2008.01.028

Leanza, L., Zoratti, M., Gulbins, E., & Szabo, I. (2014). Mitochondrial ion channels as oncological targets. *Oncogene, 33*(49), 5569–5581. doi:10.1038/onc.2013.578

Maes, M. E., Grosser, J. A., Fehrman, R. L., Schlamp, C. L., & Nickells, R. W. (2019). Completion of BAX recruitment correlates with mitochondrial fission during apoptosis. *Scientific Reports, 9*(1). doi:10.1038/s41598-019-53049-w

Mahendran, K., Romero-Ruiz, M., Schlösinger, A., Winterhalter, M., & Nussberger, S. (2012). Protein translocation through Tom40: Kinetics of peptide release. *Biophysical Journal, 102*(1), 39–47. doi:10.1016/j.bpj.2011.11.4003

Malhotra, K., Modak, A., Nangia, S., Daman, T. H., Gunsel, U., Robinson, V. L., . . . Alder, N. N. (2017). Cardiolipin mediates membrane and channel interactions of the mitochondrial TIM23 protein import complex receptor Tim50. *Science Advances, 3*(9). doi:10.1126/sciadv.1700532

Mannella, C. A., & Kinnally, K. W. (1997). Ion channels of mitochondrial membranes. *Biomembranes: A Multi-Volume Treatise Transmembrane Receptors and Channels, 377*–410. doi:10.1016/s1874-5342(96)80044-4

Marchi, S., Vitto, V. A., Danese, A., Wieckowski, M. R., Giorgi, C., & Pinton, P. (2019). Mitochondrial calcium uniporter complex modulation in cancerogenesis. *Cell Cycle, 18*(10), 1068–1083. doi:10.1080/15384101.2019.1612698

Martinou, J., & Youle, R. (2011). Mitochondria in apoptosis: Bcl-2 family members and mitochondrial dynamics. *Developmental Cell, 21*(1), 92–101. doi:10.1016/j.devcel.2011.06.017

Mishra, J., Jhun, B. S., Hurst, S., O-Uchi, J., Csordás, G., & Sheu, S. (2017). The mitochondrial $Ca^{2+}$ uniporter: Structure, function, and pharmacology. *Handbook of Experimental Pharmacology Pharmacology of Mitochondria, 129*–156. doi:10.1007/164_2017_1

Mitchell, P. (1961). Coupling of phosphorylation to electron and hydrogen transfer by a chemi-osmotic type of mechanism. *Nature, 191*(4784), 144–148. doi:10.1038/191144a0

Moran, O., Sciancalepore, M., Sandri, G., Panfili, E., Bassi, R., Ballarin, C., & Sorgato, M. (1992). Ionic permeability of the mitochondrial outer membrane. *European Biophysics Journal, 20*(6). doi:10.1007/bf00196590

Mukhopadhyay, A., Saddoughi, S. A., Song, P., Sultan, I., Ponnusamy, S., Senkal, C. E., . . . Ogretmen, B. (2008). Direct interaction between the inhibitor 2 and ceramide via sphingolipid-protein binding is involved in the regulation of protein phosphatase 2A activity and signaling. *The FASEB Journal, 23*(3), 751–763. doi:10.1096/fj.08-120550

Mulgrew-Nesbitt, A., Diraviyam, K., Wang, J., Singh, S., Murray, P., Li, Z., . . . Murray, D. (2006). The role of electrostatics in protein–membrane interactions. *Biochimica et Biophysica Acta, 1761*, 812–826.

Nury, H., Dahout-Gonzalez, C., Trézéguet, V., Lauquin, G., Brandolin, G., & Pebay-Peyroula, E. (2006). Relations between structure and function of the mitochondrial ADP/ATP carrier. *Annual Review of Biochemistry, 75*(1), 713–741. doi:10.1146/annurev.biochem.75.103004.142747

Olofsson, G., & Sparr, E. (2013). Ionization constants pKa of cardiolipin. *PLoS ONE, 8*(9). doi:10.1371/journal.pone.0073040

O'Rourke, B. (2007). Mitochondrial ion channels. *Mitochondria, 221*–238. doi:10.1007/978-0-387–69945-5_10

O'Rourke, B., Cortassa, S., & Aon, M. A. (2005). Mitochondrial ion channels: Gatekeepers of life and death. *Physiology, 20*(5), 303–315. doi:10.1152/physiol.00020.2005

Ott, C., Ross, K., Straub, S., Thiede, B., Gotz, M., Goosmann, C., . . . Kozjak-Pavlovic, V. (2012). Sam50 functions in mitochondrial intermembrane space bridging and biogenesis of respiratory complexes. *Molecular and Cellular Biology, 32*(6), 1173–1188. doi:10.1128/mcb.06388-11

Paradies, G., Paradies, V., Benedictis, V. D., Ruggiero, F. M., & Petrosillo, G. (2014). Functional role of cardiolipin in mitochondrial bioenergetics. *Biochimica Et Biophysica Acta (BBA) - Bioenergetics, 1837*(4), 408–417. doi:10.1016/j.bbabio.2013.10.006

Paschen, S. A., Waizenegger, T., Stan, T., Preuss, M., Cyrklaff, M., Hell, K., . . . Neupert, W. (2003). Evolutionary conservation of biogenesis of β-barrel membrane proteins. *Nature, 426*(6968), 862–866. doi:10.1038/nature02208

Petrosillo, G., Casanova, G., Matera, M., Ruggiero, F. M., & Paradies, G. (2006). Interaction of peroxidized cardiolipin with rat-heart mitochondrial membranes: Induction of permeability transition and cytochromecrelease. *FEBS Letters, 580*(27), 6311–6316. doi:10.1016/j.febslet.2006.10.036

Petrosillo, G., Moro, N., Ruggiero, F. M., & Paradies, G. (2009). Melatonin inhibits cardiolipin peroxidation in mitochondria and prevents the mitochondrial permeability transition and cytochrome c release. *Free Radical Biology and Medicine, 47*(7), 969–974. doi:10.1016/j.freeradbiomed.2009.06.032

Putney, J. W., & Thomas, A. P. (2006). Calcium signaling: Double duty for calcium at the mitochondrial uniporter. *Current Biology, 16*(18). doi:10.1016/j.cub.2006.08.040

Qian, S., Wang, W., Yang, L., & Huang, H. W. (2008). Structure of transmembrane pore induced by Bax-derived peptide: Evidence for lipidic pores. *Proceedings of the National Academy of Sciences, 105*(45), 17379–17383. doi:10.1073/pnas.0807764105

Qian, X., Gebert, M., Hpker, J., Yan, M., Li, J., Wiedemann, N., . . . Sha, B. (2011). Structural basis for the function of Tim50 in the mitochondrial presequence translocase. *Journal of Molecular Biology, 411*(3), 513–519. doi:10.2210/pdb3qle/pdb

Raffaello, A., Mammucari, C., Gherardi, G., & Rizzuto, R. (2016). Calcium at the center of cell signaling: Interplay between endoplasmic reticulum, mitochondria, and lysosomes. *Trends in Biochemical Sciences, 41*(12), 1035–1049. doi:10.1016/j.tibs.2016.09.001

Rizzuto, R., Simpson, A. W., Brini, M., & Pozzan, T. (1992a). Rapid changes of mitochondrial $Ca^{2+}$ revealed by specifically targeted recombinant aequorin. *Nature, 358*(6384), 325–327. doi:10.1038/358325a0

Rizzuto, R., Simpson, A. W., Brini, M., & Pozzan, T. (1992b). Correction: Rapid changes of mitochondrial $Ca^{2+}$ revealed by specifically targeted recombinant aequorin. *Nature, 360*(6406), 768–768. doi:10.1038/360768f0

Rusznák, Z., Bakondi, G., Kosztka, L., Pocsai, K., Dienes, B., Fodor, J., . . . Csernoch, L. (2007). Mitochondrial expression of the two-pore domain TASK-3 channels in malignantly transformed and non-malignant human cells. *Virchows Archiv, 452*(4), 415–426. doi:10.1007/s00428-007-0545-x

Samanta, S., Stiban, J., Maugel, T. K., & Colombini, M. (2011). Visualization of ceramide channels by transmission electron microscopy. *Biochimica Et Biophysica Acta (BBA) - Biomembranes, 1808*(4), 1196–1201. doi:10.1016/j.bbamem.2011.01.007

Sathappa, M., & Alder, N. N. (2016). The ionization properties of cardiolipin and its variants in model bilayers. *Biochimica Et Biophysica Acta (BBA) - Biomembranes, 1858*(6), 1362–1372. doi:10.1016/j.bbamem.2016.03.007

Shears, S. B., & Kirk, C. J. (1984). Determination of mitochondrial calcium content in hepatocytes by a rapid cellular fractionation technique. Vasopressin stimulates mitochondrial $Ca^{2+}$ uptake. *Biochemical Journal, 220*(2), 417–421. doi:10.1042/bj2200417

Siemen, D., Loupatatzis, C., Borecky, J., Gulbins, E., & Lang, F. (1999). $Ca^{2+}$-activated K channel of the BK-type in the inner mitochondrial membrane of a human glioma cell line. *Biochemical and Biophysical Research Communications, 257*(2), 549–554. doi:10.1006/bbrc.1999.0496

Simonyan, L., Légiot, A., Lascu, I., Durand, G., Giraud, M., Gonzalez, C., & Manon, S. (2017). The substitution of Proline 168 favors Bax oligomerization and stimulates its interaction with LUVs and mitochondria. *Biochimica Et Biophysica Acta (BBA) - Biomembranes, 1859*(6), 1144–1155. doi:10.1016/j.bbamem.2017.03.010

Siskind, L. J., Kolesnick, R. N., & Colombini, M. (2002). Ceramide channels increase the permeability of the mitochondrial outer membrane to small proteins. *Journal of Biological Chemistry, 277*(30), 26796–26803. doi:10.1074/jbc.m200754200

Siskind, L. J., Kolesnick, R. N., & Colombini, M. (2006). Ceramide forms channels in mitochondrial outer membranes at physiologically relevant concentrations. *Mitochondrion, 6*(3), 118–125. doi:10.1016/j.mito.2006.03.002

Smirnova, E., Griparic, L., Shurland, D., & Bliek, A. M. (2001). Dynamin-related protein Drp1 is required for mitochondrial division in mammalian cells. *Molecular Biology of the Cell, 12*(8), 2245–2256. doi:10.1091/mbc.12.8.2245

Sorgato, M. C., Keller, B. U., & Stühmer, W. (1987). Patch-clamping of the inner mitochondrial membrane reveals a voltage-dependent ion channel. *Nature, 330*(6147), 498–500. doi:10.1038/330498a0

Sreerama, N. (2003). Structural composition of betaI- and betaII-proteins. *Protein Science, 12*(2), 384–388. doi:10.1110/ps.0235003

Sreerama, N. (2004). On the analysis of membrane protein circular dichroism spectra. *Protein Science, 13*(1), 100–112. doi:10.1110/ps.03258404

Sreerama, N., & Woody, R. W. (2000). Estimation of protein secondary structure from circular dichroism spectra: Comparison of CONTIN, SELCON, and CDSSTR methods with an expanded reference set. *Analytical Biochemistry, 287*(2), 252–260. doi:10.1006/abio.2000.4880

Strickland, M., Yacoubi-Loueslati, B., Bouhaouala-Zahar, B., Pender, S. L., & Larbi, A. (2019). Relationships between ion channels, mitochondrial functions and inflammation in human aging. *Frontiers in Physiology, 10.* doi:10.3389/fphys.2019.00158

Suzuki, H., Kadowaki, T., Maeda, M., Sasaki, H., Nabekura, J., Sakaguchi, M., & Mihara, K. (2004). Membrane-embedded C-terminal segment of rat mitochondrial TOM40 constitutes protein-conducting pore with enriched β-structure. *Journal of Biological Chemistry, 279*(48), 50619–50629. doi:10.1074/jbc.m408604200

Szabò, I., Bock, J., Jekle, A., Soddemann, M., Adams, C., Lang, F., & Gulbins, E. (2005). A novel potassium channel in lymphocyte mitochondria. *Journal of Biological Chemistry, 280*(13), 12790–12798. doi:10.1074/jbc.m413548200

Szabò, I., Leanza, L., Gulbins, E., & Zoratti, M. (2011). Physiology of potassium channels in the inner membrane of mitochondria. *Pflügers Archiv - European Journal of Physiology, 463*(2), 231–246. doi:10.1007/s00424-011-1058-7

Szabò, I., & Zoratti, M. (1992). The mitochondrial megachannel is the permeability transition pore. *Journal of Bioenergetics and Biomembranes, 24*(1), 111–117. doi:10.1007/bf00769537

Szabò, I., & Zoratti, M. (2014). Mitochondrial channels: Ion fluxes and more. *Physiological Reviews, 94*(2), 519–608. doi:10.1152/physrev.00021.2013

Szewczyk, A. (2016). New mitochondrial potassium channels. *Biochimica Et Biophysica Acta (BBA) - Bioenergetics, 1857.* doi:10.1016/j.bbabio.2016.04.398

Szewczyk, A., Jarmuszkiewicz, W., & Kunz, W. S. (2009). Mitochondrial potassium channels. *IUBMB Life, 61*(2), 134–143. doi:10.1002/iub.155

Tait, S. W., & Green, D. R. (2010). Mitochondria and cell death: Outer membrane permeabilization and beyond. *Nature Reviews Molecular Cell Biology, 11*(9), 621–632. doi:10.1038/nrm2952

Tamura, Y., Harada, Y., Shiota, T., Yamano, K., Watanabe, K., Yokota, M., . . . Endo, T. (2009). Tim23–Tim50 pair coordinates functions of translocators and motor proteins in mitochondrial protein import. *Journal of Cell Biology, 184*(1), 129–141. doi:10.1083/jcb.200808068

Ting, S., Schilke, B. A., Hayashi, M., & Craig, E. A. (2014). Architecture of the TIM23 inner mitochondrial translocon and interactions with the matrix import motor. *Journal of Biological Chemistry, 289*(41), 28689–28696. doi:10.1074/jbc.m114.588152

Tsai, C., & Tsai, M. (2020). Mechanisms of MICU1 regulation of the mitochondrial calcium uniporter complex. *Biophysical Journal, 118*(3). doi:10.1016/j.bpj.2019.11.1060

Uren, R. T., Iyer, S., & Kluck, R. M. (2017a). Pore formation by dimeric Bak and Bax: An unusual pore? *Philosophical Transactions of the Royal Society B: Biological Sciences, 372*(1726), 20160218. doi:10.1098/rstb.2016.0218

Uren, R. T., O'Hely, M., Iyer, S., Bartolo, R., Shi, M. X., Brouwer, J. M., . . . Kluck, R. M. (2017b). Disordered clusters of Bak dimers rupture mitochondria during apoptosis. *ELife, 6.* doi:10.7554/elife.19944

Wei, M. C. (2001). Proapoptotic BAX and BAK: A requisite gateway to mitochondrial dysfunction and death. *Science, 292*(5517), 727–730. doi:10.1126/science.1059108

Westphal, D., Dewson, G., Czabotar, P. E., & Kluck, R. M. (2011). Molecular biology of Bax and Bak activation and action. *Biochimica Et Biophysica Acta (BBA) - Molecular Cell Research, 1813*(4), 521–531. doi:10.1016/j.bbamcr.2010.12.019

Westphal, D., Dewson, G., Menard, M., Frederick, P., Iyer, S., Bartolo, R., . . . Kluck, R. M. (2014). Apoptotic pore formation is associated with in-plane insertion of Bak or Bax central helices into the mitochondrial outer membrane. *Proceedings of the National Academy of Sciences, 111*(39). doi:10.1073/pnas.1415142111

Wiedemann, N., Kozjak, V., Chacinska, A., Schönfisch, B., Rospert, S., Ryan, M. T., . . . Meisinger, C. (2003). Machinery for protein sorting and assembly in the mitochondrial outer membrane. *Nature, 424*(6948), 565–571. doi:10.1038/nature01753

Xu, W., Liu, Y., Wang, S., Mcdonald, T., Eyk, J. E., Sidor, A., & O'rourke, B. (2002). Cytoprotective role of Ca$^{2+}$-activated K$^+$ channels in the cardiac inner mitochondrial membrane. *Science, 298*(5595), 1029–1033. doi:10.1126/science.1074360

Yi, M., Weaver, D., & Hajnóczky, G. (2004). Control of mitochondrial motility and distribution by the calcium signal. *Journal of Cell Biology, 167*(4), 661–672. doi:10.1083/jcb.200406038

Zhang, M., Zheng, J., Nussinov, R., & Ma, B. (2017). Release of cytochrome C from Bax pores at the mitochondrial membrane. *Scientific Reports, 7*(1). doi:10.1038/s41598-017-02825-7

Zhu, Q., Wang, Z., Ji, C., Cheng, L., Yang, Y., Ren, J., . . . Yang, Y. (2011). C6-ceramide synergistically potentiates the anti-tumor effects of histone deacetylase inhibitors via AKT dephosphorylation and α-tubulin hyperacetylation both in vitro and in vivo. *Cell Death & Disease, 2*(1). doi:10.1038/cddis.2010.96

Zoratti, M., Marchi, U. D., Gulbins, E., & Szabò, I. (2009). Novel channels of the inner mitochondrial membrane. *Biochimica Et Biophysica Acta (BBA) - Bioenergetics, 1787*(5), 351–363. doi:10.1016/j.bbabio.2008.11.015

Zorow, D. B., Kinnally, K. W., Perini, S., & Tedeschi, H. (1992). Multiple conductance levels in rat heart inner mitochondrial membranes studied by patch clamping. *Biochimica Et Biophysica Acta (BBA) - Biomembranes, 1105*(2), 263–270. doi:10.1016/0005-2736(92)90203-x

# 3 Ion Channels of the Nuclear Membrane – Physical Structures and Gating Mechanisms

The vast majority of the nuclear-encoded proteins are translated in the cytosol and may be imported into the organelle. Bidirectional cytosol↔nucleus trafficking of proteins and materials occurs continuously. Dual localization of mitochondrial and nuclear proteins is a reality. The nucleus is surrounded by a nuclear envelope (NE), and the double-layered nuclear membrane contains two lipid bilayers. Both bilayers are mostly identical in lipid composition and phospholipid mobility and are connected at annular junctions, where nucleoporin complexes (NPCs) are inserted. Nuclear membrane and NPCs work continually, independently, or through coordination to transport selective materials, including genetic substances, proteins, and electrolytes and maintain both short and long-distance bioinformatics communications between cytoplasmic and nucleoplasmic regions. Both transport and communication occur mainly through ion channels having versatile structures and selectivity. Detailed analysis of the structures and functions of these nuclear membrane channels is discussed in this chapter.

## 3.1 NUCLEAR MEMBRANE CONDUCTANCE AND CHANNELS

The nuclear membrane, the outer layer of the nucleus, is found in both animal and plant cells. The double-layered membrane of the NE are connected at annular junctions, where NPCs are inserted. NE separates the nuclear contents from its outside materials. The nuclei partition varieties of molecules from the cytosol, such as macromolecules, for example, proteins and RNA, small peptides, amino acids, sugars, and ions, for example, $K^+$, $Ca^{+2}$, and $Cl^-$. The nucleus is known to contain most of the genetic materials, so the NE is engaged in protecting them from the chemical reactions that occur outside the nucleus. The nuclear membrane controls selective unidirectional or bidirectional slow permeation and localized force-driven movement of materials and specific proteins across its double layer.

It is commonly known that NE restricts only larger molecules. NPC is assumed to contain a single diffusion cylindrical channel with a radius of ~5.35 nm and length of ~44.5 nm (Keminer and Peters, 1999), but it is flexible and can expand to accommodate particles up to ~39 nm in diameter (Panté and Kann, 2002). Does it mean that all particles smaller than this move freely across NE? Is selectivity any concern? We shall address these crucial questions in this chapter.

In a three-decade old study, in which a patch-clamp technique was applied to isolated murine pronuclei, NE was found to contain $K^+$-selective channels having multiple conductance states, with the maximal conductance being 200 pS (Mazzanti et al., 1990). These observed channels, contributing to the nuclear membrane potential, may be important in balancing the charges carried by macromolecular movements in and out of the nucleus. Similar to $K^+$-selective channels, there are varieties of channels that act in both nuclear outer membrane (NOM) and nuclear inner membrane (NIM), see Figure 3.1 (Matzke et al., 2010).

DOI: 10.1201/9781003010654-3

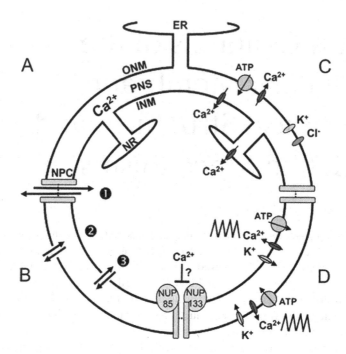

**FIGURE 3.1** The two membranes of NE, hosted ion channels and transporters, are modeled here. Various transport routes considering the structures of NE components are presented. (a) NOM or ONM, NIM or INM, PNS for perinuclear space, ER for endoplasmic reticulum, NR for nucleoplasmic reticulum, NPC for nuclear pore complex (depicted two yellow bars). (b) Three ion transport routes at the nuclear periphery (as bidirectional black arrows) are presented through NPC (1), across ONM (2), and across NIM (3). (c) Ion channel activities in mammalian cell NE. IP$_3$-regulated Ca$^{2+}$ channels in NIM (including the NR) and NOM are presumed to release Ca$^{2+}$ from the perinuclear space to the nucleoplasm and cytoplasm, respectively. A Ca$^{2+}$-ATPase identified in NOM. The direction of ion flows represented for K$^+$ and Cl$^-$ channels. (d) Nuclear membrane ion channels, transporters, and NUPs act in Nod factor signaling pathways. (For color figure see eBook figure.)

## 3.1.1 HISTORICAL PERSPECTIVES

We may wish to briefly discuss the historical progress before going into the available knowledge to date. In a six-decade old paper, Watson raised a few important questions and available evidence-based observations and conclusions (Watson, 1955). If the interphase nucleus exerted fundamental control over complex cytoplasmic activities such as protein synthesis, this control must be transmitted through the nuclear membrane, and this control must be partly mediated by large molecules that can pass between the nucleus and the cytoplasm (Anderson, 1953). The NE hosts nuclear pores which are large complexes of proteins that allow small molecules and ions to freely pass, or diffuse, into or out of the nucleus. Nuclear pores also allow necessary proteins to enter the nucleus from the cytoplasm. If the proteins have special sequences, it indicates that they belong in the nucleus.

Micromanipulation and microdissection of cells by Kite more than a century ago (Kite, 1913) and later by Chambers and Fell (Chambers and Fell, 1931) demonstrated the presence of a membrane around the nuclei of cells in interphase which resisted the penetration of the dissection needle. Once this barrier was passed, the needle could proceed through the nucleus without further interference. NEs teased from cells and examined in the electron microscope were found to consist of two parallel membranes exhibiting, on the surface, numerous ring-shaped structures which were considered to be pores, penetrating one or another of the membranes (Callan and Tomlin, 1950; Bairati and Lehmann, 1952; Harris and James, 1952; Gall, 1954). Here, we reproduce Watson's drawings on the schematic demonstration of the appearance of pores in the NE (see Figure 3.2).

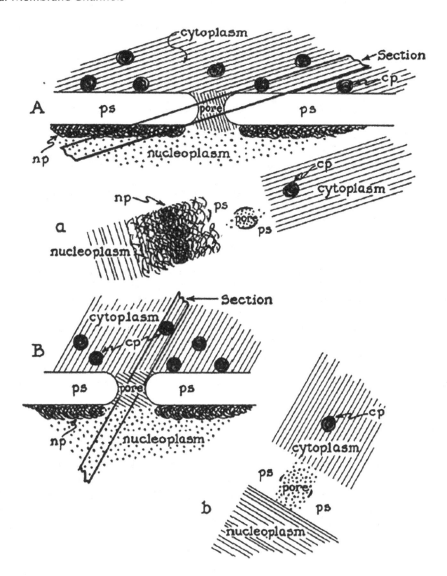

**FIGURE 3.2** Schematic demonstrating the appearance of pores in the nuclear envelope in sections oblique to the surface of the nucleus. (a) and (b) show the appearance of pores to be expected when the section is oriented. (a) The section is nearly tangential to the nuclear surface, thus, the pore appears almost in its entirety and is completely surrounded by the perinuclear space, ps. (b) On the other hand, the section is nearly normal to the nuclear surface, hence, only a small part of the circumference of the pore appears and a strip of material bridges the gap between the nucleoplasm and the cytoplasm. Symbols: ps: perinuclear space; cp: cytoplasmic particle; np: nuclear particulate matter. Reproduced from Watson (1955) to demonstrate the previous concepts about the nuclear pores and their physiological functions.

The pores used to be known as profiles of the ring-shaped structures. The inner and outer nuclear membranes are continuous with one another and enclose the perinuclear space. The pores contain a diffuse, faintly particulate material. It was already concluded by the mid-20th century that the presence of pores in the NE is a fundamental feature of all resting cells. In certain cells, the outer nuclear membrane or NOM is continuous with membranes of the endoplasmic reticulum; hence, the perinuclear space is continuous with cavities enclosed by these membranes. There were indications that this was true for all resting cells, at least in a transitory way. On the basis of these observations, a hypothesis was made that two pathways of exchange exist between the nucleus and the cytoplasm

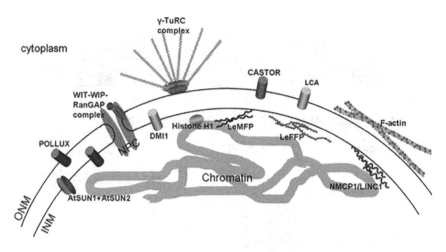

**FIGURE 3.3**  Plant NE intrinsic and associated proteins. Membrane-intrinsic plant proteins characterized to date include AtSUN1, AtSUN2, LCA, DMI1, CASTOR, POLLUX, as well as WIPs and WITs. Soluble proteins specifically localized at the nuclear periphery include RanGAP, γ-TURC, and f-actin on the cytoplasmic face, and NMCP1, LINC1, LeMFP1, LeFFPs, and histone H1 at the nucleoplasmic face. It is hypothesized that membrane-intrinsic NE membrane proteins are involved in tethering the soluble proteins to the NE; however, apart from RanGAP anchoring, the NE proteins and mechanisms involved in this process remain to be elucidated. RanGAP anchorage is mediated via WIP-WIT complexes and requires the presence of SUN proteins (Zhou et al., 2009; Graumann and Evans, 2010). ONM or NOM denote nuclear outer membrane, INM or NIM denote nuclear inner membrane.

by way of the perinuclear space and cavities of the endoplasmic reticulum and by way of the pores in the NE.

Recent investigations, based on historical progression, have produced further details about general nuclear membrane structures and specific nuclear membrane pore structures. The nuclear pore complex plays the role of a gatekeeper in traffic between the cytoplasm and the nuclear interior region. It is a large supramolecular complex composed of multiple copies of about 30 different proteins – 456 protein molecules overall. Many of the proteins described in animal and fungal systems have not been identified in plants. A simplified model on the NE is presented, see Figure 3.3 (Graumann and Evans, 2017).

### 3.1.2  CELL SIGNALING OF NUCLEAR MEMBRANES

The detailed structure of NE which is composed of the nuclear membranes, nuclear pore complex, and nuclear lamina is well known, see Figure 3.4 (Dauer and Worman, 2009). As explained earlier, NIM and NOM are separated by the perinuclear space, a continuation of the endoplasmic reticulum lumen. Consequently, proteins secreted into the endoplasmic reticulum, such as torsinA (TOR1A), can potentially reach the perinuclear space; a disease-causing torsinA variant preferentially accumulates in the perinuclear space by binding to lamina-associated protein-1. Some proteins, such as large nesprin-2 isoforms, concentrate in the outer nuclear membrane by binding to Sad1/UNC84 (SUN) domain proteins within the perinuclear space. The nuclear lamina is a meshwork of intermediate filaments on the inner aspect of the inner nuclear membrane and is composed of proteins called lamins. The lamina is associated with integral proteins of the NIM and representative examples, such as MAN1, lamina-associated polypeptide-1 (LAP1), the SUN protein lamina-associated polypeptide-2β (LAP2 2β), lamin B receptor (LBR), emerin, and a nesprin-1 isoform. The general structure of a lamin molecule is shown in the lower inset of the figure. Lamins have α-helical rod

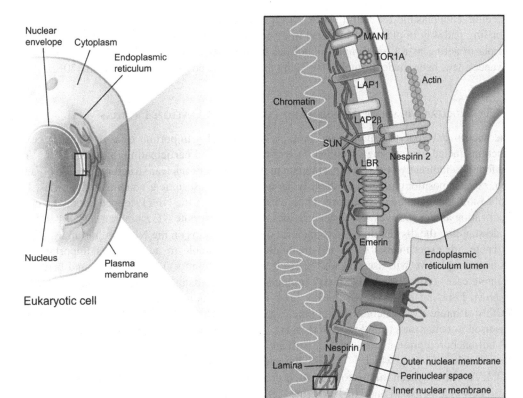

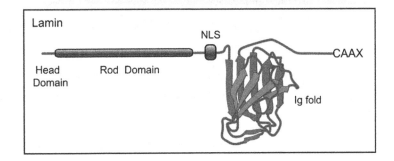

**FIGURE 3.4** Details of nuclear envelope and nuclear lamina.

domains that are highly conserved among all intermediate filament proteins and are critical for the formation of dimers and higher-ordered filaments. They have head and tail domains that vary in sequence among members of intermediate filament protein family. Within the tail domain, lamins contain a nuclear localization signal (NLS) and an immunoglobulin-like fold (Ig fold). Most lamins (not mammalian lamin C or C2) contain at their carboxyl-termini a CAAX motif that acts as a signal to trigger a series of chemical reactions leading to protein modification by fanesylation and carboxymethylation.

Numerous studies have recently uncovered new functions of NE proteins, underlying an emerging view of the NE as a critical signaling node in development and disease (see a bulk of references mentioned in the article Dauer and Worman (2009)). The first indication of a role for NE in calcium signaling was elucidated through calcium pumping ATPase of the sarcoplasmic reticulum/endoplasmic reticulum (SERCA) type called LCA (Downie et al., 1998). An NE $Ca^{2+}$ signaling pathway in plant cell involved in mycorrhizal infection and nodulation with the nuclear periplasm acting as a $Ca^{2+}$-signaling pool has been identified (Chabaud et al., 2010). Significant advances have been made through identifying the plant nucleoporins (Tamura et al., 2010).

### 3.1.3 Nuclear Membrane Translocation of Nuclear Envelope Proteins

The nuclear envelope transmembrane proteins (NETPs) play important roles in versatile cellular functions including ensuring sustained nuclear transports and participating in cell signaling. They are found to indirectly or directly participate in the creation of nuclear structure, nuclear organization, nuclear positioning, nuclear stability, as well as in transcription, splicing, epigenetics, DNA replication, and genome architecture. These vital functions of the NETPs depend upon the correct localization and/or their relative concentrations on the appropriate NE membranes NOM and NIM. Understanding the distribution and abundance of these proteins on the NE is important.

There are experimental tools and methodologies available to address this important topic. As an example, we will use one well-utilized technique "confocal microscopy" and brief on an example case study. Various landmark proteins have been identified which localize to the NOM or NIM. Labeling these known marker proteins with a fluorescent protein (e.g., mCherry) and NETPs of interest with a color fluorophore (e.g., eGFP) helps study colocalization. This approach was used to better understand how NETPs are targeted to the NIM in plants (Groves et al., 2019). ER tail-anchored proteins were tagged with an NLS and GFP and compared to the localization of the calnexin-mCherry, which is a well-characterized protein, located exclusively on the ER and NOM. As the Airyscan microscopy method does not have sufficient resolution to visually distinguish between the NOM and NIM, NE line scans were used to determine the colocalization of the two proteins, see Figure 3.5 (Tingey et al., 2019). Even line scans are limited by the overall system resolution, so several standard statistical analyses are performed on the line scan results before producing considerable conclusions (Groves et al., 2019).

Despite being simple and easy to use, this technique shows several pitfalls as follows:

- Overexpression of NETPs can result in NE mislocalization due to the leaky nature in which proteins are regulated by the NPC. This can be determined if the NOM marker, calnexin, is found on the NIM as the line scan shows false colocalizations.
- There is an uncertainty associated with the fitting of a line scan for determining the peaks.
- Line scans, a one-dimensional analysis method, do not provide information regarding diffusion or relative enrichment on the NOM or NIM.

## 3.2 NUCLEAR PORE COMPLEX TRANSPORTS AND THEIR ASSOCIATION WITH NUCLEAR ION CHANNELS

The two phospholipid bilayers of NE, NOM and NIM, meet in places where NPCs are embedded. In addition to facilitating the nucleocytoplasmic transport, NPCs are engaged in chromatin organization, regulation of gene expressions, and DNA repair mechanism.

NPC is the largest macromolecular complex of the cell. Its molecular mass ranges from ~50 MDa in yeast to ~112 MDa in vertebrates. It is composed of multiple copies, typically between 8–64 copies, of about 30 different nucleoporin proteins (Beck and Hurt, 2016). NPCs fuse the NIM and NOM to form channels across the NE. Transmembrane nucleoporins anchor the NPC in the NE, but structural nucleoporins are known to form a scaffold that shapes an hourglass passageway or pore with a diameter ranging between 35 nm (in yeast) and 50 nm (in vertebrates). The passageway lumen is filled with intrinsically disordered nucleoporins assembly harboring repeats of hydrophobic phenylalanine and glycine motifs.

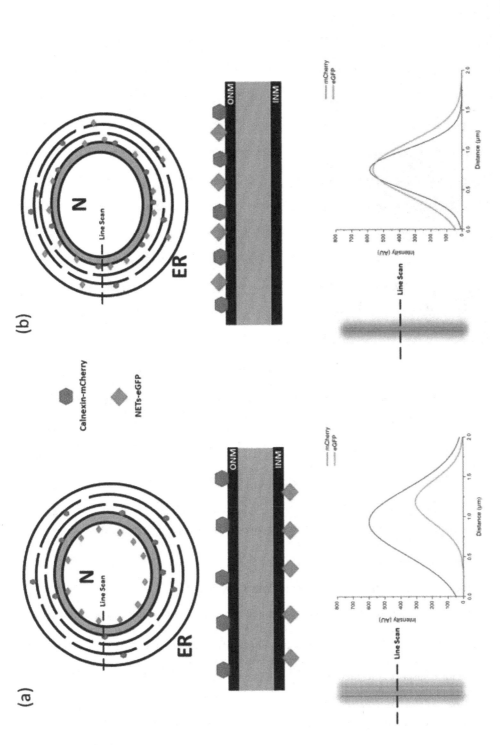

**FIGURE 3.5** A model of differential staining using Airyscan confocal microscopy. (a) Calnexin tagged with mCherry (in red color) localizes (exclusively) to the NOM and ER. The NETPs of interest, tagged with GFP (in green color), are enriched at the NIM. A line profile is generated after performing a line scan, indicating that the two fluorophores do not colocalize. (b) A line scan of NETPs, tagged with GFP (in green color), that are enriched at the NOM, and ER colocalizes with calnexin. (For color figure see eBook figure.)

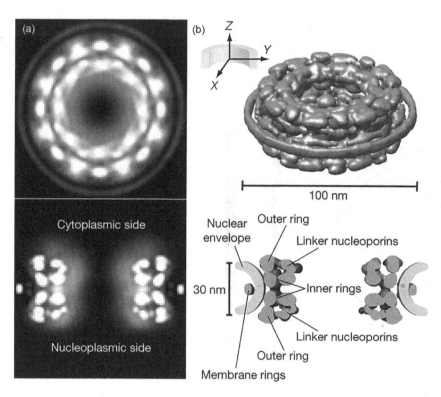

**FIGURE 3.6** (a) Projections of nucleoporin mass density (white) derived from the combined localization volumes of all structured domains and the normalized localization probability of all unstructured regions. Top, *en face* view showing a density projection along the Z axis from Z=−50 nm to Z=+50 nm. As in electron microscopy maps of the NPC, radial arms of density correspond to spokes that interconnect to form two strong concentric rings encircling a central region containing low-density unstructured material and bounded by peripheral membrane rings, giving an overall diameter of ~98 nm. Bottom, a slice along the central z axis showing a projection of density from X=−5 nm to X=+5 nm. More density can be seen on the cytoplasmic side of the NPC. The low-density unstructured material constricts the central channel to ~10 nm diameter. (b) The structured nucleoporin domains of the NPC are represented by a density contour (blue) such that the volume of the contour corresponds approximately to the combined volume of the 456 nucleoporins comprising the NPC. Top: view from a point ~30° from the equatorial plane of the NPC. Bottom: a slice along the central z axis between X=−5 nm and X=5 nm, in which the nuclear envelope is also shown (in gray). Major features of the NPC are indicated. (For color figure see eBook figure.)

### 3.2.1 Molecular Architecture of NPC and Pore Channel

Alber and coworkers investigated the molecular architecture of the NPC (Figure 3.6) of the yeast cells (Alber et al., 2007). They showed that half of the NPC is made up of a core scaffold, which is structurally analogous to vesicle-coating complexes. This scaffold forms an interlaced network that coats the entire curved surface of the NE membrane within which the complex is embedded. The selective barrier for transport is formed by a large number of proteins with disordered regions that line the inner face of the scaffold.

The major NPC structural components are the inner pore ring (which resides at the fused NIM and NOM), the nuclear and cytoplasmic rings (which are anchored by the inner pore ring), the nuclear basket, and the cytoplasmic filaments (which are peripheral elements originating from the nuclear and cytoplasmic rings) (Fahrenkrog and Aebi, 2003; Schwartz, 2016; Zhou et al., 2018). For a model demonstration of the NPC complex and associated processes, see Figure 3.7 (Beck and Hurt, 2016).

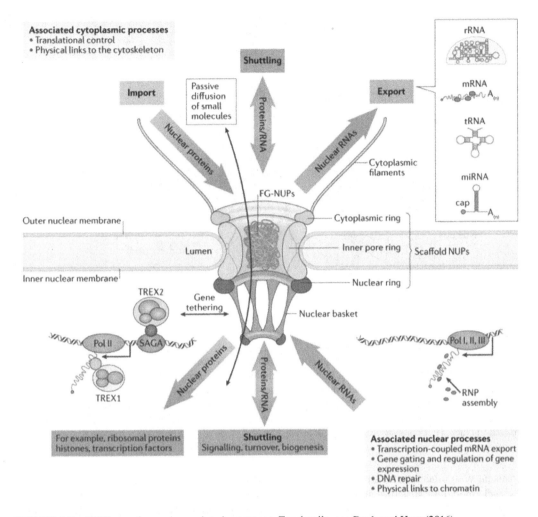

**FIGURE 3.7** NPC complex and associated processes. For details, see Beck and Hurt (2016).

NE-penetrated NPCs regulate the nucleocytoplasmic trafficking of macromolecules, and certain NPC components influence the multitude of genomic functions in a transport-independent manner. Nucleoporins (Nups) play an evolutionarily conserved role in gene expression regulation that (in metazoans) extends into the nuclear interior. In proliferative cells, Nups play a crucial role in genome integrity maintenance and mitotic progression. Some Nups located in the NPC pore perform vital roles in the gating mechanism of the NPC pores. For pore interior schematics structure, see Figure 3.8 (Ibarra and Hetzer, 2015). Three of the pore channel interior proteins, Nup62, Nup54, and Nup58, play active roles in forming the dynamic complexes (Sharma et al., 2015).

Sharma and colleagues tried to understand the molecular mechanisms active in the ordered regions of channel nucleoporins Nup62, Nup54, and Nup58, regarding the role(s) of these proteins in forming the dynamic complexes (Sharma et al., 2015). Out of ~30 nucleoporins, Nup62, Nup54, and Nup58 line the nuclear pore channel. These "channel-forming" nucleoporins contain an ordered region of ~150–200 residues, which is segmented into 3–4 α-helical regions with ~40–80 residues. The segmentations are evolutionarily conserved between uni- and multicellular eukaryotes.

The analysis by utilizing the entire ordered regions of pore interior-aligned Nup62, Nup54, and Nup58 demonstrates that they participate in forming a dynamic "triple complex," heterogeneously formed from Nup54·Nup58 and Nup54·Nup62 interactomes. These data and interpretations are

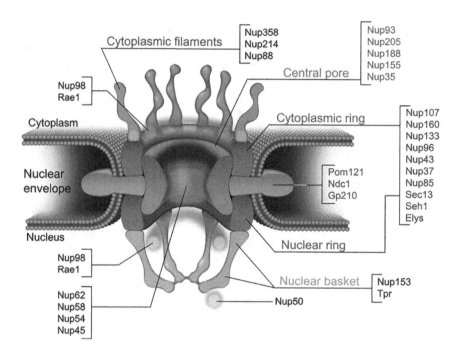

**FIGURE 3.8** NPC structure and molecular composition. Representation and predicted molecular composition of the vertebrate NPC. For the corresponding orthologs in other model organisms, see Rothballer and Kutay (2012). The NPC core scaffold and its components are represented in dark blue and light blue, the transmembrane subunits are shown in green, the cytoplasmic structures are represented in magenta, and the basket is shown in orange. Peripheral Nups showing higher mobility by FRAP and other approaches are depicted in black. (For color figure see eBook figure.)

consistent with other crystal structure-deduced copy numbers and stoichiometries, as well as with ring cycle model for structure and dynamics of the nuclear pore channel. Piecing together the crystal structures of channel nucleoporin segments into an understandable model for a nuclear pore channel is presented in Figure 3.9 (Sharma et al., 2015).

### 3.2.2 ENERGETICS OF THE NPC PORE TRANSPORT

NPCs are known to both import and export macromolecular cargoes into and out of the nucleus against the concentration gradients of the cargoes. The NPC provides two types of nucleocytoplasmic transport, namely, the passive diffusion of small molecules and the active chaperon-mediated translocation of large molecules. Many biophysical features of NPC transport have been demonstrated (Ghavami et al., 2016; Jovanovic-Talisman and Zilman, 2017).

From the thermodynamic perspective, the required energy inputs are provided by Guanosine-5′-triphosphate (GTP) hydrolysis during the transport cycle (Cautain et al., 2014). Unlike other types of transporters that generally require flip/flop-type conformational changes, NPCs do not possess the back-and-forth open↔close transitions of gates between states, see Figure 3.10 (Jovanovic-Talisman and Zilman, 2017).

Translocation of individual transport protein–cargo complexes is not directly coupled to GTP hydrolysis and occurs by thermal diffusion. Transport protein–cargo complexes can transport in both directions through the NPC, often resulting in abortive translocations. This is observed using single-molecule fluorescence microscopy and measurements of bulk fluxes (Kopito and Elbaum, 2007). The vital function of the transport proteins is to ferry cargo, but they can also translocate

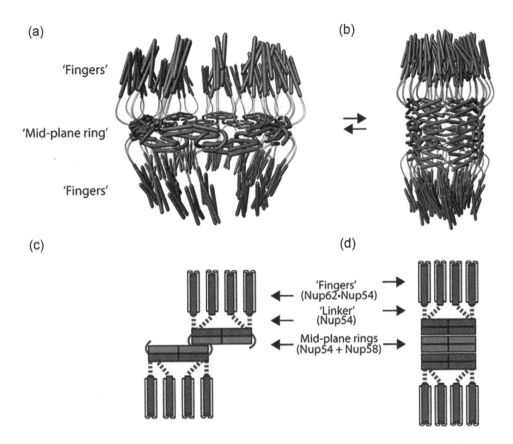

**FIGURE 3.9** Piecing together crystal structures of channel nucleoporin segments into a model for a nuclear pore channel. Interconvertible dilated (a) and constricted (b) channel states, respectively. The segment paths of ordered regions of channel Nups (Nup62 (gray), Nup54 (blue), Nup58 (red) through two principal structural elements of the channel, mid-plane ring, and attached fingers are shown. The latter project to cytoplasm and nucleoplasm. (c, d) A single module (out of all eight) is represented, each for the dilated (c) or constricted (d) channel states. The "linker region" of Nup54 is indicated either by curved lines (see a and b) or by dotted lines (see c and d). For details, see Sharma et al. (2015). (For color figure see eBook figure.)

through NPCs without cargo. Cargo-bound and cargo-free transport proteins, multiple copies, are found in NPCs. They play a role in making or shaping the NPC structures (Cautain et al., 2014).

The interaction between intrinsically disordered proteins lining the central channel of the NPC and the transporting cargoes is an important determining factor. Coarse-grained molecular dynamics simulations have been performed to quantify the energy barrier active against molecules passing through the NPC (Ghavami et al., 2016).

Two aspects of the transport are discussed and briefly explained in this study.

First, the passive transport of a few model cargo molecules with different sizes has been studied, and the NPC size selectivity feature has been investigated. This study suggests that the transport probability of cargoes gets significantly reduced when their size (diameter) surpasses ~5 nm. Figure 3.11 shows the correlation of the energy barrier and the cargo diameter (Ghavami et al., 2016). The energy barrier is defined as the difference between the mean value of the potential of mean force at $-5.0 \text{ nm} < z < 5.0 \text{ nm}$ and $20 \text{ nm} < z < 27 \text{ nm}$, and is an indication of the work required to translocate cargoes from the cytoplasm to the NPC core region.

Second, simulations show that incorporating hydrophobic binding spots on the surface of the cargo effectively decreases the energy barrier of the pore. This computational finding is consistent

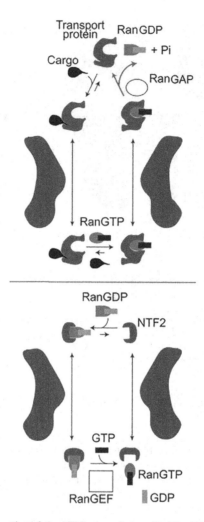

**FIGURE 3.10**  Simplified schematics of the NPC operating cycle. The NPC import process has two inter-linked cycles. The first one (upper panel) uses the energy that is released by the hydrolysis of one GTP molecule, catalyzed by RanGAP, to import one molecular cargo into the nucleus, relying on the higher RanGTP concentration in the nucleus, relative to the cytoplasm. The red line is the only nonequilibrium step of the import cycle, which depends on the metabolic energy release in GTP hydrolysis form. The nucleocytoplasmic RanGTP/RanGDP gradient, maintained by the second cycle, is shown in the lower panel, which relies on the high RanGEF concentration in the nucleus due to its association with chromatin (Jovanovic-Talisman and Zilman, 2017).

with the experimental findings of Naim et al. (2009), who reported that large inert cargoes were able to transport through the pore when hydrophobic amino acid side chains were attached to the cargo's surface.

Finally, a simple transport model is proposed which characterizes the energy barrier of the NPC as a function of diameter and hydrophobicity of the transporting particles. The attachment of hydrophobic binding spots to the cargo complex surface lowers the energy barrier below $k_B T$, facilitating the transport of large cargo molecules, see Figure 3.12 (Ghavami et al., 2016). The energy map suggests that the efficient transport occurs in a strip confined between two iso-lines of $+k_B T$ and $-k_B T$ (gray area in Figure 3.12). The region below the $+k_B T$ line holds a small number of hydrophobic bindings spots to reduce the free energy barrier enough for possible transporting. The area above

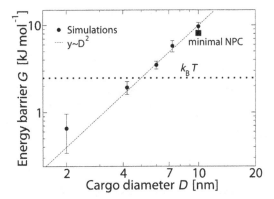

**FIGURE 3.11** The energy barrier G plotted against the cargo diameter D. The dashed line is a quadratic fit to the data and the error bars indicate the standard deviation of the data for the interval −5.0 nm < z < 5.0 nm and 20 nm < z < 27 nm.

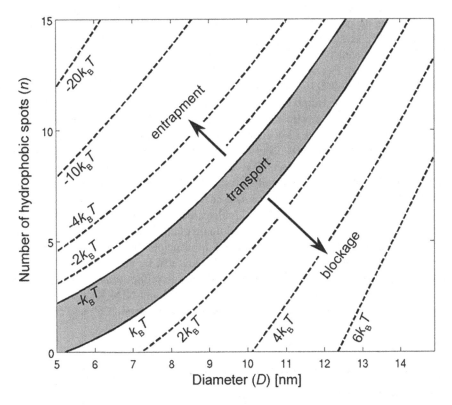

**FIGURE 3.12** Contour plot of the energy barrier G of the NPC as a function of cargo diameter D and the number of hydrophobic binding spots n.

the −$k_B$T line holds a large number of hydrophobic binding spots on the cargo. These spots have a considerably high affinity to the FG-Nups resulting in the entrapment of the cargo inside the NPC. Thus, we may conclude that the hydrophobicity and spacing between binding spots are crucial in controlling the active transport. The number and spacing of binding spots may even determine whether a cargo can be expelled from, transported through, or trapped inside the NPC.

## 3.3 NUCLEAR OUTER MEMBRANE CHANNELS

The presence of ion channels in the NOM of NE was confirmed about three decades ago in various investigations (Mazzanti et al., 1990; 1991; Tabares et al., 1991; Matzke et al., 1992; Bustamante, 1992; 1994). The patch-clamp technique characterized the channel conductance in different cell nuclei, namely, large and the low-conducting Cl⁻ channels, as well as cation-selective channels (Bustamante, 1994).

$Ca^{2+}$, $K^+$, and Cl⁻ are among the most important ions known to cross NOM. That is, both cations and anions cross across NOM. The NIM is enriched with processes allowing cations (explained later in details). The aspects of NOM transport events or NOM-hosted ion channels are discussed here.

### 3.3.1 CATION CHANNELS OF THE NUCLEAR OUTER MEMBRANE

Ion gradients can be generated and maintained across either the NOM or NIM, with ions stored in and released from the perinuclear space. Besides their identification in proteomics analyses (Matzke et al., 2010), NOM and NIM ion channels and transporters have been discovered using versatile techniques and approaches, namely, genetic, immunological, pharmacological, and electrophysiological. The most comprehensive results concern $Ca^{2+}$ ions, and there is now ample evidence, particularly from animal cells, favoring the presence of $Ca^{2+}$ channels and transporters in nuclear membranes.

Major intracellular channels releasing $Ca^{2+}$ in animal cells are inositol (1,4,5)-triphosphate receptors (IP₃Rs) and the related ryanodine receptors (RyRs) (Gerasimenko and Gerasimenko, 2004; Bootman et al., 2009). IP₃R has been detected in both the NOM and NIM of animal cells (Humbert et al., 1996; Bootman et al., 2009). The NIM channels enable $Ca^{2+}$ release from perinuclear stores directly into the nucleoplasm (Humbert et al., 1996; Bootman et al., 2009). Functioning RyRs have been localized to the NIM (Gerasimenko and Gerasimenko, 2004). $Ca^{2+}$-ATPases and inositol 1,3,4,5-tetrakisphosphate-operated $Ca^{2+}$ channels replenishing the $Ca^{2+}$ store in the perinuclear space have been detected in the NOM, but not yet in the NIM of animal cells (Humbert et al., 1996; Bootman et al., 2009). In addition, IP₃Rs and RyRs are also observed in the nucleoplasmic reticulum allowing the subnuclear control of $Ca^{2+}$ signaling (Echevarría et al., 2003; Bootman et al., 2009). Figure 3.13 presents pathways that regulate nuclear $Ca^{2+}$ levels (Bootman et al., 2009).

Nuclear patch clamping has been successful in detecting the different types of $K^+$ channels with varying conductance in the NE of animal cells (Guihard et al., 2000; Mazzanti et al., 2001; Bkaily et al., 2009). Patch clamping of intact nuclei should, in principle, primarily detect channels in the NOM (Bkaily et al., 2009), although NIM channels may also be accessible to the patch pipette when an isolated nuclear membrane is incorporated into proteoliposomes (Matzke et al., 1990; Guihard et al., 2000). Membrane potentials created by NOM and NIM $K^+$ channels can regulate other nuclear membrane voltage-dependent ion channels, such as certain $Ca^{+2}$ and chloride channels (Bkaily et al., 2009).

### 3.3.2 ANION CHANNELS OF THE NUCLEAR OUTER MEMBRANE

Anion channels in nuclear membrane have been extensively investigated. For example, chloride channels in the NOM were demonstrated in early studies (Tabares et al., 1991). Ongoing studies keep producing new discoveries (Fedorenko et al., 2007; Marchenko and Fedorenko, 2011; Fedorenko and Marchenko, 2014).

By applying patch-clamp technique on isolated nuclei of pyramidal neurons, from the hippocampal CA1 area, the biophysical properties of the spontaneously active ion channels in the nuclear membranes was made in 2007 (Fedorenko et al., 2007). In the NOM, anion channels were found to have a unitary conductance of 156 pS and very rapid fluctuation kinetics, see Figure 3.14 (Fedorenko et al., 2007). In the NIM, cationic channels were recorded with a unitary conductance of 248 pS and very slow kinetics (not shown here).

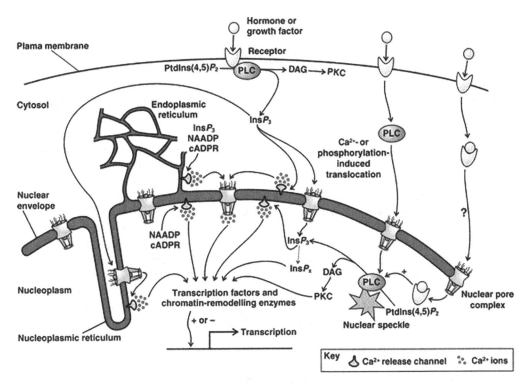

**FIGURE 3.13** Pathways known to regulate nuclear Ca²⁺ levels. Mechanisms of Ca²⁺ release in the nucleus and cytosol are different. Ca²⁺ release and the production of Ca²⁺-releasing messengers may be localized in the individual compartments, but nuclear and cytosolic Ca²⁺ levels may also be coordinated via diffusion of messengers through NPCs. There are some controversial pathways. The translocation of growth factor receptors into the nucleus, in particular, has not been observed in all of the studies ("?"marked). + and − indicate a positive or negative effect on the downstream process, respectively (Bootman et al., 2009)

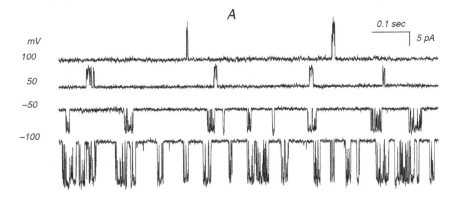

**FIGURE 3.14** Functioning of a spontaneously active Cl⁻ channel in the NOM of the NE of hippocampal pyramidal neurons. Original traces at different potentials indicated at the left (mV).

Channels of both anions and cations (Fedorenko et al., 2007) have demonstrated clearly observable voltage dependences. These channels are assumed to be important in maintaining the function of the intermembrane space of the NE of these cells, forming a considerable calcium store. These nuclear membrane channels also play important roles in the maintenance of the ion balance between the cytoplasm and perinuclear space, and between the latter and karyoplasm, as well as in the neutralization of voltage shifts during Ca²⁺ release.

Fedorenko and colleagues extended their investigations to understand the varieties of calcium channel types that might exist in the membranes of isolated nuclei of pyramidal neurons of the hippocampal CA1 area. In the NIM of these cells, they, for the first time, found inositol trisphosphate receptors (IP$_3$Rs) activated by inositol trisphosphate applied at a concentration of ≥0.1 µM. The approximate conductivity of single channels was 366 pS, and these channels were found to be permeable for both monovalent and bivalent cations. The data indicate that the NE of pyramidal neurons of the hippocampal CA1 area can play the role of a calcium store from which Ca$^{2+}$ might enter the cell nucleus directly (Fedorenko et al., 2008).

Fedorenko and colleagues then explored the NE ion channels of the nuclear membrane of hippocampal neurons (Fedorenko and Marchenko, 2014). They found that the nuclear membrane of CA1 pyramidal and dentate gyrus granule (DG), but not CA3 pyramidal neurons, was enriched in functional inositol 1,4,5-trisphosphate receptors/Ca(2+)-release channels (IP3 Rs) localized mainly in the NIM. They showed that the hippocampal neurons' nuclear membranes contain distinct sets of neuron-specific ion channels; IP3 Rs, but not RyRs, are targeted to the NIM of CA1 pyramidal and DG granule, but they were absent in the nuclear membranes of CA3 pyramidal neurons; the NE of these neurons is specialized to release Ca$^{2+}$ into the nucleoplasm, which may amplify Ca$^{2+}$ signals entering the nucleus from the cytoplasm or should generate Ca$^{2+}$ transients on its own. Table 3.1 lists a group of ion channels found in the nuclear membranes of hippocampal neurons. Both anion and cation-selective channels are observed (though with varied densities) in both NOM and NIM. The value of the Cl$^-$-conducting channel conductance is the lowest while Ca$^{2+}$-conducting channel conductance is the highest.

## 3.4   NUCLEAR INNER MEMBRANE CHANNELS

NIM protein trafficking is an important aspect determining (partially or fully depending on the protein type) both general diffusion and specific ion channel transport of materials and information across the NIM. The knowledge regarding traffic to the NIM is crucial for understanding the NIM channel functions. Table 3.2 lists the sorting for the NIM insertion of the following three classes of proteins and machineries: membrane proteins (MPs), cytosolic proteins, and nuclear import machineries (Laba et al., 2014). The sorting has found the following two major predictions or conclusions:

**TABLE 3.1**

**NOM and NIM-Hosted Cation and Anion Channels (Fedorenko and Marchenko, 2014)**

Ion Channels in the Nuclear Membrane of Hippocampal Neurons

| Cell Type | Conductance[a] | Selectivity | Localization | Identity | Density |
|---|---|---|---|---|---|
| CA1 pyramidal neurons | 248±6 pS | K$^+$ | Mainly the inner nuclear membrane | Unknown | High |
| | 366±5 pS | Ca$^{2+}$ | Mainly the inner nuclear membrane | IP$_3$Rs | High |
| | 156±4 pS | Cl$^-$ | The outer nuclear membrane | Unknown | Low |
| CM pyramidal neurons | 210±6 pS | K$^+$ | Both the inner and the outer nuclear membranes | Unknown | Very low |
| DG granule neurons | 179±15 pS | K$^+$ | Mainly the inner nuclear membrane | Unknown | High |
| | 378±11 pS | Ca$^{2+}$ | Mainly the inner nuclear membrane | IP$_3$Rs | High |
| | 719 pS | Ca$^{2+}$ | The outer nuclear membrane | RyRs | Very low |
| | 305±16 pS | Cl$^-$ | The outer nuclear membrane | Unknown | Low |
| | 80±8 pS and 163±9 pS | Cl$^-$ | Inner nuclear membrane | CIC | Low |

[a]   slope conductance in symmetric KCl solution.

**TABLE 3.2**

**Sorting Signals in Integral NIM Proteins**

| Protein | MP Insertion | Cytosolic Protein Sorting | Nuclear Import Machinery | Unclassified | Refs. |
|---|---|---|---|---|---|
| **Yeast** | | | | | |
| Mps3 | Cotranslational Sec61 system | | • Indirect dependence on Kap123, Kap95, and RanGTP–RanGDP gradient; piggyback mechanism via binding to histone H2Z.A<br>• Nuclear retention | | Gardner et al. (2011) |
| Heh1/Src1 | • Src1 small: cotranslational Sec61<br>• Full length Heh1: cotranslational Sec61 | | • NLS, RanGTP-RanGDP gradient, Kap60, Kap95, Nup170, Nup2<br>• Nuclear retention | | King et al. (2006) |
| Heh2 | • Cotranslational Sec61<br>• We interpret "INM sorting motif" as a topology indicator | | • NLS, RanGTP-RanGDP gradient, Kap60, Kap95, Nup170, Nup2, GLFG domains of Nup100, Nup57, Nup145<br>• Nuclear retention | | Liu et al. (2010) |
| **Human** | | | | | |
| SUN1 | Cotranslational Sec61 | | • Nuclear retention | Localization depends on farnesylated prelamin A | Haque et al. (2009) |
| SUN2 | Cotranslational Sec61 | Golgi retrieval signal | • NLS, importin-α, importin-β, RanGTP–RanGDP gradient,<br>• Nuclear retention | SUN2 mobility requires ATP | Turgay et al. (2010) |
| Emerin | Tail-anchored protein, possibly posttranslational insertion by GET pathway | Subpopulation in plasma membrane in heart tissue from human, rat, and mouse (sorting signals unknown) | • Nuclear retention | Emerin mobility requires ATP | Berk et al. (2013) |
| LAP2β | Tail-anchored protein, possibly posttranslational insertion by GET pathway | | • Nuclear retention | | Ohba et al. (2004) |
| LEM2 | Cotranslational Sec61 | | • Nuclear retention | | Brachner (2005) |
| MAN1 | Cotranslational Sec61 | | • Nuclear retention | | Pan et al. (2005) |
| LBR | • Cotranslational Sec61<br>• N terminal domain probably codefines topology;<br>• "INM sorting motif" | Distinct functions at ER and NE (sorting signals unknown) | • RanGTP-dependent interaction with Importinβ (not importin-α-dependent)<br>• Nuclear retention | Mobility of LBR is dependent on RanGTP and Nup35 | Clayton et al. (2010) |
| Nurim | • Cotranslational Sec61<br>• "INM sorting motif" | | • Nuclear retention (but not to DNA and lamins) | | Rolls et al. (1999) |

i. sorting signals and various molecular mechanisms in MP biogenesis, MP traffic, and nuclear transports are relevant with respect to NIM traffic; and

ii. the interplay of the effects among these signals and molecular mechanisms determines the rates of traffic to the NIM.

For detailed understanding, see the references quoted inside Table 3.2 for each observation.

Ample evidence favors the existence of facilitated transport (a process in which materials diffuse across the membrane with the help of MPs) of integral inner membrane proteins similar to that of soluble proteins. As an example, for evidence regarding *S. cerevisiae* NIM, see Laba et al. (2014). The study also argues in favor of finding evidence suggesting that the facilitated transport is present in both yeast and humans.

Specific domains in MP structure may appear important in their NIM recognition (early), insertion, and trafficking. Many of the NIM proteins have a relatively large N-terminal extralumenal domains, often containing regions proven to be relevant for membrane trafficking. The early recognition (including as early as during translation) of NIM proteins was first predicted for viral peptides, and later for native NIM proteins (Braunagel et al., 2007). A large amount of data demonstrate this feature, see Figure 3.15 (Laba et al., 2014).

### 3.4.1 Voltage-gated Potassium Channels in the Nuclear Inner Membrane

A decade ago, Chen and coworkers observed the voltage-gated potassium channel KV10.1 in NIM (Chen et al., 2011). KV10.1 was found to be expressed at the NIM in both human and rat models. KV10.1 channels at the NIM are not all transported, directly from the ER, but rather have been exposed to the extracellular milieu. Various imaging and patch-clamp experiments helped discover the existence of this potassium channels. The imaging data and figures are not presented here (readers may find them in the original article). The results helped hypothesize that KV10.1 channels at the NE might participate in the homeostasis of nuclear $K^+$, or indirectly interact with heterochromatin, perhaps affecting gene expressions.

The existence of KV10.1 channels in the NIM was revealed based on both optical and electron microscope data, as well as biochemical evidence, in both native and heterologous systems (Chen et al., 2011). Careful inspection of the data shows that all the criteria of a transmembrane protein, being localized to NIM, are not only fulfilled but are also functional in NIM as an ion channel that conducts potassium ions. The electrophysiological experiments recorded ion channel current traces (see Figure 3.16), indicating protein orientation; the extracellular loops of the channel facing the pipette favor localization at the NIM (Chen et al., 2011). The identity of the channel as KV10.1 is supported not only by the fact that it is not found in wild-type nuclei but also by its voltage dependence, particularly dependence on the prepulse potential-single-channel conductance and pharmacology, most importantly by the inhibition measured in the presence of mAb56, which is the most selective blocker available.

The channel openings were compatible with KV10.1 in terms of the measured conductance ($8.1 \pm 0.4$ pS; Figure 3.16a, b) and voltage dependence, with the highest open probability found at $-60$ mV and the lowest at $+60$ mV in the pipette, which would represent a membrane potential of opposite sign in an inside-out configuration, that is, if the channel's extracellular side faces the pipette and the intracellular domains are located toward the nucleoplasm.

### 3.4.2 Voltage-independent Monovalent Cationic Channels in the Nuclear Inner Membrane

A nuclear monovalent cationic channel (NMCC) is, for the first time, found prominently expressed in the NIM, isolated from flexor digitorum brevis skeletal muscle fibers of adult mice (Yarotskyy

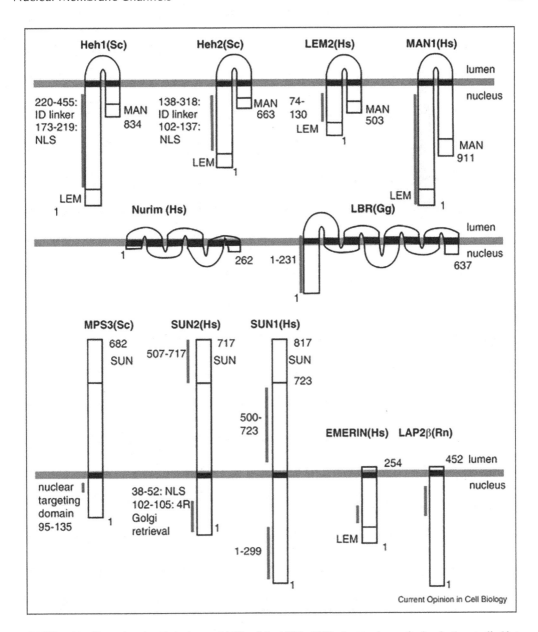

**FIGURE 3.15** Targeting signals in integral MPs of the NIM of NE, the topology of a few better-studied integral MPs. The red bar represents part of the sequence shown experimentally to be important for NIM localization of proteins. LEM for Lap1-emerin-MAN1, MAN for Heh/Man1 carboxy-terminal homology domain, and CTHD and SUN for Sad-Unc-84 homology domains are indicated. Hs: Homo sapiens, Sc: *Saccharomyces cerevisiae*, Rn: *Rattus norvegicus*, Gg: *Gallus gallus*.

and Dirksen, 2014). The activity of a smaller conductance channel was detected in isolated nuclei (~64% of the patches). These prolonged channel openings were observed across a wide voltage range (from −50 to 50 mV; Figure 3.17a and c), having a unitary average channel conductance 158±7 pS. In addition, these channels exhibited a markedly high, voltage-independent mean open probability ($P_o$) (Figure 3.17 c) (Yarotskyy and Dirksen, 2014).

In isotonic 140 mM KCl, the skeletal muscle NMCC was found to exhibit a unitary conductance of ~160 pS and high voltage-independent open probability. Based on single-channel

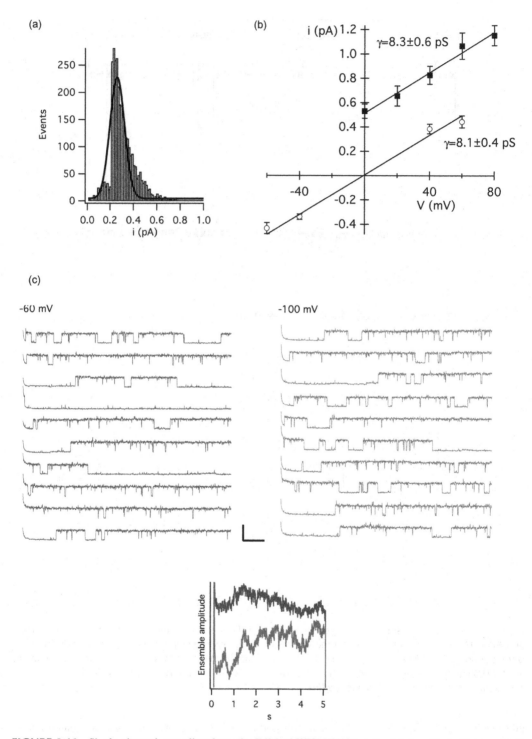

**FIGURE 3.16**   Single-channel recording from the INM of HEK-KV10.1 cells. (a) Amplitude histogram of events recorded at +40 mV in symmetrical potassium (pooled data from five independent recordings). (b) Single-channel amplitudes versus voltage in symmetrical (circles) or asymmetrical potassium (squares). Error bars represent SEM. The solid line represents a linear fit of the data that gives a slope conductance of 8.1 and 8.3 pS. (c) Traces recorded at +60 mV from a holding potential of −60 (blue) or −100 mV (red). The latency time before the first opening is increased when the holding potential is more negative, as clearly seen in the ensemble currents depicted in the inset. Scale bars: 2 pA, 500 ms. (For color figure see eBook figure.)

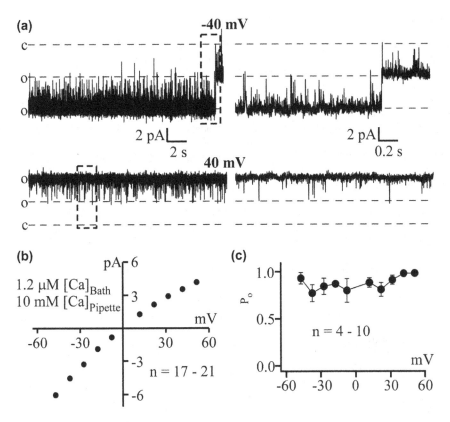

**FIGURE 3.17** Channel activity in the NIM (electrophysiology records). (a) Representative traces of the channel activity recorded at −40 mV (top) and 40 mV (bottom) pipette potentials. Traces shown on the right side of the panel are expanded regions of the traces on the left (of dashed boxes). (b) Average (±SE) i–V relationships represents the plot of the unitary channel current amplitude against pipette voltage. (c) Average (±SE) open probability ($P_o$) of channels recorded from the NIM.

reversal potential measurements, NMCCs were found to be slightly more permeable to potassium ions over sodium ($P_K/P_{Na} = 2.68 \pm 0.21$) and cesium ($P_K/P_{Cs} = 1.39 \pm 0.03$) ions. NMCCs were found not to permeate divalent cations, so are inhibited by calcium ions, and demonstrate weak rectification in the asymmetric $Ca^{2+}$-containing solutions. These results (Yarotskyy and Dirksen, 2014) together characterize a voltage-independent NMCC in skeletal muscle, and that its properties are ideally suited to serve as one of the countercurrent mechanisms during calcium release of NE.

Yarotskyy and Dirksen observed, in additional investigations, similar NMCC channel activity in nuclei isolated from tibialis anterior skeletal muscle (data not shown), in addition to those of flexor digitorum brevis skeletal muscle fibers (Yarotskyy and Dirksen, 2014). The NMCC characterized here is likely to be an important nuclear ion channel that is widely expressed in skeletal muscle. However, NMCC expression may not be specific to skeletal muscle because a channel with similar properties has been reported in Purkinje neurons (Marchenko et al., 2005). This group reported that the nuclei from Purkinje neurons exhibit a large conductance monovalent cationic channel with high open probability, and with a preference for K+ over Na+ ions, as well as inhibition by $Ca^{2+}$ ions. A second larger conductance NMCC channel was later recorded from the nuclei of Purkinje neurons (Fedorenko et al., 2010). However, this channel was distinctive as it exhibited a prominent voltage-dependent change in channel $P_o$, which is not observed in NMCC, recorded from the NIM of skeletal muscle cells.

## REFERENCES

Alber, F., Dokudovskaya, S., Veenhoff, L. M., Zhang, W., Kipper, J., Devos, D., Suprapto, A., Karni-Schmidt, O., Williams, R., Chait B. T., Rout, M. P., & Sali, A. (2007). Determining the architectures of macromolecular assemblies. *Nature, 450*, 683–694.

Anderson, N. G. (1953). On the nuclear envelope. *Science, 117*, 517–521.

Bairati, A., & Lehmann, F.E. (1952). Über die submikroskopische Struktur der Kernmembran bei*Amoeba proteus*. *Experientia 8*, 60–61. doi:10.1007/BF02139020

Beck, M., & Hurt, E. (2016). The nuclear pore complex: Understanding its function through structural insight. *Nature Reviews Molecular Cell Biology, 18*(2), 73–89. doi:10.1038/nrm.2016.147

Berk, J. M., Tifft, K. E., & Wilson, K. L. (2013). The nuclear envelope LEM-domain protein emerin. *Nucleus, 4*(4), 298–314. doi:10.4161/nucl.25751

Bkaily, G., Avedanian, L., & Jacques, D. (2009). Nuclear membrane receptors and channels as targets for drug development in cardiovascular diseases This article is one of a selection of papers from the NATO Advanced Research Workshop on Translational Knowledge for Heart Health (published in part 1 of a 2-part Special Issue). *Canadian Journal of Physiology and Pharmacology, 87*(2), 108–119. doi:10.1139/y08-115

Bootman, M. D., Fearnley, C., Smyrnias, I., MacDonald, F., & Roderick, H. L. (2009). An update on nuclear calcium signalling. *Journal of Cell Science, 122*(14), 2337–2350. doi:10.1242/jcs.028100

Brachner, A. (2005). LEM2 is a novel MAN1-related inner nuclear membrane protein associated with A-type lamins. *Journal of Cell Science, 118*(24), 5797–5810. doi:10.1242/jcs.02701

Braunagel, S. C., Williamson, S. T., Ding, Q., Wu, X., & Summers, M. D. (2007). Early sorting of inner nuclear membrane proteins is conserved. *Proceedings of the National Academy of Sciences, 104*(22), 9307–9312. doi:10.1073/pnas.0703186104

Bustamante, J. O. (1992). Nuclear ion channels in cardiac myocytes. *European Journal of Physiology, 421*(5), 473–485. doi:10.1007/bf00370259

Bustamante, J. O. 1994. Topical review: Nuclear electrophysiology. *The Journal Membrane Biology, 138*, 105–112.

Callan, H. G., & Tomlin, S. G. (1950). Experimental studies on amphibian oocyte nuclei I. Investigation of the structure of the nuclear membrane by means of the electron microscope. (1950). *Proceedings of the Royal Society of London. Series B - Biological Sciences, 137*(888), 367–378. doi:10.1098/rspb.1950.0047

Cautain, B., Hill, R., Pedro, N. D., & Link, W. (2014). Components and regulation of nuclear transport processes. *FEBS Journal, 282*(3), 445–462. doi:10.1111/febs.13163

Chabaud, M., Genre, A., Sieberer, B. J., Faccio, A., Fournier, J., Novero, M., . . . Bonfante, P. (2010). Arbuscular mycorrhizal hyphopodia and germinated spore exudates trigger $Ca^{2+}$ spiking in the legume and nonlegume root epidermis. *New Phytologist, 189*(1), 347–355. doi:10.1111/j.1469-8137.2010.03464.x

Chambers, R., & Fell, B. (1931). *Proceedings of the Royal Society London Series B, 109*, 380.

Chen, Y., Sánchez, A., Rubio, M. E., Kohl, T., Pardo, L. A., & Stühmer, W. (2011). Functional KV10.1 channels localize to the inner nuclear membrane. *PLoS ONE, 6*(5). doi:10.1371/journal.pone.0019257

Clayton, P., Fischer, B., Mann, A., Mansour, S., Rossier, E., Veen, M., . . . Hoffmann, K. (2010). Mutations causing Greenberg dysplasia but not Pelger anomaly uncouple enzymatic from structural functions of a nuclear membrane protein. *Nucleus, 1*(4), 354–366. doi:10.4161/nucl.1.4.12435

Dauer, W. T., & Worman, H. J. (2009). The nuclear envelope as a signaling node in development and disease. *Developmental Cell, 17*(5), 626–638. doi:10.1016/j.devcel.2009.10.016

Downie, L., Priddle, J., Hawes, C., & Evans, D. E. (1998). A calcium pump at the higher plant nuclear envelope? *FEBS Letters, 429*(1), 44–48. doi:10.1016/s0014-5793(98)00564-x

Echevarría, W., Leite, M. F., Guerra, M. T., Zipfel, W. R., & Nathanson, M. H. (2003). Regulation of calcium signals in the nucleus by a nucleoplasmic reticulum. *Nature Cell Biology, 5*(5), 440–446. doi:10.1038/ncb980

Fahrenkrog, B., & Aebi, U. (2003). The nuclear pore complex: Nucleocytoplasmic transport and beyond. *Nature Reviews Molecular Cell Biology, 4*(10), 757–766. doi:10.1038/nrm1230

Fedorenko, E. A., Duzhii, D. E., & Marchenko, S. M. (2007). Spontaneously active ion channels of membranes of the nuclear envelope of hippocampal pyramidal neurons. *Neurophysiology, 39*(1), 1–6. doi:10.1007/s11062-007-0001-1

Fedorenko, E. A., Duzhii, D. E., & Marchenko, S. M. (2008). Calcium channels in the nuclear envelope of pyramidal neurons of the rat hippocampus. *Neurophysiology, 40*(4), 238–242. doi:10.1007/s11062-009-9047-6

Fedorenko, O. A., & Marchenko, S. M. (2014). Ion channels of the nuclear membrane of hippocampal neurons. *Hippocampus, 24*(7), 869–876. doi:10.1002/hipo.22276

Fedorenko, O., Yarotskyy, V., Duzhyy, D., & Marchenko, S. (2010). The large-conductance ion channels in the nuclear envelope of central neurons. *Pflügers Archiv - European Journal of Physiology, 460*(6), 1045–1050. doi:10.1007/s00424-010-0882-5

Gall, J. G. (1954). *Experimental Cell Research, 7,* 197.

Gardner, J. M., Smoyer, C. J., Stensrud, E. S., Alexander, R., Gogol, M., Wiegraebe, W., & Jaspersen, S. L. (2011). Targeting of the SUN protein Mps3 to the inner nuclear membrane by the histone variant H2A.Z. *The Journal of Cell Biology, 193*(3), 489–507. doi:10.1083/jcb.201011017

Gerasimenko, O., & Gerasimenko. J. (2004). New aspects of nuclear calcium signalling. *Journal of Cell Science, 117*(15), 3087–3094. doi:10.1242/jcs.01295

Ghavami, A., van der Giessen, E., & Onck, P. R. (2016). Energetics of transport through the nuclear pore complex. *PLoS ONE, 11*(2), e0148876. doi:10.1371/journal.pone.0148876

Graumann, K., & Evans, D. E. (2010). The plant nuclear envelope in focus. *Biochemical Society Transactions, 38*(1), 307–311. doi:10.1042/bst0380307

Graumann, K., & Evans, D. E. (2017). The nuclear envelope - structure and protein interactions. *Annual Plant Reviews Online, 19*–56. doi:10.1002/9781119312994.apr0498

Groves, N. R., Mckenna, J. F., Evans, D. E., Graumann, K., & Meier, I. (2019). A nuclear localization signal targets tail-anchored membrane proteins to the inner nuclear envelope in plants. *Journal of Cell Science, 132*(7). doi:10.1242/jcs.226134

Guihard, G., Proteau, S., Payet, M. D., Escande, D., & Rousseau, E. (2000). Patch-clamp study of liver nuclear ionic channels reconstituted into giant proteoliposomes. *FEBS Letters, 476*(3), 234–239. doi:10.1016/s0014-5793(00)01752-x

Haque, F., Mazzeo, D., Patel, J. T., Smallwood, D. T., Ellis, J. A., Shanahan, C. M., & Shackleton, S. (2009). Mammalian SUN protein interaction networks at the inner nuclear membrane and their role in laminopathy disease processes. *Journal of Biological Chemistry, 285*(5), 3487–3498. doi:10.1074/jbc.m109.071910

Harris P., & James T. (1952). *Gxperientia, 8,* 384.

Humbert, J. P., Matter, N., Artault, J. C., Koppler, P., & Malviya, A. N. (1996). Inositol 1,4,5-trisphosphate receptor is located to the inner nuclear membrane vindicating regulation of nuclear calcium signaling by inositol 1,4,5-trisphosphate. Discrete distribution of inositol phosphate receptors to inner and outer nuclear membranes. *Journal of Biological Chemistry, 271,* 478–485.

Ibarra, A., & Hetzer, M. W. (2015). Nuclear pore proteins and the control of genome functions. *Genes & Development, 29*(4), 337–349. doi:10.1101/gad.256495.114

Jovanovic-Talisman, T., & Zilman, A. (2017). Protein transport by the nuclear pore complex: Simple biophysics of a complex biomachine. *Biophysical Journal, 113*(1), 6–14. doi:10.1016/j.bpj.2017.05.024

Keminer, O., & Peters, R. (1999). Permeability of single nuclear pores. *Biophysical Journal, 77*(1), 217–228. doi:10.1016/s0006-3495(99)76883-9

King, M. C., Lusk, C., & Blobel, G. (2006). Karyopherin-mediated import of integral inner nuclear membrane proteins. *Nature, 442*(7106), 1003–1007. doi:10.1038/nature05075

Kite, G. L. (1913). *American Journal of Physiology-Heart and Circulatory Physiology.* 32###146

Kopito, R. B., & Elbaum, M. (2007). Reversibility in nucleocytoplasmic transport. *Proceedings of the National Academy of Sciences, 104*(31), 12743–12748. doi:10.1073/pnas.0702690104

Laba, J. K., Steen, A., & Veenhoff, L. M. (2014). Traffic to the inner membrane of the nuclear envelope. *Current Opinion in Cell Biology, 28,* 36–45. doi:10.1016/j.ceb.2014.01.006

Liu, D., Wu, X., Summers, M. D., Lee, A., Ryan, K. J., & Braunagel, S. C. (2010). Truncated isoforms of Kap60 facilitate trafficking of Heh2 to the nuclear envelope. *Traffic, 11*(12), 1506–1518. doi:10.1111/j.1600-0854.2010.01119.x

Marchenko, S. M., & Fedorenko, O. A. (2011). Spontaneously active ion channels of the nuclear envelope membrane. *International Journal of Physiology and Pathophysiology, 2*(2), 185–197. doi:10.1615/intjphyspathophys.v2.i2.90

Marchenko, S. M., Yarotskyy, V. V., Kovalenko, T. N., Kostyuk, P. G., & Thomas, R. C. (2005). Spontaneously active and InsP3-activated ion channels in cell nuclei from rat cerebellar Purkinje and granule neurones. *The Journal of Physiology, 565*(3), 897–910. doi:10.1113/jphysiol.2004.081299

Matzke, A. J., Behensky, C., Weiger, T., & Matzke, M. (1992). A large conductance ion channel in the nuclear envelope of a higher plant cell. *FEBS Letters, 302*(1), 81–85. doi:10.1016/0014-5793(92)80290-w

Matzke, A. J., Weiger, T. M., & Matzke, M. (1990). Detection of a large cation-selective channel in nuclear envelopes of avian erythrocytes. *FEBS Letters, 271*(1–2), 161–164. doi:10.1016/0014-5793(90)80397-2

Matzke, A. J., Weiger, T. M., & Matzke, M. (2010). Ion channels at the nucleus: Electrophysiology meets the genome. *Molecular Plant, 3*(4), 642–652. doi:10.1093/mp/ssq013

Mazzanti, M., Bustamante, J. O., & Oberleithner, H. (2001). Electrical dimension of the nuclear envelope. *Physiological Reviews, 81*(1), 1–19. doi:10.1152/physrev.2001.81.1.1

Mazzanti, M., Defelice, L. J., Cohen, J., & Malter, H. (1990). Ion channels in the nuclear envelope. *Nature, 343*(6260), 764–767. doi:10.1038/343764a0

Mazzanti, M., Defelice, L. J., & Smith, E. F. (1991). Ion channels in murine nuclei during early development and in fully differentiated adult cells. *The Journal of Membrane Biology, 121*(2), 189–198. doi:10.1007/bf01870532

Naim, B, Zbaida, D, Dagan, S, Kapon, R, & Reich, Z. (2009). Cargo surface hydrophobicity is sufficient to overcome the nuclear pore complex selectivity barrier. *The EMBO Journal, 28*(18), 2697–2705. pmid:19680225

Ohba, T., Schirmer, E. C., Nishimoto, T., & Gerace, L. (2004). Energy- and temperature-dependent transport of integral proteins to the inner nuclear membrane via the nuclear pore. *Journal of Cell Biology, 167*(6), 1051–1062. doi:10.1083/jcb.200409149

Pan, D., Estévez-Salmerón, L. D., Stroschein, S. L., Zhu, X., He, J., Zhou, S., & Luo, K. (2005). The integral inner nuclear membrane protein MAN1 physically interacts with the R-smad proteins to repress signaling by the transforming growth factor-β superfamily of cytokines. *Journal of Biological Chemistry, 280*(16), 15992–16001. doi:10.1074/jbc.m411234200

Panté, N., & Kann, M. (2002). Nuclear pore complex is able to transport macromolecules with diameters of ~39 nm. *Molecular Biology of the Cell, 13*(2), 425–434. doi:10.1091/mbc.01-06-0308

Rolls, M. M., Stein, P. A., Taylor, S. S., Ha, E., Mckeon, F., & Rapoport, T. A. (1999). A visual screen of a Gfp-fusion library identifies a new type of nuclear envelope membrane protein. *The Journal of Cell Biology, 146*(1), 29–44. doi:10.1083/jcb.146.1.29

Rothballer, A., & Kutay, U. (2012). SnapShot: The Nuclear Envelope I. *Cell, 150*(4). doi:10.1016/j.cell.2012.07.024

Schwartz, T. U. (2016). The structure inventory of the nuclear pore complex. *Journal of Molecular Biology, 428*(10), 1986–2000. doi:10.1016/j.jmb.2016.03.015

Sharma, A., Solmaz, S. R., Blobel, G., & Melčák, I. (2015). Ordered regions of channel nucleoporins Nup62, Nup54, and Nup58 form dynamic complexes in solution. *Journal of Biological Chemistry, 290*(30), 18370–18378. doi:10.1074/jbc.m115.663500

Tabares, L., Mazzanti, M., & Clapham, D. E. (1991). Chloride channels in the nuclear membrane. *The Journal of Membrane Biology, 123*(1), 49–54. doi:10.1007/bf01993962

Tamura, K., Fukao, Y., Iwamoto, M., Haraguchi, T., & Hara-Nishimura, I. (2010). Identification and characterization of nuclear pore complex components in Arabidopsis thaliana. *The Plant Cell, 22*(12), 4084–4097. doi:10.1105/tpc.110.079947

Tingey, M., Mudumbi, K. C., Schirmer, E. C., & Yang, W. (2019). Casting a wider net: Differentiating between inner nuclear envelope and outer nuclear envelope transmembrane proteins. *International Journal of Molecular Sciences, 20*(21), 5248. doi:10.3390/ijms20215248

Turgay, Y., Ungricht, R., Rothballer, A., Kiss, A., Csucs, G., Horvath, P., & Kutay, U. (2010). A classical NLS and the SUN domain contribute to the targeting of SUN2 to the inner nuclear membrane. *The EMBO Journal, 29*(14), 2262–2275. doi:10.1038/emboj.2010.119

Watson, M. L. (1955). The nuclear envelope. *The Journal of Biophysical and Biochemical Cytology, I*(3).

Yarotskyy, V., & Dirksen, R. (2014). Monovalent cationic channel activity in the inner membrane of nuclei from skeletal muscle fibers. *Biophysical Journal, 107*(9), 2027–2036. doi:10.1016/j.bpj.2014.09.030

Zhou, K., Rolls, M. M., Hall, D. H., Malone, C. J., & Hanna-Rose, W. (2009). A ZYG-12–dynein interaction at the nuclear envelope defines cytoskeletal architecture in the C. elegans gonad. *The Journal of Cell Biology, 186*(2), 229–241. doi:10.1083/jcb.200902101

Zhou, X., Boruc, J., & Meier, I. (2018). The plant nuclear pore complex - The nucleocytoplasmic barrier and beyond. *Annual Plant Reviews Online*, 57–91. doi:10.1002/9781119312994.apr0499

# 4 Artificial Ion Channels

Biological ion channels are made up of proteins that have matured over about 4 billion years in living systems. During this period, they have perfected the precision chemical modification of the interior of the pore wall through chemical selection processes. Thus, those pores possess specific selectivity criteria. Therefore, biological channels act together with their host cell membranes to determine which electrolytes, materials, and information are to be transported across the cell membrane barrier in a precisely self-controlled manner. On the other hand, artificial channels or pores, constructed using artificial/synthetic agents, have recently evolved. As these channels are constructed using synthetic materials, they are also commonly called "synthetic ion channels." Artificial channels or pores, at their current stages, have been going through continuous developments in achieving several goals, for example, synthesis of bioadaptable agents as the channel building blocks that migrate easily to hydrophilic/hydrophobic interfaces and locate at "soft condensed matter" cell membrane and stabilize there in the hydrophobic region spanning over two hydrophilic surfaces. The constructed pores must be compatible with the size and stability of biological cell membranes so that they can comfortably host without a severe energetic penalty. Above all the synthetic pores must be selective for agents that are biologically demanding, such as various drugs we try to push through cell membranes for delivery into various target regions inside cells. Achieving these goals and the precise physical and chemical modifications remain elusive, but superior techniques are expected to help us in finding superior materials than the existing ones to let us reach the goal.

## 4.1 ARTIFICIAL ION CHANNELS – CURRENT STATUS

Convergent multidimensional self-assembly strategies have been used over decades for the synthesis of unimolecular ion channels or noncovalent self-organized channels. The use of unimolecular agents has a vital advantage as they can rearrange themselves in biological environments to construct different structures having diverse biophysical properties. These channels are designed to mimic natural ion channel proteins, for which a rich array of interconverting or adaptive channel conductance states can be observed. Recently, a review provided an overview on the development of various self-assembled artificial channels, especially regarding their design, self-assembly behavior, bilayer transport phenomena, underlying mechanisms of transport, and ultimately the comparison of these properties with their natural ion channel counterparts (Zheng et al., 2020).

Nanopores as sensors for DNA (Deamer and Akeson, 2000) received attention long ago. As soon as the biological pore α-hemolysin (Kasianowicz et al., 1996) appeared as a model protein to serve in understanding the nanopore transport of materials that are not usually considered for general biological channels, the research received significant attention in searching for artificial nanopores. Later, the discovery of nonbiological solid-state nanopores opened up a wider range of new research areas (Dekker, 2007; Yin et al., 2020).

Comparable investigations among amphiphiles and peptide channels about two decades ago also led to the discovery of artificial channels using synthetic materials and various derivatives. Among these, we recommend readers to consult a few key papers of Fyles (2007; 2013). A brief analysis has been presented later.

As the discovery of artificial channels progresses, computational scientists are increasingly stepping into the field and proposing various models based on versatile *in silico* assay studies including molecular dynamic (MD) simulations. These computational studies help us understand the artificial channel structures and energetics, and above all their statistical behavior in constructed membranes

DOI: 10.1201/9781003010654-4

in controlled environments mimicking biological systems. Crucial biophysical parameters related to channel structures and functions are being cross-checked. We have presented a few example cases later to address them.

The optimal goal is to construct any biological channel similar to an artificial or synthetic channel that might help in drug delivery and delivery of various other nano, micro, and even macromolecules in a controlled fashion. This chapter briefly discuss the latest signs of progress and our understanding of both scientific discoveries and technological innovations.

## 4.2 ARTIFICIAL POTASSIUM CHANNELS

The possibility of constructing self-organized channels based on heteroditopic alkylureido crown ethers has been tried for almost two decades. They are among the best K⁺-selective artificial channels, showing considerable activities mimicking regular channels (Barboiu et al., 2003; Barboiu, 2004; Gilles and Barboiu, 2015; Sun et al., 2015; 2016; Schneider et al., 2017). The supramolecular channels emerge from H-bonded self-assemblies of crown ethers, aligned around a central pore. Monomers as carrier-like transporters are combined (Barboiu et al., 2003) with supramolecular aggregates showing potential channel functions, see Figure 4.1 (Barboiu et al., 2004).

A few crown ether molecules have been tested and found to be successful molecular carriers or supramolecular channels that span lipid bilayer membranes (Sun et al., 2016; Schneider et al., 2017). Their dynamics and transport properties are similar to natural channels active under membrane potentials. The modification of the main structural and functional moieties of the ionophores may be predicted to induce controlled recognition and improved transport behaviors. Of special interest are structure-controlled functions for constructing the channels and for building their superstructures from the directional self-assembly of optimal units (Barboiu, 2018):

  i. the cationic binding groups are rigid benzo-15-crown-5 and benzo-18-crown-6 of flexible 15-crown-5 macrocycles,
 ii. the guiding interactions are the urea orbis (acylhydrazone) H-bonds, and
iii. the nature of the hydrophobic tail determines the dynamics of the channel superstructures at the interface with the bilayer membranes.

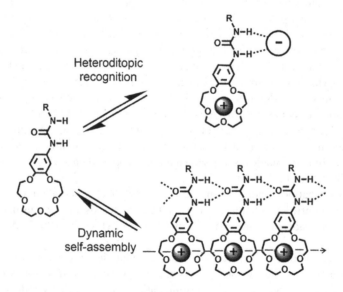

**FIGURE 4.1**  Heteroditopic ion-pair recognition and H-bonding self-assembly of macrocyclic supramolecular ion channels (Barboiu et al., 2004).

A total of 24 macrocyclic ionophores (see 1–24) were prepared for the construction of the ion channel systems, see Figure 4.2 (Barboiu, 2018). Compounds 1–12 contain rigid benzo-15-crown-5 (1–9) or benzo-18-crown-6 (9–12) macrocyclic ion-binding sites and are decorated with linear or branched alkyl tails of different lengths. Here, linear alkyl groups proved to be important to promote the creation/formation of self-assembled channels, but the crowded bulky tails may disturb the binding and self-assembly by favoring the emergence of small aggregates. Flexible 15-crown-5 macrocycles with variable alkyl chain lengths were synthesized in a similar manner. They led to the adaptive binding of Na$^+$ and K$^+$ cations as a function of their concentration in the membrane.

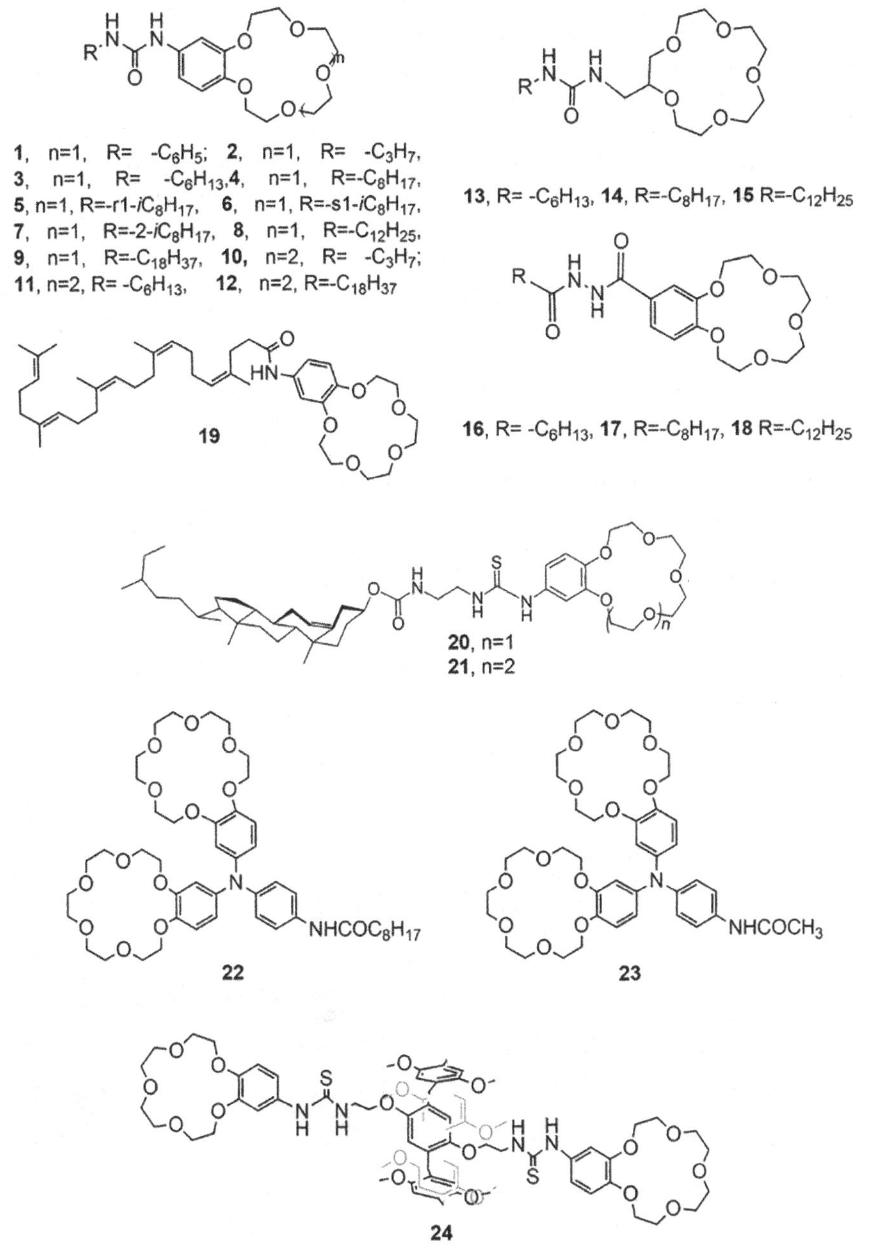

**1**, n=1, R= -C$_6$H$_5$; **2**, n=1, R= -C$_3$H$_7$,
**3**, n=1, R= -C$_6$H$_{13}$,**4**, n=1, R=-C$_8$H$_{17}$,
**5**, n=1, R=-r1-iC$_8$H$_{17}$, **6**, n=1, R=-s1-iC$_8$H$_{17}$,
**7**, n=1, R=-2-iC$_8$H$_{17}$, **8**, n=1, R=-C$_{12}$H$_{25}$,
**9**, n=1, R=-C$_{18}$H$_{37}$, **10**, n=2, R= -C$_3$H$_7$;
**11**, n=2, R= -C$_6$H$_{13}$, **12**, n=2, R=-C$_{18}$H$_{37}$

**13**, R= -C$_6$H$_{13}$, **14**, R=-C$_8$H$_{17}$, **15** R=-C$_{12}$H$_{25}$

**16**, R= -C$_6$H$_{13}$, **17**, R=-C$_8$H$_{17}$, **18** R=-C$_{12}$H$_{25}$

**19**

**20**, n=1
**21**, n=2

**22**

**23**

**24**

**FIGURE 4.2** Crown ether compounds 1–24 studied as K$^+$-channel-forming components. (Reprinted (adapted) with permission from Barboiu (2018). Copyright (2018) American Chemical Society.)

Acyl hydrazide-substituted benzo-15-crown-5 ethers show low transport rates, in contrast to their stronger self-association via multiple H-bonding (Li et al., 2018). A series of squalyl (19) and cholesteryl (20–21) 15-crown-5 ethers were then designed that form H-bonded channels. They stabilize via aggregation of bulky anchoring arms, resulting in the formation of preorganized clusters of the macrocycles in the lipid bilayers. Multivalent macrocyclic systems have been designed for the generation of directional ion-transporting channels based on triarylamine pillars (22,23) (Schneider et al., 2017) or a pillar [5] and a central platform (Barboiu et al., 2004).

As discussed here, the latest developments of artificial supramolecular carriers or channels that selectively transport $K^+$ cations against other cations are discussed in Barboiu (2018). Synthetic methods for the design of alternative biomimetic artificial $K^+$ channel systems are of tremendous interest due to increased interest in potassium channels.

The channel selectivity is mainly determined by the perfect coordination of $K^+$ cations together with the orientation of the dipolar carbonyl oxygens that surround permeating cations in the pore filter, rather than the pore size (see Figure 4.3a). Channel conformational motion adaptively determines that there are eight carbonyl oxygens that compensate for the dehydration energy, and there are four such coordination sites along the filter. Molecular recognition involving synthetic receptors is governed by the positioning of the coordinating groups of the receptor, replacing the hydration water molecules around the $K^+$ cations, as in the active gate of KcsA $K^+$ channels. The selective recognition of the optimal over the imperfect coordination features is crucial for both the recognition and transport functions. Most of the pioneering examples include crown ethers (see Figure 4.3a–c) Crown ether receptors equatorially bind the cations, but they do not complete all of the hydration sites, so dehydration is not compensated. This allows anionic counterions, water molecules, or aromatic groups to occupy the vacant apical positions (see Figure 4.3d, e). Moreover, cation–$\pi$ interactions are known, and many examples are of particular biological significance.

Detailed analysis (not presented here) suggests that these systems (Barboiu, 2018) can be considered as adaptive self-instructed ion channels, where the $K^+$ solute drives the selection and formation of specific superstructures for its selective transport. The artificial $K^+$-selective channels described here may be regarded as biomimetic alternatives to KcsA channels.

## 4.3 CONSTRUCTING SOLID-SATE NANOPORES IN MEMBRANES

Three decades ago, the scientific community started to think of the possibility of using nanopores as sensors for DNA (Deamer and Akeson, 2000). If DNA could be transferred through a nanopore in a

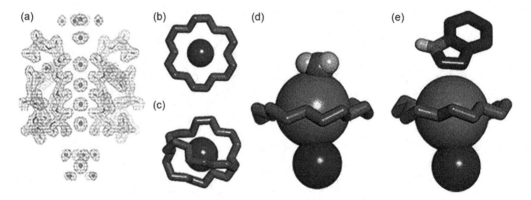

**FIGURE 4.3** (a) $K^+$ cations in the KcsA $K^+$ channel filter or outside the channel surrounded by carbonyl groups or water, respectively. (b, c) $K^+$ recognition by (b) 18-crown-6 and (c) 222 cryptand. (d, e) Examples in which 18-crown-6 equatorially coordinates the $K^+$ cations while (d) anionic counterions and water molecules or (e) aromatic indoles bind the vacant apical coordination sites. (Reproduced with permission from Barboiu (2018) and associated articles. Copyright 2005 Royal Society of Chemistry and 2001 Nature, respectively.)

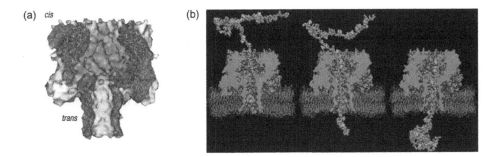

**FIGURE 4.4**   DNA translocation through a biological nanopore. α-hemolysin is a transmembrane protein that contains a pore that is approximately 1.4 nm wide at its narrowest point, which allows single-stranded DNA or RNA to move in and out of cells. (a) The cross-sectional structure of α-hemolysin. The trans side is located toward the cytosol in the cell, whereas the cis side points outwards. (b) Snapshots of an ssDNA molecule passing through such a nanopore. The DNA molecule enters the cavity at the cis side (left). Subsequently, the membrane voltage drives the DNA through the pore toward the trans side (middle, right). These snapshots are the result of a molecular dynamics simulation.

linear fashion, this might serve as a device to read the DNA sequence rapidly. This must be of obvious interest for genomics applications. Early experimental results were reported on the biological pore α-hemolysin (Kasianowicz et al., 1996), which has become the model protein for this type of research. Recently, nonbiological solid-state nanopores have opened up an even wider range of new research areas. Figure 4.4 shows DNA translocation through a biological nanopore (Dekker, 2007).

α-hemolysin is a protein secreted by *Staphylococcus aureus* as a toxin. It forms nanopores that spontaneously insert themselves into a lipid membrane, featuring a transmembrane ion channel with a width of 1.4 nm at its narrowest point (Song et al., 1996). The pore maintains high concentration ionic conductance of about 1 nS for typical conditions of one molar salt. An applied transmembrane voltage of 100 mV leads to a current of ~100 pA. A DNA can thread such pores, which would perhaps allow the DNA sequence to be read. The pore, due to its size, allows the passage of single-strand DNA (ssDNA), but not double-strand DNA (dsDNA) because, with a diameter of 2.2 nm, it is too wide. DNA is highly charged, so it can be driven through the nanopore in a linear head-to-tail fashion by an electric field. As it enters the nanopore, the ionic current is reduced simply because part of the liquid volume that carries the ionic current is occupied by the DNA.

Use of nanosensors via utilization of constructed nanopores in membranes for rapid electrical detection and characterization of biomolecules occurs with the help of versatile nanopores: α-hemolysin protein nanopores in lipid membranes (Kasianowicz et al., 1996) and solid-state nanopores (Li et al., 2001) in $Si_3N_4$. Subsequently, a new technique was reported for fabricating silicon oxide nanopores with single-nanometer precision and direct visual feedback using state-of-the-art silicon technology and transmission electron microscopy (TEM) (Storm et al., 2003). First, a pore of 20 nm is opened in a silicon membrane using electron-beam lithography and anisotropic etching. The membrane is thermally oxidized on both sides with a 40-nm thick $SiO_2$ layer. Using electron-beam lithography and reactive-ion etching, opening of squares occurs with dimensions up to 500 nm in the $SiO_2$ mask layer at the top. Subsequently, pyramid-shaped holes are etched using anisotropic KOH wet etching. Stripping the 40-nm oxide in buffered hydrogen fluoride opens up the nanopore in the silicon membrane, see Figure 4.5 (Storm et al., 2003).

After thermal oxidation, the pore is reduced to a single nanometer when it is exposed to a high-energy electron beam. This fluidizes the silicon oxide leading to a shrinking of the small hole due to surface tension. When the electron beam is switched off, the material quenches and retains its shape.

The pore size tuning technique was tested on holes fabricated using a different process. A focused electron beam with a spot size of a few nanometers can be used to drill holes in thin free-standing

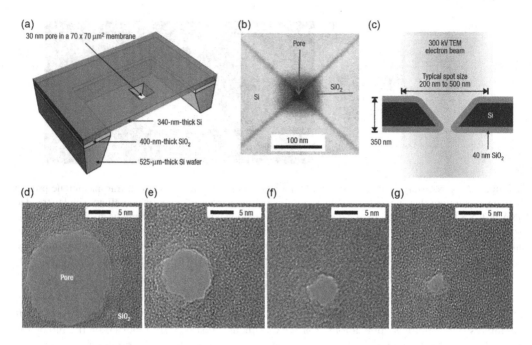

**FIGURE 4.5** Fabrication of silicon oxide nanometer-sized pores. (a) Cross-section view of our device. It consists of a 340-nm-thick free-standing single-crystalline silicon membrane supported by a KOH-etched wafer 525 μm thick. The membrane contains one or more submicrometer, pyramid-shaped pores, anisotropically etched with KOH from the top. (b) Top views canning electron micrograph of a nanofabricated pore after thermal oxidation. The pore is approximately $20 \times 20 \, nm^2$ in size, and is surrounded by a $SiO_2$ layer of approximately 40 nm thickness. (c) Cross-sectional view of the pore inside the electron microscope. (d–g) Sequence of micrographs obtained during imaging of a silicon oxide pore in a TEM microscope. The electron irradiation causes the pore to shrink gradually to approximately 3 nm.

$SiO_2$ membranes, with an estimated thickness of 10 nm, see Figure 4.6 (Storm et al., 2003). Using a pore that has been drilled using an electron-beam intensity above $1 \times 10^8 \, Am^{-2}$, as small as 6 nm holes can be obtained.

A rigorous analysis on synthetic channels was reported by Kasianowicz et al. (2008). Here, synthetic silicon nitride nanopore (Kim et al., 2006), carbon multiwall nanotube in an epoxy matrix (Ito et al., 2003), and nanopores formed in track-etched polyimide, polycarbonate, or poly(ethylene terephthalate) membranes (Li et al., 2004) have been presented. Due to copyright issues, we cannot present them here. The readers may consult the study for further information.

Solid-state nanopores in silicon nitride have been developed to take advantage of the potentially improved stability offered by semiconductor materials. Solid-state nanopores were initially used to detect individual double-stranded DNA (dsDNA) molecules, which are too large to transport through many ion channels. Other nanopores fabricated from carbon nanotubes or by heavy ion bombardment combined with chemical etching are found to have also made inroads into nanoscale pore sensor elements.

Continued developments have been made in designing solid-state nanopores. Very recently, a study addressed nanopore fabrication via transient high electric field controlled breakdown method which was used for detection of single RNA molecules, see Figure 4.7 (Yin et al., 2020). The fabrication of nanopores through a dielectric breakdown method, achieved by simple, low-cost desktop setups, has promoted the research of solid-state nanopore sensing. It reports a method for fabricating nanopores using transient high electric field controlled breakdown (THCBD) to form electric-field-dependent nanopores with different diameters of the order of milliseconds. By manipulating a micropipette with high electric field to establish the meniscus contact with the SiNx membrane, nanopores can be formed using an "auto-brake" fabrication process.

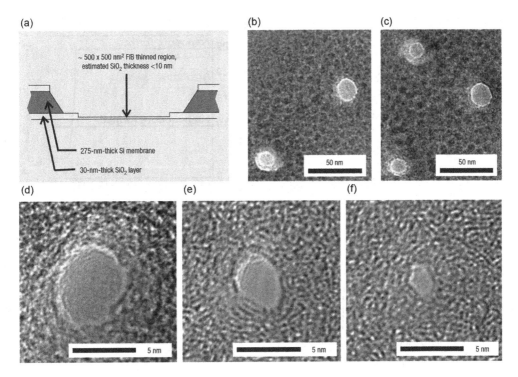

**FIGURE 4.6** TEM-drilled nanopores in thin free-standing $SiO_2$ membranes. (a) Cross-section of a thin $SiO_2$ membrane within a silicon-based membrane. Fabrication of this structure starts with a bare silicon membrane, which is oxidized with about 30 nm of $SiO_2$ on both sides. Using electron-beam lithography and reactive-ion etching, we open up $1 \times 1\,\mu m^2$ squares in the oxide layer. After a KOH wet etch, we obtain 30-nm-thick $SiO_2$ membranes. Subsequently, these are thinned further in a focused ion-beam microscope (FEI Strata DB235) to a final estimated thickness of less than 10 nm. (b) TEM micrograph of a part of a membrane with two holes that were drilled by a finely focused electron beam inside the TEM microscope. (c) TEM micrograph after drilling a third hole in the membrane depicted in (b). (d–f) Sequence of TEM images obtained on a shrinking nanopore with an initial diameter of about 6 nm and a final diameter of only 2 nm.

The nanopores formed by this method are useful in detecting two types of RNA molecules: transfer RNA from yeast extract and synthetic RNA oligonucleotide fragment (rArArArArArArArArArArArArArA), see Figure 4.8 (Yin et al., 2020).

## 4.4 ARTIFICIAL DNA CHANNELS

DNA was used as a building material (Burns et al., 2013) to create an atomistically determined molecular valve that is expected to control when and which cargo is transported across a bilayer. The valve, made from seven concatenated DNA strands, can bind a specific ligand, and, in response, undergoes a nanomechanical change to open up the membrane-spanning channel.

DNA is known to fold into predetermined structures (Zheng et al., 2009); hence, may meet an important criterion required for creating synthetic channels (Langecker et al., 2013).

The first DNA channel is claimed to have been designed to mimic a biological membrane channel α-hemolysin, a channel that had been extensively used in biosensing applications (Kasianowicz et al., 2008; Howorka and Siwy, 2009). However, the channel was constructed using the DNA origami method and was larger (~30 nm in each dimension) than a typical biological protein channel. Hence, further development was inevitable.

Membrane-spanning DNA nanopores have been built featuring a central hollow barrel that is open at both ends (Burns et al., 2013). The barrel is composed of six hexagonally arranged,

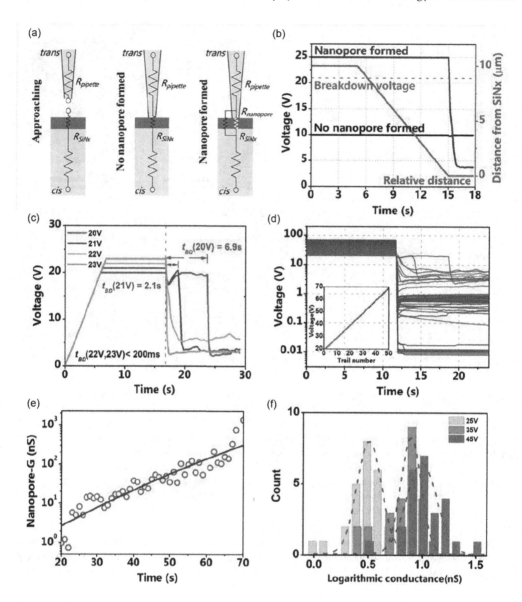

**FIGURE 4.7** Formation of nanopores in THCBD is an electric-field-dependent process. (a) The equivalent circuit shows three statuses of fabrication in a resistance model: (i) before contact is made, the total resistance is dominated by Rair; (ii) if contact is established but no nanopore is formed, the total resistance is dominated by RSiNx; (iii) if nanopores are successfully formed, the total resistance is dominated by Rnanopore. (b) Feedback voltage curves correspond to the fabrication process with voltages above (blue line) and below (black line) the breakdown voltage, and the relative distance is represented with a red curve. (c) Feedback voltage curves from 20 to 23 V show the fabrication time decreasing with increasing voltage and eventually exceeding the sampling period (200 ms), tBD (20V)=6.9 s, tBD (21 V)=2.1 s, tBD (22, 23 V)<200 ms. (d) Feedback voltage curves of fabrication experiments (voltage ranging from 20 to 70 V) show most of the breakdown events occurred transiently. For clarity, only the same period before and after the contact time-point is plotted. Inset: A gradient color legend indicating trail number and fabrication voltage. (e) Nanopore conductance versus fabrication voltage. (f) Nanopore conductance frequency histogram (n=25), fabricated with a voltage of 25, 35, 45 V, is fitted by Gaussian distribution. For details, see Yin et al. (2020). (Reprinted (adapted) with permission from Yin et al. (2020). Copyright (2020) American Chemical Society.)

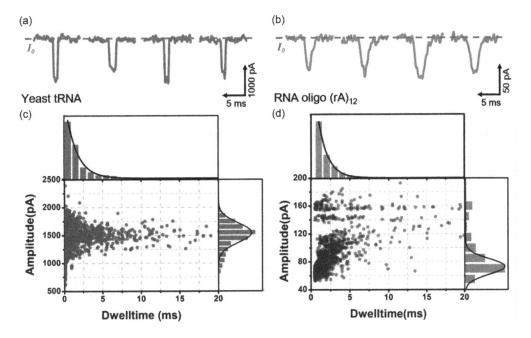

**FIGURE 4.8** RNA molecule analysis of THCBD nanopores. (a) Translocation events of yeast tRNA molecule through a nanopore in 1 M LiCl, pH 8 with an applied voltage of 200 mV. (b) Translocation events of synthesis RNA oligo $(rA)_{12}$ through a nanopore in 1 M LiCl, pH 8 with an applied voltage of 80 mV. (c) Yeast tRNA translocation event of dwell-time versus amplitude is shown in scatter diagram, and frequency count histogram of dwell-time (top) and amplitude (right) is shown. (d) The same for RNA oligo $(rA)_{12}$. (Reprinted (adapted) with permission from Yin et al. (2020). Copyright (2020) American Chemical Society.)

interconnected DNA duplexes enclosing a 2-nm-wide lumen with a tunable height ranging from 17 to 42 nm. Hydrophobic anchors have been included (Burns et al., 2013) to insert the negatively charged pores into the hydrophobic bilayer membrane. However, the barrels do not exploit the full design scope offered by DNA nanotechnology and do not exhibit the higher-order functions of general ion channels that can bind ligands, respond by nanomechanical opening, and select cargo for transport.

Burns and colleagues used the simple geometric shape of an open barrel as a starting point to rationally design a nanodevice that can regulate the flux of materials across a bilayer membrane. They first aimed at reducing the pore height to the order of the bilayer thickness, and thereby avoided structural flexibility and potential leakiness. A height of 9 nm was achieved with nanopore NP. Figure 4.9 shows a DNA nanopore featuring a nanomechanical and sequence-specific gate to regulate transmembrane flux (Burns et al., 2016). The six-helix-bundle architecture features six concatenated DNA strands, each of which are seen to connect two neighboring duplexes at their termini via single-stranded loops. The DNA origami plate might be structurally too flexible and leaky to form a considerable tight seal to the pore. This was overcome by designing a nanodevice that, in its closed state NP-C, features a simple "lock" strand that is bound closely to the entrance by hybridization to two docking sites to form a duplex across the channel opening. The docking sites are formed by the extension of two duplex staves. A "key" can hybridize to the lock and physically remove it to render the device in the open state, NP-O. The 2-nm-wide channel is expected to form a device that can regulate the flow of small organic molecules, including drugs. The design further equipped the nanodevices with hydrophobic cholesterol groups to anchor the hydrophilic nanostructures into the bilayer.

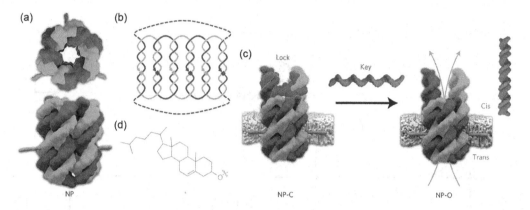

**FIGURE 4.9** A rationally designed DNA nanopore features a nanomechanical and sequence-specific gate to regulate transmembrane flux. (a) Structural model of pore NP composed of six DNA strands alternately in dark and pale blue. On the outside, the pore carries cholesterol-based membrane anchors (orange). (b) Two-dimensional map illustrating the connectivity of the six DNA strands of pore NP. The arrows indicate the 3′-termini of the DNA strands. (c) The "lock" DNA (red) of the closed nanopore NP-C is hybridizing to "key" DNA (green) to release open-channel NP-O. (d) Chemical structure of the cholesterol membrane anchor which is attached to the 3′-end of DNA oligonucleotides. (For color figure see eBook figure.)

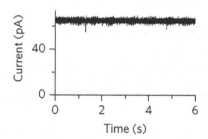

**FIGURE 4.10** Ionic current trace recorded of a single NP pore. 1 M KCl, 10 mM HEPES, pH 8.0, and at +40 mV relative to the cis side of the membrane.

Single-channel current recordings were used to see whether NP is membrane-spanning and structurally stable in the bilayer. Pore NP was found to form a stable channel across the membrane, a steady current (in a representative current trace) at +40 mV, and a narrow distribution of conductances (mean of 1.62 ± 0.09 nS, n = 100). The pore is also found Ohmic in behavior, reflecting its vertical symmetry. See Figure 4.10 (Burns et al., 2016).

The channels can distinguish with high selectivity the transport of small organic molecules that differ by the presence of a positively or negatively charged group. The DNA device could be used for controlled drug release and building synthetic cell-like or logic ionic networks. In contrast to ion channels of the biological systems, the synthetic channel is made from DNA, which folds into predictable structures and relative to proteins is easier to control. The channel is held in the lipid bilayer membrane via hydrophobic anchors. In the presence of a specific DNA "key," it maintains selective transports of small organic molecules with a positive charge across the bilayer. DNA channels are comparatively cheap and fast to produce and may be found with potential uses in drug delivery and synthetic biology.

## 4.5 AMPHIPHILIC MOLECULES INDUCE ION CHANNELS IN MEMBRANES

Let us start with a comparison of channels formed by well-studied peptide gramicidin A and less-studied polyene antibiotic amphotericin. The peptide gramicidin is known to form a head-to-head helical dimer that opens a tubular channel within a bilayer membrane (D-amino acids are

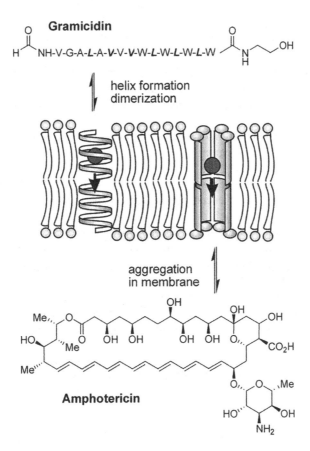

**FIGURE 4.11** Examples of ion channels formed by naturally occurring compounds. The peptide gramicidin forms a head-to-head helical dimer that opens a tubular channel within a bilayer membrane (D-amino acids are indicated in italic). The amphiphile polyene antibiotic amphotericin forms an aggregate channel surrounding an aqueous pore. (Reprinted (adapted) with permission from Fyles (2007). Copyright (2007) American Chemical Society.)

indicated in italic), while amphiphile antifungal agent polyene antibiotic amphotericin is known to form an aggregate channel surrounding an aqueous pore, see Figure 4.11 (Fyles, 2007).

The polyene antibiotic amphotericin forms channels in sterol-containing bilayer membranes and provides an alternative small-molecule model to relate channel structure to function. The amphotericin monomer has dual amphiphilic character: the mycosamine head group orients the molecule in a bilayer membrane with this polar group in contact with the aqueous phase and the polyene tail in contact with the lipid hydrophobic region. Several monomers experiencing correct orientation then aggregate to create a water-filled tube lined by the hydroxyl groups on the edge of the amphotericin macrocycle. A single amphotericin molecule is just half the thickness of a bilayer membrane. Two such aggregates geometrically associate in an end-to-end fashion and create a membrane-spanning artificial ion channel.

Gramicidin and amphotericin exhibit a membrane-spanning tube of roughly 2.5 nm in length with an internal diameter of 0.3 nm or more. These channel structural paradigms may provide the basic design criteria for artificial ion channels, as follows (Fyles, 2007):

i. A membrane (~4 nm thick)-spanning structure is required.
ii. The active membrane-spanning structure must enclose a significant volume for the passage of the ion. Consequently, the active structures have molecular weights in excess of 3–4 kDa.

Both the dimensions and molecular weights are large by the standards of small-molecule synthesis.

iii. The interior of the channel is hydrophilic. In gramicidin channels, the ions are largely desolvated during passage, but the larger amphotericin aggregate allows the passage of hydrated ions. An artificial structure must provide stabilizing contacts for a transiting ion. These can be direct transporter–ion interactions, or more simply transporter–solvated ion interactions.

iv. The channel must embed itself into a bilayer membrane through hydrophobic contacts with the lipid hydrocarbon chain regions.

Both gramicidin and amphotericin channels are too short to cover the bilayer thickness and ion passing through them are meant to encounter the hydrophilic lipid head groups unless some sort of accommodation is created through the membrane thickness change might happen due to the elastic properties of bilayers (Ashrafuzzaman and Tuszynski, 2012). To overcome this bilayer thickness-channel length mismatch and other required properties for the synthetic agents to consider including membrane-spanning bola-amphiphilic character, inward facing polar functionality to bind water and transiting ions, and hydrophobic outward facing functionality to stabilize and orient the structure within a bilayer membrane, Fyles and colleagues synthesized linear oligomers, see Figure 4.12 (Fyles, 2013). Compounds proven to be active ion channels with the bilayer clamp technique showed remarkable results that the channels had conductance within the range of natural ion channels and gramicidin.

Fyles and colleagues continued to exploring alternative macrocyclic "wall units," see Figure 4.13 (Fyles, 2013). The parent compound (6, Figure 4.13) was active but was essentially insoluble;

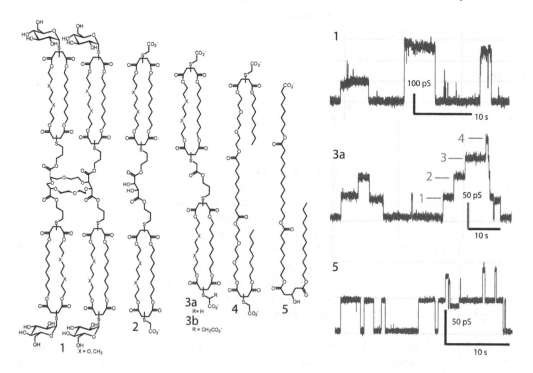

**FIGURE 4.12** Structural evolution of linear oligoesters and their corresponding bilayer conductance. Time profiles are under comparable conditions: 1 M CsCl electrolyte, diphytanoylphosphatidyl choline (diPhyPC) membrane, and 100 mV applied potential. The amphiphile polyene antibiotic amphotericin forms an aggregate channel surrounding an aqueous pore. (Reprinted (adapted) with permission from Fyles (2013). Copyright (2013) American Chemical Society.)

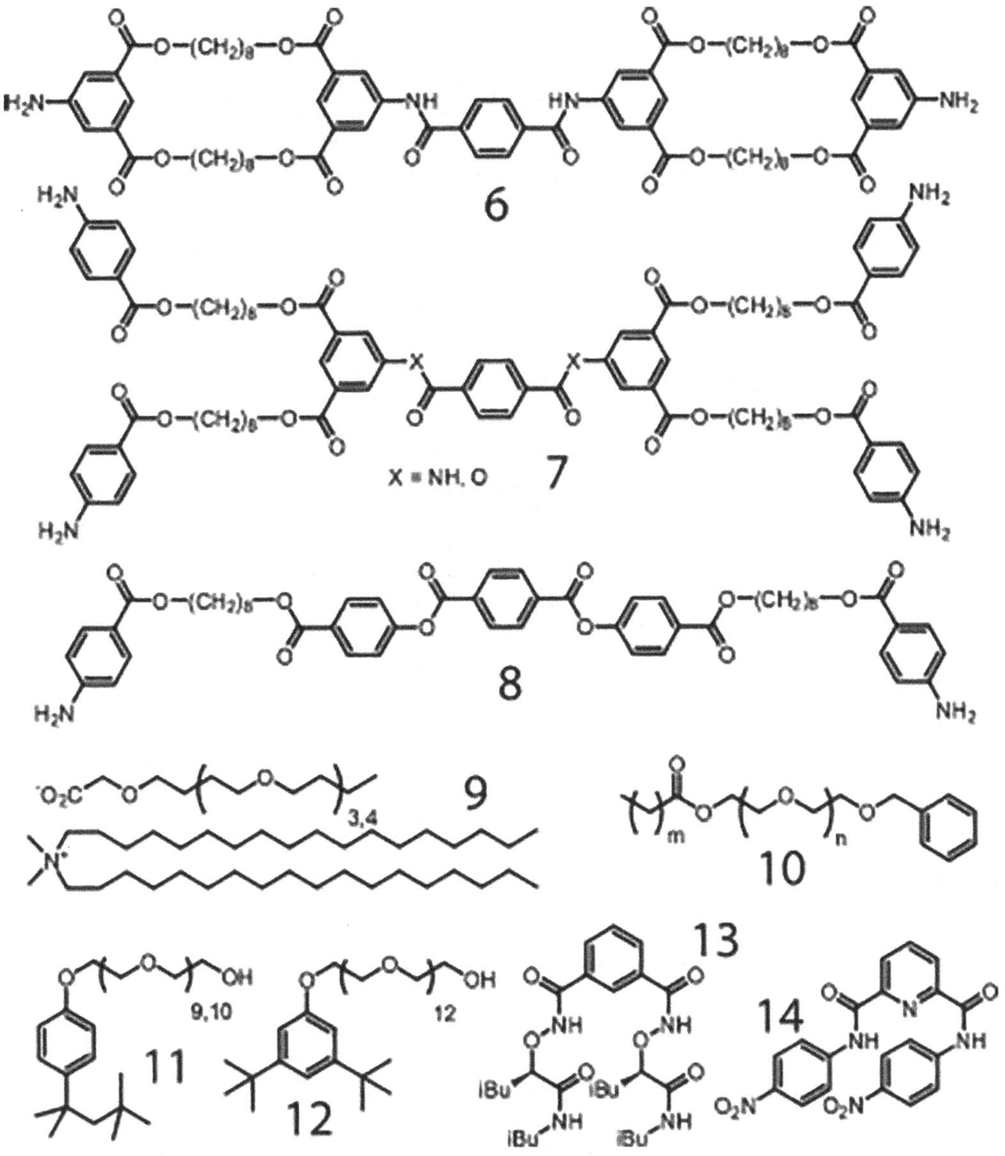

**FIGURE 4.13** Simple compounds that form transporting structures. (Reprinted (adapted) with permission from Fyles (2013). Copyright (2013) American Chemical Society.)

acyclic versions (7) were much more tractable and demonstrated sustained regular on-off openings (40 pS under conditions comparable to earlier presented Figure 4.12) with good cation selectivity. Compound 8 uncovered a rare exponential voltage dependence of the mean current carried by these channels. This series of compounds behaves as the earlier series; simplified compounds retain the activity of more complex parents.

The synthetic ion transporter mechanisms invoke transmembrane structures that provide a continuous pathway from one face of the bilayer barrier to the opposite. To access such structures, the transporter must penetrate the bilayer, following a process similar to that of lipid flip-flop and is occasionally described as a rate-limiting process. A mechanistic framework consistent with various observations is shown in Figure 4.14.

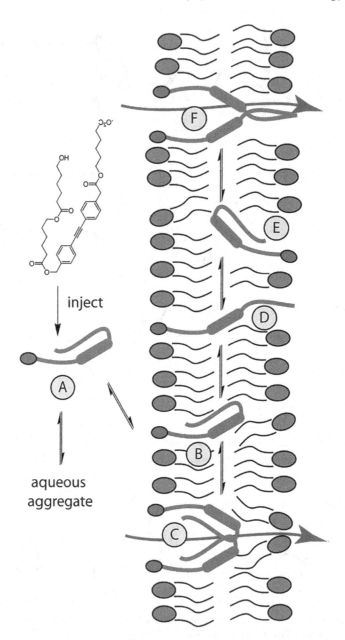

**FIGURE 4.14** A proposed mechanistic framework for transport by oligoester channel-forming compounds. (Reprinted (adapted) with permission from Fyles (2013). Copyright (2013) American Chemical Society.)

The injection of a compound results immediately in the formation of some monomer state (Figure 4.14A) in water. The monomer can associate with the bilayer in various ways, but a flexible amphiphile would perhaps rapidly partition to form a U-shaped insert (Figure 4.14B). Once inserted, the object would perturb the membrane environment to an extent that aggregation and eventually microphase separation would become a driving force. These loosely structured aggregates (Figure 4.14C) would produce the kind of short lifetime excimer emission, and we see these as good candidates for one of the conducting structures formed. Figure 4.14D and E represent the naturally expected unfolded and refolded (a complete flip-flop) conditions while spanning with the other surface of the membrane. Then the final stage Figure 4.14F might give shorter-lived or less regular conducting states albeit with potentially much higher conductance. These are all predictions by Fyles (2013), but require structural studies to get verified.

## 4.6 MOLECULAR DYNAMICS SIMULATIONS TO ADDRESS THE ARTIFICIAL ION CHANNEL ASSEMBLY IN LIPID BILAYERS

We have understood, though in brief, about various ion channels that are experimentally addressed. MD simulations provide further details into the structure and function of channels that are not occasionally addressed in experiments. Here, we shall explain two cases as examples to enrich our understanding of artificial channels.

### 4.6.1 SIMULATING POROUS NANOCAPSULE AS AN ION CHANNEL

Biological ion channel-type functionality is observed in porous polyoxometalate Keplerate nanocapsules. Coarse-grained molecular dynamics (CGMD) simulations were used to demonstrate a route for embedding negatively charged nanocapsules into lipid bilayer membranes via self-assembly (Carr et al., 2008). High negative charge of the capsules temper their use as artificial ion channels. A homogeneous mixture of water, cationic detergent, and phospholipid was reported to spontaneously self-assemble around the nanocapsule into a layered, liposome-like structure, where the nanocapsule was enveloped by a layer of cationic detergents followed by a layer of phospholipids, see Figure 4.15 (Carr et al., 2008). A video on the self-assembly process may be watched at the following link: https://pubs.acs.org/doi/abs/10.1021/nl802366k.

Fusion of the layered liposome with a lipid bilayer membrane was reported to embed the nanocapsule into the lipid bilayer membrane. The resulting assembly was found to remain stable even after the surface of the capsule was exposed to electrolyte.

Water was observed to flow across (into and out of) the capsule as Na$^+$ cations entered, suggesting that a polyoxometalate nanocapsule can form a functional synthetic ion channel in a lipid bilayer membrane, see Figure 4.16 (Carr et al., 2008). At the end of the simulation of fusion, the capsule remained trapped inside the bilayer, not directly connected to the solvent. A model was created by removing POPC and DODA above and below the capsule and the empty space was filled with water and ions, which would address whether the capsule might be stable in a configuration which is open to water on either side of the membrane. Following a brief period, where the capsule, DODA, and POPC were constrained, while the water and ions were allowed to assume a new equilibrium conformation, the system was simulated free of restraints under constant pressure conditions. Figure 4.17 illustrates one system that was obtained at the end of a 351 ns simulation. In equilibrium configuration, water was observed to flow across the capsule as Na$^+$ cations entered, which shows that the capsule can be both stable in this conformation and transport ions across the membrane, even when the capsule's charge is reduced by the presence of the cationic DODA.

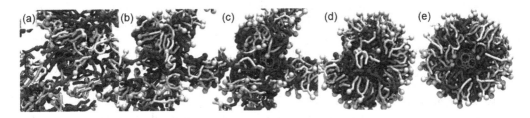

**FIGURE 4.15** Self-assembly of a capsule liposome. The molecular dynamics snapshots illustrate the molecular configuration obtained after 0 (a), 3.36 (b), 7.36 (c), 41.68 (d), and 100.24 ns (e) of CGMD simulation. The porous nanocapsule is shown in red, dimethyldioctadecylammonium (DODA) in blue, and palmitoyloleoylphosphatidylcholine (POPC) in white; water and ions are not shown. (Reprinted (adapted) with permission from Carr et al. (2008). Copyright (2008) American Chemical Society.) (For color figure see eBook figure.)

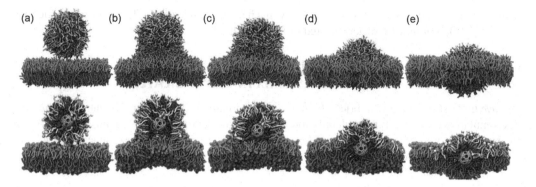

**FIGURE 4.16** Fusion of a liposome with a lipid bilayer. Snapshots illustrate the state of the CGMD simulation after 0 (a), 125.2 (b), 431.6 (c), 607.36 (d), and 940.4 ns (e). The porous nanocapsule is shown in yellow, DODA in blue, POPC in white, and the top and bottom leaflets of the POPC bilayer in green and red, respectively. For each panel, the image in the bottom row is a cut-away view of the same configuration as shown in the top row. An animation illustrating this simulation is available in Supporting Information in the published article. (Reprinted (adapted) with permission from Carr et al. (2008). Copyright (2008) American Chemical Society.) (For color figure see eBook figure.)

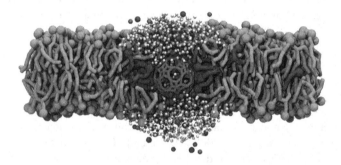

**FIGURE 4.17** Porous nanocapsule as an ion channel. The snapshot illustrates the outcome of a CGMD simulation of a porous nanocapsule (red) held in a POPC bilayer membrane (cyan) by DODA (blue). The top of the capsule is open to water (white) and ions (blue $Na^+$, red $Cl^-$). Note the ions inside and around the capsule. (Reprinted (adapted) with permission from Carr et al. (2008). Copyright (2008) American Chemical Society.) (For color figure see eBook figure.)

### 4.6.2 Simulating DNA Channels

MD simulations were performed to characterize the biophysical properties of DNA channels (previously explained) with atomic precision (Yoo and Aksimentiev, 2015). Two variants of the six-helix DNA channel were simulated to determine the microscopic structure in a lipid bilayer membrane and elucidate the ion conductance mechanism, including its dependence on the degree of hydrophobicity of the transmembrane pore. DNA channels were unexpectedly found to exhibit mechanosensitive gating and transport charged solutes against the electrostatic gradient by electro-osmosis.

A DNA nanochannel was constructed consisting of six 42-basepair dsDNA helices (system I) using the caDNAno program. Neighboring helices were connected to each other via DNA origami-like crossovers, Figure 4.18a; six base pairs of each helix were covalently modified by attaching ethyl groups to the DNA phosphate groups, forming a transmembrane domain. The addition of an ethyl group to the backbone of a DNA nucleotide makes the DNA nucleotide electrically neutral under physiological (pH 7.5) conditions and more hydrophobic. The MD simulation data of six-helix DNA membrane channels are presented in Figure 4.18.

The local structures of the channels were found to undergo considerable fluctuations, departing from the idealized design. The transmembrane ionic current flows both through the central pore of

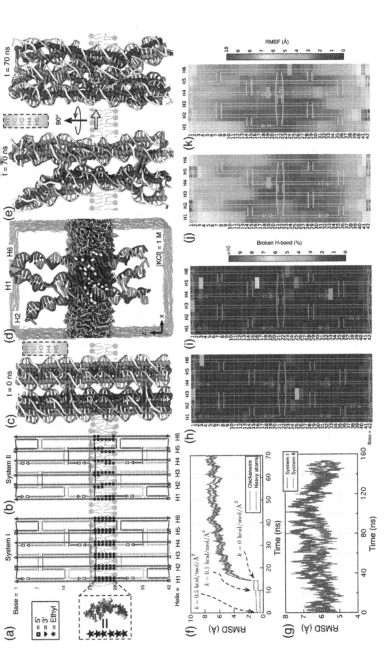

**FIGURE 4.18** MD simulation of six-helix DNA membrane channels. (a) CaDNAno design of system I. System I consists of six interconnected dsDNA helices (H1–H6); each helix contains 42 base pairs. To enable placement of the channel in a lipid bilayer, 72 backbone phosphate groups (stars) were neutralized by covalently attaching ethyl groups (Burns et al., 2013). The molecular image in the inset shows a six-nucleotide DNA fragment (pink) containing six ethyl groups (gray). (b) CaDNAno design of system II. System II has the same structure as system I except that the DNA phosphates located less than 20 Å away from the pore axis were not modified by the ethyl groups. (c) An idealized initial structure of system I shown using a cartoon representation (gray) overlaid with a custom chickenwire representation (colors). In the chickenwire representation, beads indicate the locations of the centers of mass of individual base pairs; horizontal connections between pairs of beads indicate interhelical crossovers. (d) A fully assembled system I at the end of a 70 ns equilibration simulation. For clarity, this cross-sectional view shows only three DNA helices: H1, H2, and H6. The DNA strands are shown in cartoon representation with each strand having the same color as in the caDNAno design; the ethyl groups are shown as white spheres, the lipid molecules as molecular bonds, and the 1 M KCl solution as a semitransparent surface. (e) Cartoon (gray) and chickenwire (color) representations of system I after the 70 ns equilibration. (f) Root-mean-square deviations (RMSDs) of the DNA's coordinates from the ideal design (t=0) during the 70 ns equilibration. Blue and red curves indicate the RMSD values computed for nonhydrogen atoms and chickenwire beads, respectively. (g) RMSD of the DNA's chickenwire coordinates with respect to the ideal design during the 150 ns production simulations of systems I (red) and II (blue). (h,i) Probability of observing a broken base pair during the 150 ns production simulations of systems I (h) and II (i). Each block indicates a unique base pair of the channel; the probability is color-coded for each block. The caDNAno design of the channel is overlaid in yellow. (j,k) Root-mean-square fluctuations of each base pair's center of mass during the production simulations of systems I (j) and II (k). (Reprinted (adapted) with permission from Yoo and Aksimentiev (2015). Copyright (2015) American Chemical Society.) (For color figure see eBook figure.)

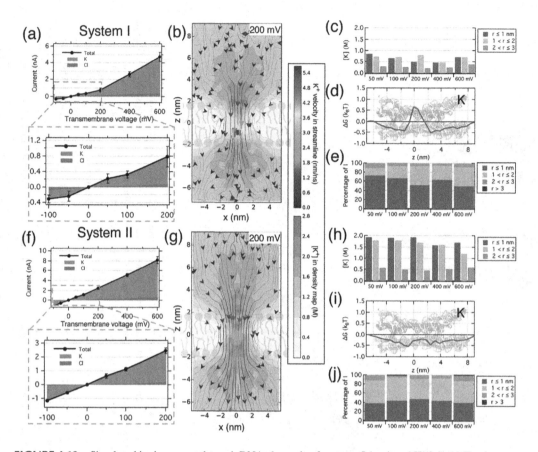

**FIGURE 4.19** Simulated ionic current through DNA channels of systems I (a–e) and II (f–j). (a) Total current (black) and the currents of K$^+$ (blue) and Cl$^-$ (red) ions versus transmembrane voltage for system I. The error bars indicate the standard deviation of a 10 ns block average of the corresponding total ionic current trace. (b) Local density (grayscale) and local velocity (streamlines) of K$^+$ ions in system I at 200 mV. The maps show the x–z cross-sections of the corresponding three-dimensional density and velocity fields. (c) Local density of K$^+$ ions at the constriction (|z|< 5 Å) of the channel for three cylindrical segments concentric with the channel's axis; r denotes the radial distance from the axis. (d) Free energy change (ΔG) of K$^+$ ions along the channel axis. ΔG was computed by taking a Boltzmann inversion of the K$^+$ concentration profile along the channel axis. The concentration profile was computed from the 150 ns production simulation of system I by counting the number of ions in cylindrical bins of 1.0 nm radius and 0.7 nm spacing along the z axis. (e) Fraction of the total ionic current flowing through cylindrical segments concentric with the channel's axis. The calculation was done for the |z|< 5 Å cross-section of the channel. (f–j) Same as in panels (a–e) but for system II. (Reprinted (adapted) with permission from Yoo and Aksimentiev (2015). Copyright (2015) American Chemical Society.) (For color figure see eBook figure.)

the channel as well as along the DNA walls and through the gaps in the DNA structure. Figure 4.19 demonstrates these features in detail.

The following conclusions can be drawn from the simulations:

- chemical modifications of the channel's surface can have a considerable effect on the transmembrane transport of water and ions.
- the DNA channels exhibit a mechanosensitive response. It opens up the possibility of using such channels as force sensors. DNA channels' conductance was found to depend on the membrane tension, making them potentially suitable for force-sensing applications.
- the electro-osmosis governs the transport of drug-like molecules through the DNA channels.

## REFERENCES

Ashrafuzzaman, M., & Tuszynski, J. A. (2012). *Membrane Biophysics*. Berlin, Heidelberg: Springer-Verlag. doi:10.1007/978-3-642-16105-6

Barboiu, M. (2004). Supramolecular polymeric macrocyclic receptors – Hybrid carrier versus channel transporters in bulk liquid membranes. *Journal of Inclusion Phenomena, 49*(1/2), 133–137. doi:10.1023/b:jiph.0000031126.58442.f4

Barboiu, M. (2018). Encapsulation versus self-aggregation toward highly selective artificial K+ channels. *Accounts of Chemical Research, 51*(11), 2711–2718. doi:10.1021/acs.accounts.8b00311

Barboiu, M., Cerneaux, S., Lee, A. V., & Vaughan, G. (2004). Ion-driven ATP pump by self-organized hybrid membrane materials. *Journal of the American Chemical Society, 126*(11), 3545–3550. doi:10.1021/ja039146z

Barboiu, M., Vaughan, G., & Lee, A. V. (2003). Self-organized heteroditopic macrocyclic superstructures. *Organic Letters, 5*(17), 3073–3076. doi:10.1021/ol035096r

Burns, J. R., Seifert, A., Fertig, N., & Howorka, S. (2016). A biomimetic DNA-based channel for the ligand-controlled transport of charged molecular cargo across a biological membrane. *Nature Nanotechnology, 11*(2), 152–156. doi:10.1038/nnano.2015.279

Burns, J. R., Stulz, E., & Howorka, S. (2013). Self-assembled DNA nanopores that span lipid bilayers. *Nano Letters, 13*(6), 2351–2356. doi:10.1021/nl304147f

Carr, R., Weinstock, I. A., Sivaprasadarao, A., Müller, A., & Aksimentiev, A. (2008). Synthetic ion channels via self-assembly: A Route for embedding porous polyoxometalate nanocapsules in lipid bilayer membranes. *Nano Letters, 8*(11), 3916–3921. doi:10.1021/nl802366k

Deamer, D. W., & Akeson, M. (2000). Nanopores and nucleic acids: Prospects for ultrarapid sequencing. *Trends in Biotechnology, 18*(4), 147–151. doi:10.1016/s0167-7799(00)01426-8

Dekker, C. (2007). Solid-state nanopores. *Nature Nanotechnology, 2*(4), 209–215. doi:10.1038/nnano.2007.27

Fyles, T. M. (2007). Synthetic ion channels in bilayer membranes. *Chemical Society Reviews, 36*(2), 335–347. doi:10.1039/b603256g

Fyles, T. M. (2013). How do amphiphiles form ion-conducting channels in membranes? Lessons from linear oligoesters. *Accounts of Chemical Research, 46*(12), 2847–2855. doi:10.1021/ar4000295

Gilles, A., & Barboiu, M. (2015). Highly selective artificial K+ channels: An example of selectivity-induced transmembrane potential. *Journal of the American Chemical Society, 138*(1), 426–432. doi:10.1021/jacs.5b11743

Howorka, S., & Siwy, Z. (2009). Nanopore analytics: Sensing of single molecules. *Chemical Society Reviews, 38*(8), 2360. doi:10.1039/b813796j

Ito, T., Sun, L., & Crooks, R. M. (2003). Simultaneous determination of the size and surface charge of individual nanoparticles using a carbon nanotube-based coulter counter. *Analytical Chemistry, 75*(10), 2399–2406. doi:10.1021/ac034072v

Kasianowicz, J. J., Brandin, E., Branton, D., & Deamer, D. W. (1996). Characterization of individual polynucleotide molecules using a membrane channel. *Proceedings of the National Academy of Sciences, 93*(24), 13770–13773. doi:10.1073/pnas.93.24.13770

Kasianowicz, J. J., Robertson, J. W., Chan, E. R., Reiner, J. E., & Stanford, V. M. (2008). Nanoscopic porous sensors. *Annual Review of Analytical Chemistry, 1*(1), 737–766. doi:10.1146/annurev.anchem.1.031207.112818

Kim, M., Wanunu, M., Bell, D., & Meller, A. (2006). Rapid fabrication of uniformly sized nanopores and nanopore arrays for parallel DNA analysis. *Advanced Materials, 18*(23), 3149–3153. doi:10.1002/adma.200601191

Langecker, M., Arnaut, V., Martin, T. G., List, J., Renner, S., Mayer, M., . . . Simmel, F. C. (2013). Synthetic lipid membrane channels formed by designed DNA nanostructures. *Biophysical Journal, 104*(2). doi:10.1016/j.bpj.2012.11.3022

Li, J., Stein, D., Mcmullan, C., Branton, D., Aziz, M. J., & Golovchenko, J. A. (2001). Ion-beam sculpting at nanometre length scales. *Nature, 412*(6843), 166–169. doi:10.1038/35084037

Li, N., Yu, S., Harrell, C. C., & Martin, C. R. (2004). Conical nanopore membranes. Preparation and transport properties. *Analytical Chemistry, 76*(7), 2025–2030. doi:10.1021/ac035402e

Li, Y., Zheng, S., Legrand, Y., Gilles, A., Van Der Lee, A., & Barboiu, M. (2018). Structure-driven selection of adaptive transmembrane Na+carriers or K+channels. *Angewandte Chemie International Edition, 57*(33), 10520–10524. doi:10.1002/anie.201802570

Schneider, S., Licsandru, E., Kocsis, I., Gilles, A., Dumitru, F., Moulin, E., . . . Barboiu, M. (2017). Columnar self-assemblies of triarylamines as scaffolds for artificial biomimetic channels for ion and for water transport. *Journal of the American Chemical Society, 139*(10), 3721–3727. doi:10.1021/jacs.6b12094

Song, L., Hobaugh, M. R., Shustak, C., Cheley, S., Bayley, H., & Gouaux, J. E. (1996). Structure of staphy-lococcal alpha-hemolysin, a heptameric transmembrane pore. *Science, 274*(5294), 1859–1865. doi:10.1126/science.274.5294.1859

Storm, A. J., Chen, J. H., Ling, X. S., Zandbergen, H. W., & Dekker, C. (2003). Fabrication of solid-state nano-pores with single-nanometre precision. *Nature Materials, 2*(8), 537–540. doi:10.1038/nmat941

Sun, Z., Barboiu, M., Legrand, Y., Petit, E., & Rotaru, A. (2015). Highly selective artificial choles-teryl crown ether K+-channels. *Angewandte Chemie International Edition, 54*(48), 14473–14477. doi:10.1002/anie.201506430

Sun, Z., Gilles, A., Kocsis, I., Legrand, Y., Petit, E., & Barboiu, M. (2016). Squalyl crown ether self-assembled conjugates: An example of highly selective artificial K⁺ channels. *Chemistry - A European Journal, 22*(6), 2158–2164. doi:10.1002/chem.201503979

Yin, B., Fang, S., Zhou, D., & Yuan, J. (2020). Nanopore fabrication via transient high electric field controlled breakdown and detection of single RNA molecules. doi:10.1021/acsabm.0c00812.s001

Yoo, J., & Aksimentiev, A. (2015). Molecular dynamics of membrane-spanning DNA channels: Conductance mechanism, electro-osmotic transport, and mechanical gating. *The Journal of Physical Chemistry Letters, 6*(23), 4680–4687. doi:10.1021/acs.jpclett.5b01964

Zheng, J., Birktoft, J. J., Chen, Y., Wang, T., Sha, R., Constantinou, P. E., . . . Seeman, N. C. (2009). From molecular to macroscopic via the rational design of a self-assembled 3D DNA crystal. *Nature, 461*(7260), 74–77. doi:10.1038/nature08274

Zheng, S., Huang, L., Sun, Z., & Barboiu, M. (2020). Self-assembled artificial ion-channels toward natural selection of functions. *Angewandte Chemie*. doi:10.1002/ange.201915287

# 5 Nonchannel Membrane Gating

Cellular membrane, the semi-permeable compartment, maintains transmembrane transport of materials and information. This transport may be both "controlled" and "uncontrolled". Natural and artificial agents that are meant to naturally travel or artificially forced to get delivered back-and-forth or unidirectionally across various membranes, such as plasma, nuclear, and mitochondrial membranes follow mainly two routes. They are membrane-hosted versatile ion channel and nonchannel routes. Phenomenology and mechanisms of their delivery through ion channels have been rigorously addressed in earlier chapters of this book. However, nonchannel route(s) across the cell membrane is largely unaddressed. Ion channel transports mostly refer to the controlled transport phenomena, while nonchannel transports may be both controlled and uncontrolled. The main difference between these two types of transports involving the role(s) of membranes may lie in the relative involvement of the physicochemical states of membranes, that is, the former (controlled transport) is largely regulated by membrane physical properties, while the latter (uncontrolled transport) mainly relies on primarily the natural chemical composition and secondarily the physical properties of the membrane. Moreover, controlled transport occurs through a physical state which may undergo structural changes, while the latter is mostly naturally inbuilt process that may not be considerably influenced due to changes in localized events or structures. Moreover, the membrane dynamics may have different quantitative and qualitative effects on both controlled and uncontrolled membrane transport phenomena. This chapter discusses the existing mechanisms in natural systems, as well as the novel artificial mechanisms that may eventually help create breakthroughs in medical applications.

## 5.1 CHANNEL VERSUS NONCHANNEL TRANSPORTS

Membrane transport occurs mainly through the following three ways:

- Transport through carrier (transporter) transport proteins of membrane,
- Transport through channel proteins of membranes,
- Transport without the use of transporters or channel. This transport occurs without the use of any transport proteins. This way, membranes would be permeable to only some gases and small molecules.

The channel protein transports of membrane are already addressed in earlier chapters. In this chapter, we aim to address the other two transport phenomena and mechanisms that are covered in the nonchannel transport category.

There are a few intriguing questions related to the cell membrane transport phenomena that biophysicists and chemists have been exploring for answers. These are as follows: how different types/ classes of molecules, namely, small molecules, peptides, and proteins cross the mammalian plasma membrane, how such permeation events are measured experimentally, and which technologies are being developed to deliver impermeable molecules to the cytoplasm, see Figure 5.1.

Experimental methods have been used to determine, qualitatively or quantitatively, whether a molecule has successfully traversed the cell membrane. Various molecules, including ions, small solutes, and metabolites, along with bacterial toxins and viruses, are thought to traverse the cell membrane. There are engineering strategies that have been proposed to improve the permeability of small molecules, peptides, and proteins. We shall address a few of the transport phenomena that do not concern the involvement of ion channel transport, instead we wish to pinpoint nonchannel transport of the cell membrane.

DOI: 10.1201/9781003010654-5

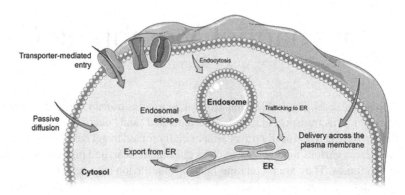

**FIGURE 5.1**  Possible routes of cytosolic entry. Molecules may passively diffuse across the cell membrane, or be shuttled in via natural or artificial delivery mechanisms. Membrane transporters allow the passage of various ions and metabolites. Protein toxins and viruses have evolved complex translocation mechanisms, hijacking the host's ER transporters in some instances. Engineering approaches to improve a payload's permeability may involve physically disrupting the membrane, chemically modifying the payload, or attaching the payload – covalently or non-covalently – to an intracellular delivery system that can disrupt cell membranes. In any case, the translocation event can occur across the plasma membrane, or across internal cellular membranes following endocytosis (termed "endosomal escape"). (Images were adapted from Servier Medical Art. Details in Yang and Hinner (2014).) (For color figure see eBook figure.)

The term "gated" implies that the ion channel is controlled by a "gate" that must be opened for ions to pass through. The gates are usually opened by the binding of an incoming signal "ligand" to the receptor, which allows almost instantaneous passage of millions of ions from one side of the membrane to the other. This type of gating is specific for specific channels, termed "channel-gating." However, the cell membrane behaves like a partition having instantaneous- or slow-diffusing gates, ensuring membrane-gating (nonchannel gating), for specific substances, mentioned earlier, which do not pass through conventional ion channel gates. Subsequent sections explain the nonchannel membrane-gating phenomena and mechanisms using a few example cases.

## 5.2  GENERAL DIFFUSION OF MEMBRANES: TRANSPORT WITHOUT THE USE OF TRANSPORTERS OR CHANNELS

This transport occurs without the use of any transport proteins. Without the use of transport proteins, membranes are permeable to some gases and small molecules. The cell membrane's lipophilicity defines the metabolites that can freely move in and out of the cell, a process called "simple diffusion." Simple diffusion is directed from a region of high solute concentration to a region of low solute concentration. Various factors determine the net rate of diffusion, including the concentration difference of the solute, pressure difference between the cell and the environment, membrane electric potential, and osmosis (Guyton and Hall, 2000).

Generally, it is expected that the cross-membrane movement of the materials also depends on membrane composition and dynamics-related parameters that determine the fluidity of the membrane. This biophysical status of the membrane is responsible for in-membrane diffusion. The membrane dynamics and diffusion of agents inside it were observed by measuring various diffusion-related parameters for a marker, the sphingolipid-binding domain (SBD) derived from the amyloid peptide Aβ, on live neuroblastoma cells (Sankaran et al., 2009). Investigating the organization and dynamics of SBD-bound lipid microdomains under the conditions of cholesterol removal and cytoskeleton disruption was performed. Although it is considered symmetric, the cell membrane may host, on a localized scale, various physical domains, disorders, and varied pressure profiles that must have considerable effects on membrane fluidity, so on the transmembrane diffusivity. The

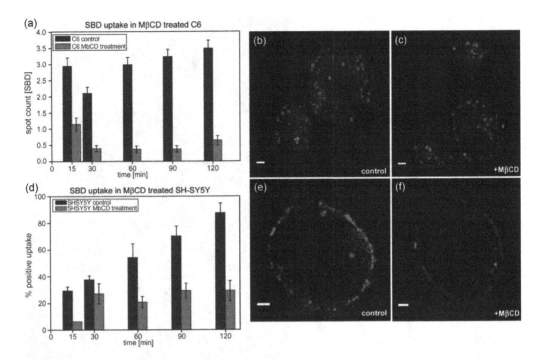

**FIGURE 5.2** SBD uptake at the plasma membrane is cholesterol-dependent. (a) Average number of SBD-tetramethyl rhodamine (TMR)-positive vesicles per cell in live c6 neurons untreated (black bars) or treated (red bars) with 10 mM methyl-β-cyclodextrin (MβCD). Cholesterol depletion (due to the effects of MβCD) significantly decreases the uptake of SBD over the indicated time course. (b,c) Fixed c6 neurons showing filip in staining of cholesterol (blue) and SBD (red) after 30 min of SBD incubation. In cholesterol-depleted neurons (C), SBD remains predominantly at the plasma membrane. (d) Average percentages of live cells with internalized SBD in SH-SY5Y cells untreated (black bars) or treated with 10 mM MβCD (red bars). In contrast to the gradual increase of SBD uptake in controls, cholesterol depletion prevents SBD uptake. (e,f) Live SH-SY5Y cells showing reduced incorporation and uptake at the plasma membrane of SBD, as measured by the presence or absence of internalized SBD-positive vesicles within the cell, at 60 min after incubation with 10 mM MβCD. Bars = 2 µm. Details in Hebbar et al. (2008). (For color figure see eBook figure.)

fluidity of a cell membrane may get compromised as the cell is treated with adsorbing drugs, but may get back to normal (predrug-treatment state) fluidity condition after the span of a time period (Sankaran et al., 2009). However, in this investigation, the change in membrane fluidity causes the rate of the cellular internalization of agents to be altered, although such transmembrane diffusion is also dependent on the timescale when the membrane fluidity state remains compromised due to the effects of drug treatments.

An important investigation suggested that the SBD is taken up by neuronal cells in a cholesterol- and sphingolipid-dependent manner via detergent-resistant microdomains and that the drug treatments have considerable roles in the cellular uptake mechanisms. Figures 5.2 and 5.3 clarify the matter in detail (Hebbar et al., 2008).

More than two decades ago, there was a clear cut demonstration on the membrane diffusion of peroxynitrite (Denicola et al., 1998). Peroxynitrite anion (ONOO−) is a reactive species of increasingly recognized biological relevance that contributes to oxidative tissue damage. In this study, the diffusion of peroxynitrite across erythrocyte membranes was studied. Peroxynitrite is found to cross the erythrocyte membrane by two different mechanisms: in the anionic form through the DIDS-inhibitable anion channel, and in the protonated form by passive diffusion (Denicola et al., 1998; Ferrer-Sueta and Radi, 2009).

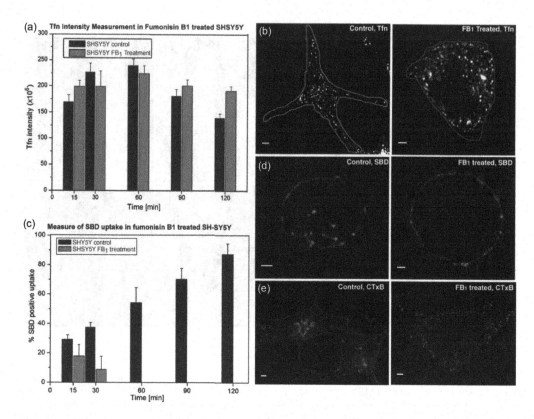

**FIGURE 5.3** SBD uptake is dependent on sphingolipid levels. SH-SY5Y cells were treated with 10 mM fumonisin B1 (FB1) for 2 h and then incubated with transferrin-Alexa-594 (Tfn), SBD-TMR, or Alexa Fluor-594 CtxB, and subsequently imaged in a medium containing 10 mM FB1. (a) FB1 treatment does not significantly alter the uptake of Tfn. The graph shows the intensity measurement after Tfn uptake at specific time points in treated (red bars) and untreated (black bars) cells. (b) Representative cells used for the analysis of Tfn uptake, with the cell boundary outlined. The intensity of spots within the cell was measured and compared between treated and control cells. (c) SBD uptake is reduced as indicated by spot count analysis. The graph indicates the percentage of SBD-positive spots in FB1-treated (red bars) and control cells (black bars). Spot count was used as a direct measurement of uptake instead of intensity because of the relatively lower uptake of SBD into SH-SY5Y cells compared with that of SBD into c6 cells (see supplementary Fig. II in the article). (d) Representative SH-SY5Y cells before and after FB1 treatment incubated with SBD-TMR. Fewer internal vesicles, carrying internalized SBD, are seen in the treated cells, where SBD remains at the plasma membrane. (e) Control and FB1-treated cells show differences in CtxB trafficking. CtxB is ordinarily trafficked to the Golgi body, seen near the center of neuroblastoma cells, but in FB1-treated cells, this is altered to a vesicular distribution throughout the cell. Bars = 2 μm. Details in Hebbar et al. (2008). (For color figure see eBook figure.)

Another example of passive diffusion in the biological system is that of the membrane transport of steroid hormones. Steroid hormones are unique among signaling molecules as they must travel to their target to trigger biological responses. After passing the plasma membrane, steroid hormones bind to nuclear receptors, which are ligand-dependent transcription factors that directly regulate gene expression. In this manner, steroid hormones influence nearly all aspects of animal development and physiology, including immunity, metabolism, growth, and sexual maturation. Steroids are small, lipophilic molecules, just "greasy" enough to enter phospholipid bilayers but also just "wet" enough to avoid becoming trapped within membranes. Given these characteristics, it seemed likely that steroid hormones could enter cells by passive diffusion across the plasma membrane (Plagemann and Erbe, 1976). However, many other studies suggested that steroid hormones require protein transporters for import into target cells, but no specific transporters were ever identified

(Milgrom et al., 1973). Thus, in the absence of a bona fide steroid hormone transporter, the simpler "passive diffusion" model prevailed for more than four decades, so much so that many current biology and medical textbooks state this model as fact. However, Okamoto and colleagues have identified and characterized a steroid hormone transporter protein in the fruit fly *Drosophila melanogaster* (Okamoto et al., 2018). This provides evidence that the prevailing dogma of steroid hormone entry by passive diffusion needs to be reconsidered.

Passive transport is classified into diffusion, facilitated diffusion, and bulk flow. The simple diffusion is a process described by the random movement of solutes that results in the net transport (flux) along a concentration gradient. This process, first, may be described by the first law of Fick (1855a; 1855b). More detailed models take into account the partitioning of solutes between the aqueous media and the barrier itself, as well as the flux resistance that arises from water layers adherent to the barrier, which results in the second law of Fick. The actual extent of the diffusion of molecules across biological membranes is caused by the interaction between the composition and the structural arrangement of the membrane and by the physicochemical characteristics of the permeant. For simplification, the primary structural element of a phospholipid biomembrane bilayer can be considered a continuous, lipophilic phase between two aqueous compartments. A complete understanding of this membrane diffusion of materials requires to adopt permeability screening approaches including *in silico* modeling, liposome-based experimental models, cell culture models, and *in situ* perfusion. In this chapter, we do not discuss this in detail, but readers may find a lot of articles to read and enrich their knowledge. Moreover, I have two chapters in my previous book *Nanoscale Biophysics of the Cell* to explain the cell membrane diffusion, theoretically, computationally and experimentally using my own research data and data of others published in various references therein. All relevant theories discovered over many decades have also been chronologically listed and explained here. Therefore, this book can be consulted for deeper understanding (Ashrafuzzaman, 2018).

## 5.3 MEMBRANE TRANSPORT THROUGH TRANSPORTERS

Carrier proteins are involved in membrane transport of materials and information. Although ion channels are also carrier proteins, but the transporters do have different ion channel structural moieties inside membrane. We explain this using a few example cases.

Generally, the hydrophobic substances can easily cross the lipid cell membrane by simple diffusion, while hydrophilic substances cannot. There are two basic modes of cellular transport for hydrophilic substances, that is, active and passive transport. These basic mechanisms can operate as follows: without a carrier protein (simple diffusion), with a carrier protein (facilitated diffusion), and with the expenditure of energy (primary and secondary active). The various modes of transport are shown in Figure 5.4 (Sahoo et al., 2014).

Membrane transport of a single solute at a time is a process referred to as a uniport. For example, "facilitated diffusion" is an example of uniport transport (Lodish et al., 2000). In facilitated diffusion (or carrier diffusion), the cargo molecule itself causes a conformational change in the carrier protein, which opens up a channel for the cargo to cross the cell membrane. The capacity of this bidirectional transport mechanism is limited by the time needed to change the conformation back and forth (Guyton and Hall, 2000).

Active transport allows the movement of molecules against their concentration and electrochemical gradients. This transport process requires ATP energy or other high energy phosphate bonds. There are three types of active transport mechanisms (Alberts et al., 2002):

- *Primary active transport.* ATP hydrolysis is directly connected to this type of transport.
- *Secondary active transport.* ATP hydrolysis is generated as an electrochemical gradient.
- *Tertiary active transport.* In this case, a secondary active transport process is further coupled to another distinct exchange mechanism. One example is the coupling of amino

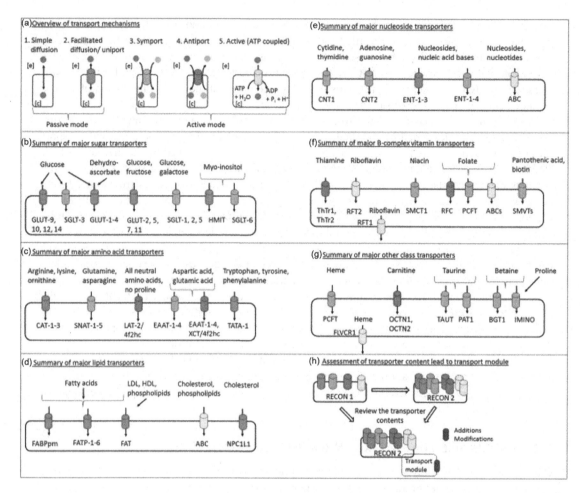

**FIGURE 5.4** Overview of transport mechanisms and major transport proteins of the various metabolite classes. (a) The basic modes for metabolite transport across the plasma membrane are shown. Based on the energy association, transport processes can be categorized into active and passive modes based on the energy used. The active mode can be further classified into primary and secondary mechanisms, while the metabolites can also be transported mainly via simple diffusion or facilitated diffusion driven by an increase in entropy. Specialized transport mechanisms (e.g., receptor-mediated endocytosis and tertiary active processes) are not shown. (b-g) Highlights major transport proteins involved in the transport of various substrates belonging to the sugar, amino acid, lipid, nucleoside, vitamin, and other classes mentioned in the text. (h) The present work accesses the coverage and gain in membrane transport systems with reference to the global human metabolic reconstruction, Recon 2 over Recon 1. The review of the relevant scientific literature led to the generation of transport reaction module that contained the proposed additions and modifications, discussed throughout the text. Refer to the article for a further explanation of these transport processes. The color coding for the transport mechanism as shown in (a) has been maintained in the other panels. For details, see Sahoo et al. (2014). (For color figure see eBook figure.)

acid transport system A (SNAT2) with system L (LAT1/4f2hc) for leucine uptake. SNAT2 utilizes the electrochemical gradient established by the $Na^+/K^+$ ATPase pump to drive its substrate into the cell (Baird et al., 2009).

Another transport mechanism is worth mentioning. Symport is the transport of multiple solutes across the cell membrane at the same time and in the same direction. If the inward transport of one solute is connected to the outward transport of another solute, the process is referred to as antiport (Alberts et al., 2002).

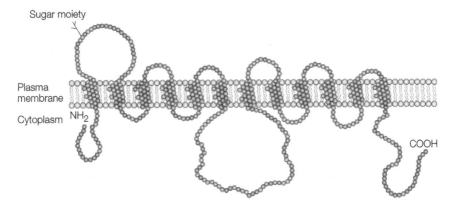

**FIGURE 5.5** Schematic representation of the GLUT family of proteins. The GLUT family comprises 13 members at present, which are predicted to span the membrane 12 times with both amino- and carboxyl-termini located in the cytosol. On the basis of sequence homology and structural similarity, three subclasses of sugar transporters have been defined: Class I (GLUTs 1–4) are glucose transporters; Class II (GLUTs 5, 7, 9 and 11) are fructose transporters; and Class III (GLUTs 6, 8, 10,12 and HMIT1) are structurally atypical members of the GLUT family, which are poorly defined at present. The diagram shows a homology plot between GLUT1 and GLUT4. Residues that are unique to GLUT4 are shown in red. Details in Bryant et al. (2002). (For color figure see eBook figure.)

As an example, we shall address the glucose transport across membranes. For most living cells, glucose is a key energy source. Due to its polar nature and large size, glucose molecules cannot traverse the lipid membrane of the cell by simple diffusion. Instead, glucose entry into the cells is affected by a large family of structurally related transport proteins known as glucose transporters. Two main types of glucose transporters have been identified:

- *Sodium–glucose linked transporters (SGLTs).* Sodium–glucose linked transporter-1 (SGLT1) was the first among SGLTs to be discovered and extensively studied. It comprises 14 transmembrane helices, both having the COOH and $NH_2$ terminals face the extracellular space.
- *Facilitated diffusion glucose transporters (GLUTs).* GLUTs comprise 12 membrane-spanning regions with intracellularly located amino and carboxyl terminals. Three subclasses, class I, II and III, of facilitative transporters have been identified. GLUT is a uniporter-type transporter protein.

As an example, we may consider a case study on regulated transport of the glucose transporter GLUT4 (Bryant et al., 2002). In muscle and fat cells, insulin stimulates the delivery of the glucose transporter GLUT4 from an intracellular location to the cell surface, where it facilitates the reduction of plasma glucose levels. The molecular mechanisms that mediate this translocation event involve integrating two fundamental processes – the signal transduction pathways that are triggered when insulin binds to its receptor and the membrane transport events that need to be modified to divert GLUT4 from intracellular storage to an active plasma membrane shuttle service. Figure 5.5 shows the schematic representation of the GLUT family of proteins engaging plasma membrane, and Figure 5.6 shows a model that depicts the transport of GLUT4 in insulin-responsive cells (Bryant et al., 2002).

Although we have picked a case study and have presented the available models, many more developments have been made since this study was published in 2002 (Bryant et al., 2002). The readers should consult them for further insights into understanding the GLUT transport of glucose molecules. We do not discuss them here due to limited space allocated for this subject in the book.

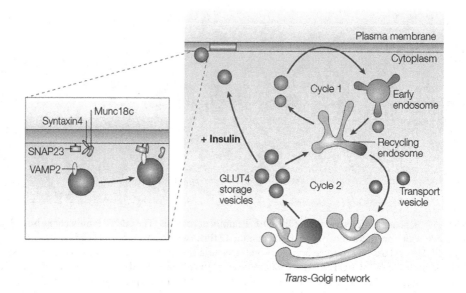

**FIGURE 5.6**  A model depicting the transport of GLUT4 in insulin-responsive cells. The model depicts two main intracellular-recycling pathways: cycle 1, between the cell surface and endosomes; and cycle 2, between the trans-Golgi network (TGN) and endosomes. GLUT4 transport is intricately controlled at several points along these cycles. On entry into the endosomal system, GLUT4 is selectively retained at the expense of other recycling transport, such as the transfer in receptor that constitutively moves through cycle 1. This retention mechanism might predispose GLUT4 for sorting into transport vesicles that bud slowly from the endosome and are targeted to the TGN. GLUT4 is sorted into a secretory pathway in the TGN. This sorting step probably involves a specialized population of secretory vesicles that excludes other secretory cargo, and that does not fuse constitutively with the plasma membrane. Vesicles that emerge from this sorting step, which we have previously referred to as GLUT4 storage vesicles or GSVs, might constitute most of the GLUT4 that is excluded from the endosomal system. In the absence of insulin, GSVs might slowly fuse with endosomes, thereby accounting for the presence of a significant but small pool of GLUT4 in endosomes, even in the absence of insulin. Insulin would then shift GLUT4 from this TGN–endosome cycle to a pathway that takes GLUT4 directly to the cell surface. The inset shows the SNARE proteins that are thought to regulate docking and fusion of GSVs with the cell surface. The t-SNAREs Syntaxin 4 and SNAP23 in the plasma membrane of fat and muscle cells form a ternary complex with the v-SNARE VAMP2, which is present on GSVs. Munc18c has been identified as the SM (Sec1-like/Munc18 family) protein (BOX 3) that controls the formation of this ternary complex. Details in Bryant et al. (2002).

## 5.4   MEMBRANE TRANSPORT THROUGH PEPTIDE-INDUCED DEFECTS

Gramicidin S and D are found to have high potential of almost 100% killing mechanisms against most of the Gram-positive and few Gram-negative bacteria and fungi, which can be used as an alternative antibiotic in the near future (Pavithrra and Rajasekaran, 2018). Considerable evidence has accumulated that the major target of gramicidin S (GS) is the lipid bilayer of bacterial or eukaryote membranes, and that GS kills bacteria by permeabilizing their inner membranes (Prenner et al., 1999).

Yonezawa et al. (1986) studied the binding of $^{14}$C-GS and various radioactive GS derivatives to intact cells and to cell wall-less protoplasts of several species of bacteria with different susceptibilities to this antimicrobial peptide. In general, GS absorbed more readily to GS-susceptible bacteria such as *Bacillus subtilis* and *Staphylococcus aureus* than to GS-resistant bacteria such as *E. coli*. In both cases, GS binding increased linearly with increasingly peptide concentration. At the minimal inhibitory concentration (MIC), about $1.3 \times 10^6$ molecules of GS were bound per bacterial cell, enough to cover the cell surface. At the MIC, protoplasts of *B. subtilis* bound about 80% of the

GS bound by intact cells, indicating that the majority of added peptide is bound to the bacterial inner membrane and not the extracellular cell wall. The absorption of GS in all cases was shown to increase the permeability of cells. This suggests that the mode of action of GS involves binding to the bacterial inner membrane and the resultant perturbation of inner membrane structure and function.

We inspected the membrane effect of GS to address its effects in model membrane system. We observe them to induce defects, for the first time, instead of inducing channels, and that the defect formation mechanism is phenomenologically highly GS concentration and transmembrane voltage-dependent. Figure 5.7 presents representative EP traces (Ashrafuzzaman et al., 2008).

We have addressed that the effects of GS on lipid bilayers depend both qualitatively and quantitatively on the type and charge of the lipid and to a lesser extent on the presence of cholesterol, but is almost insensitive to bilayer thickness. The concentration- and voltage-dependent permeabilization of phospholipid bilayers by GS does not occur by the formation of discrete, long-lived, peptide-induced channel structures or by overt bilayer solubilization, but by the formation of a wide distribution of short-lived, GS-induced defects in the host phospholipid bilayer.

Although we claim that the membrane current is due to defects in the membrane caused or created by GS molecules or their interactions with membrane components, we certainly do not know the pattern of their structures involving lipids, cholesterols, etc. However, certainly we can predict with confidence that the GS and membrane components do not make any stable structure in the membrane which might be able to create some stable leaks, equivalent to channels, but transient

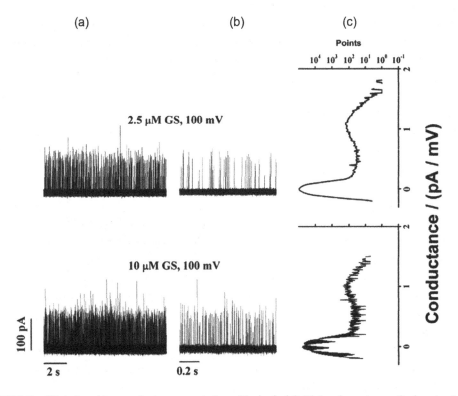

**FIGURE 5.7** GS-induced ion conductance events in zwitterionic 1,2-Dioleoyl-sn-glycero-3-phosphocholine/ n-decane bilayers at different concentrations of GS. (a) and (b) show long-time (11 s) and short-time (1 s) current traces of GS-induced ion conductance events, respectively. (c) All point conductance-level histograms constructed from the long-time traces (a). Two peaks (C) at 0 pA/mV and around 1 pA/mV respectively represent the baseline conductance of the "unperturbed" bilayer and the conductance levels of the GS-induced ion conductance events. 1.0 M, NaCl, pH 7.0.

leaks which are far off in regard to stability required for stable current events. For details, readers may consider to go through the paper (Ashrafuzzaman et al., 2008).

## 5.5  TOPOLOGICAL NANODEFECTS OF RBC MEMBRANES AND RELATED MEMBRANE TRANSPORTS

Defects in the RBC membrane in disease condition (bipolar depressive disease, commonly known as manic-depressive illness) was first predicted in 1982 (Merz, 1982). Intact living lymphocytes, RBCs, and fibroblasts were analyzed within an hour of sampling by either technique, namely, fluorescent spectroscopy or nuclear magnetic resonance. Both procedures yielded similar findings: distinct abnormalities in molecular movement within the surface membranes of cells taken from patients with bipolar depression.

Using atomic force microscopy (AFM), the transformation of cell morphology and the appearance of topological nanodefects of packed red blood cell (PRBC) membranes were addressed recently (Kozlova et al., 2017). The study concluded that there is a transition period of 20–26 days, in which an increase in the Young's modulus of the membranes 1.6–2 times and transition of cells into irreversible forms, which was preceded by the appearance of topological nanodefects of membranes. The AFM data can be understood in Figure 5.8.

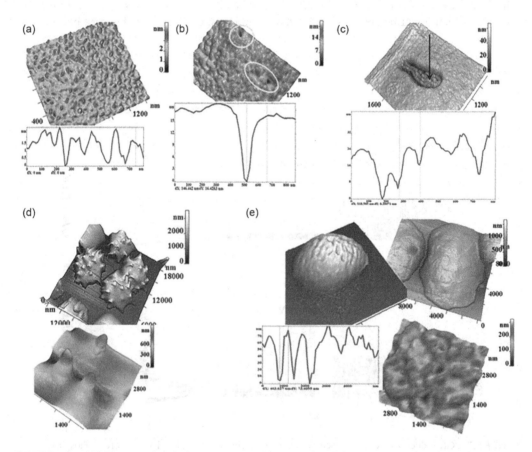

**FIGURE 5.8**  AFM images of typical nanosurfaces of II order of PRBC membranes and corresponding nanosurface profiles. (a) For discocyte on the fifth day of PRBC storage. (b) Single topological defects (highlighted by yellow circles) on the surface of membranes on the 16th day of PRBC storage. (c) Domains with a grain-like structure.(d) Spheroehinocyte on the 33th day of storage and its nanosurface. (e) Swelled cells e on the 40th day of storage and its nanosurface. Details in Kozlova et al. (2017).

Quality of PRBC is determined largely by the shape of cells and their membrane structures. Because the RBC membrane itself is the main target for pathological influence, cell shape and the cell membrane integrities are the main determinants of the clinical effects of PRBC storage. Changes in the forms of erythrocytes and disturbances in the structure of their membranes can lead to decrease in the deformability of PRBC, to worsening of rheological properties of blood, to weakening of gas transport function (Ciccoli et al., 2012). Abnormal RBC shape is found in the study by Ciccoli and colleagues to be related to oxidative membrane damage and hypoxia. Other types of issues related to general transports of RBC membranes, due to nanodefects, may severely emerge (Petkova-Kirova et al., 2019).

Congenital hemolytic anemias are inherited disorders caused partially due to the RBC membrane property alternation (such as the creation of nanodefects), in addition to cytoskeletal protein defects, deviant hemoglobin synthesis, and metabolic enzyme deficiencies. Occasionally, although the causing mutation might be known, the pathophysiology and the connection between the particular mutation and the symptoms of the disease are not completely understood. In many cases, abnormal RBC cation content and cation leaks go along with the disease. Therefore, the RBC membrane conductance assessment may help in understanding if there is any link between the alterations in membrane properties, including the loss of integrity of the membrane, the rise of nanodefects, etc., and the diseases. Recent electrophysiological measurements of the general conductance of RBCs will help us understand this clearly (Kirova et al., 2019).The whole-cell currents from 29 patients with different types of congenital hemolytic anemias were recorded. Fourteen of the patients with hereditary spherocytosis (HS) due to mutations in α-spectrin, β-spectrin, ankyrin, and band three protein; six patients with hereditary xerocytosis due to mutations in Piezo1; six patients with enzymatic disorders (three patients with glucose-six-phosphate dehydrogenase deficiency; one patient with pyruvate kinase deficiency; one patient with glutamate-cysteine ligase deficiency and one patient with glutathione reductase deficiency); one patient with β-thalassemia; and two patients were carriers of several mutations and a complex genotype. Various mutation sites are visible in Figure 5.9 (Mohandas and Gallagher, 2008).

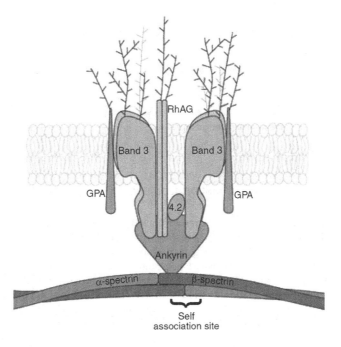

**FIGURE 5.9** Membrane defects in hereditary spherocytosis (HS) affect the "vertical" interactions anchoring the membrane skeleton to the lipid bilayer. Deficiency in any one of the protein components (Band 3, RhAG, ankyrin, protein 4.2 and spectrin) involved in the anchoring process leads to HS (Mohandas and Gallagher, 2008).

While the patients with β-thalassemia and metabolic enzyme deficiencies showed no changes in their membrane conductance, the patients with HS (not shown here) and hereditary xerocytosis (as a representative demonstration of all records, see Figure 5.10) showed largely variable results depending on the underlying mutation (Kirova et al., 2019).

P50.2 (Family 3), although showing no difference with its transportation control (Figure 5.10Aa), demonstrated increased conductance compared to the general, pooled, control (Figure 5.10Ab). There was also a family (Family 4 with P52.1) in which the Piezo1 mutation (c.7483_7488dupCTGGAG p.2495_2496dupLeuGlu) was associated with a decreased conductance compared both to the transportation and to the general, pooled, control (Figure 5.10Ba, b, respectively).

Based on varied conductance results of Korova and colleagues, this study concluded that changes in conductance are incurred by certain α-spectrin [c.2755G>T (p.Glu919) and c.678G>A p.(Glu227fs) whenever an $α^{LELY}$ allele is present], band three protein [c.2348T>A p.(Ile783Asn)], and Piezo1 (c.7483_7488dupCTGGAG p.2495_2496dupLeuGlu) mutations as a difference is observed with both the general and the transportation control. This study could not come up with

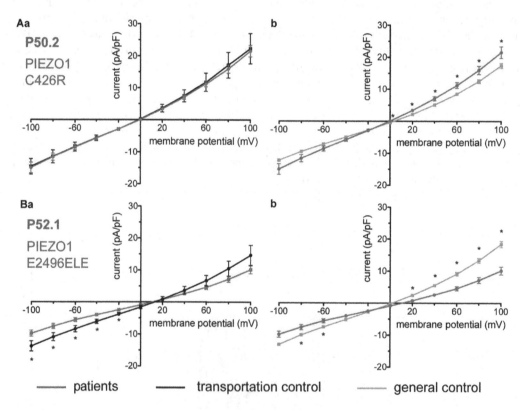

**FIGURE 5.10** Whole-cell recordings of ion currents from RBCs of healthy donors and hereditary xerocytosis patients. Compared are the I/V curves of P50.2 (n=14) with its own transportation control C50 (n=7) (**Aa**) as well as with a general control (n=175) (**Ab**). Comparison of capacitance of P50.2 with control capacitances gave no difference with the general or the transportation control (0.73 pA/pF P50.2 vs. 0.80 pA/pF transportation control; p>0.05; 0.73 pA/pF P50.2 vs. 0.69 pA/pF general control; p>0.05). Compared are the I/V curves of P52.1 (n=20) with its own transportation control C52 (n=11) (**Ba**) as well as with a general control (n=175) (**Bb**), where n denotes the number of cells. No changes were observed in capacitance either with the transportation (0.75 pA/pF P52.1 vs. 0.62 pA/pF C52; p>0.05) or with the general control (0.75 pA/pF P52.1 vs. 0.69 pA/pF general control; p>0.05). Currents were elicited by voltage steps from −100 to 100 mV for 500 ms in 20 mV increments at $V_h$=−30 mV. Data are expressed as mean current density±SEMs. Significant differences are determined based on an unpaired t-test with * representing p<0.05. Mutations below patients numbers are designated as amino acid substitutions in the respective protein. Details in Kirova et al. (2019).

a common channel increased activity/dysfunction or just an effect (decrease or increase in conductance) accompanying RBC ion disbalance disorders as large variability in results was identified. Identification of the specific channels that underlie the changed conductance demands future investigation. However, clear evidence of the creation of nanodefects in the RBC membrane may hint toward some defect-induced changes (Kozlova et al., 2017).

## REFERENCES

Alberts, B., Johnson, A., & Lewis, J. (2002). *Molecular Biology of the Cell*. New York, NY: Garland Science.

Ashrafuzzaman, M. (2018). *Nanosclae Biophysics of the Cell*. Springer International Publishing Switzerland AG, part of Springer Nature. doi:10.1007/978-3-319-77465-7

Ashrafuzzaman, M., Andersen, O., & McElhaney, R. (2008). The antimicrobial peptide gramicidin S permeabilizes phospholipid bilayer membranes without forming discrete ion channels. *Biochimica Et Biophysica Acta (BBA) - Biomembranes, 1778*(12), 2814–2822. doi:10.1016/j.bbamem.2008.08.017

Baird, F. E., Bett, K. J., Maclean, C., Tee, A. R., Hundal, H. S., & Taylor, P. M. (2009). Tertiary active transport of amino acids reconstituted by coexpression of System A and L transporters in Xenopus oocytes. *American Journal of Physiology-Endocrinology and Metabolism, 297*, E822–E829. doi:10.1152/ajpendo.00330.2009

Bryant, N., Govers, R., & James, D. (2002). Regulated transport of the glucose transporter GLUT4. *Nature Reviews Molecular Cell Biology, 3*, 267–277. doi:10.1038/nrm782

Ciccoli, L., De Felice, C., Paccagnini, E., Leoncini, S., Pecorelli, A., Signorini, C., Belmonte, G., Valacchi, G., Rossi, M., & Hayek, J. (2012). Morphological changes and oxidative damage in Rett Syndrome erythrocytes. *Biochimica Et Biophysica Acta (BBA) - General Subjects, 1820*(4), 511–520. doi:10.1016/j.bbagen.2011.12.002

Denicola, A., Souza, J. M., & Radi, R. (1998). Diffusion of peroxynitrite across erythrocyte membranes. *Proceedings of the National Academy of Sciences, 95*(7), 3566–3571. doi:10.1073/pnas.95.7.3566

Ferrer-Sueta, G., & Radi, R. (2009). Chemical biology of peroxynitrite: Kinetics, diffusion, and radicals. *ACS Chemical Biology, 4*(3), 161–177. doi:10.1021/cb800279q

Fick, A. (1855a). Ueber diffusion. *Annalen der Physik (in German), 94*(1), 59–86. Bibcode:1855AnP...170...59F. doi:10.1002/andp.18551700105

Fick, A. (1855b). V. on liquid diffusion. *Philosophical Magazine, 10*(63), 30–39. doi:10.1080/14786445508641925

Guyton, A. C., & Hall, J. E. (2000). *Textbook of Medical Physiology*. Philadelphia, PA: W. B. Saunders Company.

Hebbar, S., Lee, E., Manna, M., Steinert, S., Kumar, G. S., Wenk, M., . . . Kraut, R. (2008). A fluorescent sphingolipid binding domain peptide probe interacts with sphingolipids and cholesterol-dependent raft domains. *Journal of Lipid Research, 49*(5), 1077–1089. doi:10.1194/jlr.m700543-jlr200

Kozlova, E., Chernysh, A., Moroz, V., Sergunova, V., Gudkova, O., & Manchenko, E. (2017). Morphology, membrane nanostructure and stiffness for quality assessment of packed red blood cells. *Scientific Reports, 7*(1). doi:10.1038/s41598-017-08255-9

Lodish, H., Berk, A., & Zipursky, S. L. (2000). *Molecular Cell Biology: Uniporter-Catalyzed Transport*. New York, NY: W. H. Freeman and Company.

Merz, B. (1982). Cell membrane defects in mental illness. *JAMA: The Journal of the American Medical Association, 248*(6), 633. doi:10.1001/jama.1982.03330060011007

Milgrom, E., Atger, M., & Baulieu, E. (1973). Studies on estrogen entry into uterine cells and on estradiol-receptor complex attachment to the nucleus — Is the entry of estrogen into uterine cells a protein-mediated process? *Biochimica Et Biophysica Acta (BBA) - General Subjects, 320*(2), 267–283. doi:10.1016/0304-4165(73)90307-3

Mohandas, N., & Gallagher, P. G. (2008). Red cell membrane: Past, present, and future. *Blood, 112*(10), 3939–3948. doi:10.1182/blood-2008-07-161166

Okamoto, N., Viswanatha, R., Bittar, R., Li, Z., Haga-Yamanaka, S., Perrimon, N., & Yamanaka, N. (2018). A membrane transporter is required for steroid hormone uptake in Drosophila. *Developmental Cell, 47*(3). doi:10.1016/j.devcel.2018.09.012

Pavithrra, G., & Rajasekaran, R. (2018). Identification of effective dimeric gramicidin-D peptide as antimicrobial therapeutics over drug resistance: in-silico approach. *Interdisciplinary Sciences: Computational Life Sciences, 11*(4), 575–583. doi:10.1007/s12539-018-0304-5

Petkova-Kirova, P., Hertz, L., Danielczok, J., Huisjes, R., Makhro, A., Bogdanova, A., Mañú-Pereira, M. del, Vives Corrons, J.-L., van Wijk, R., & Kaestner, L. (2019). Red blood cell membrane conductance in hereditary haemolytic anaemias. *Frontiers in Physiology, 10*. doi:10.3389/fphys.2019.00386

Plagemann, P. G., & Erbe, J. (1976). Glucocorticoids—uptake by simple diffusion by cultured reuber and novikoff rat hepatoma cells. *Biochemical Pharmacology, 25*(13), 1489–1494. doi:10.1016/0006-2952(76)90066-6

Prenner, E. J., Lewis, R. N., & Mcelhaney, R. N. (1999). The interaction of the antimicrobial peptide gramicidin S with lipid bilayer model and biological membranes. *Biochimica Et Biophysica Acta (BBA) - Biomembranes, 1462*(1–2), 201–221. doi:10.1016/s0005-2736(99)00207-2

Sahoo, S., Aurich, M. K., Jonsson, J. J., & Thiele, I. (2014). Membrane transporters in a human genome-scale metabolic knowledgebase and their implications for disease. *Frontiers in Physiology, 5.* doi:10.3389/fphys.2014.00091

Sankaran, J., Manna, M., Guo, L., Kraut, R., & Wohland, T. (2009). Diffusion, transport, and cell membrane organization investigated by imaging fluorescence cross-correlation spectroscopy. *Biophysical Journal, 97*(9), 2630–2639. doi:10.1016/j.bpj.2009.08.025

Yang, N. J., & Hinner, M. J. (2014). Getting across the cell membrane: An overview for small molecules, peptides, and proteins. *Site-Specific Protein Labeling Methods in Molecular Biology,* 29–53. doi:10.1007/978-1-4939-2272-7_3

Yonezawa, H., Okamoto, K., Tomokiyo, K., & Izumiya, N. (1986). Mode of antibacterial action by gramicidin S. *The Journal of Biochemistry, 100*(5), 1253–1259. doi:10.1093/oxfordjournals.jbchem.a121831

# 6 Ion Channel Energetics

Ion channel conduction of current across the bilayer membrane relies on the energetics of the channel coupled to the membrane constituents. The channel subunits and membrane materials are associated with each other differently in different regions (hydrophobic or hydrophilic) and course of time. The ion channel energetics is mainly dependent on the strength of its self-assembly and its assembly involving other membrane components. Due to the dynamic nature of both channel and membrane structures, these soft condensed materials' inclusive assemblies as well as the strength of mutual association are subject to change with time. These phenomena determining the existence, energetics, and stability of ion channels belong to the areas covered partly in classical mechanics. However, there are quantum phenomena that might tune the channel energetics at a different scale which is much lower than those treated in classical mechanics. This chapter aims to explain a few crucial factors that are involved in determining and modulating ion channel energetics. A universal conduction model will be proposed based on the existing knowledge and relevant predictions.

## 6.1 PHYSICAL STATES OF ION CHANNELS

The majority of statistical mechanics problems may be addressed by the famous theory of Boltzmann. Most biological soft matter are materials that are easily deformable due to the influence of thermal fluctuations and external forces. General statistical mechanics principles including Boltzmann statistics are often applicable in understanding certain aspects of the physical states of soft materials in biological systems. Various physics principles are applied to understand a multitude of physical complexity involved in living systems. Soft matter or soft condensed matter biological systems comprise various physical systems that are locally deformable or structurally modifiable due to modest applications of thermal or mechanical stresses. These changes are of the magnitude of thermal fluctuations. The ion channel is a many-body system/structure constructed in a soft condensed matter phase. The channel integrity and function follow principles of statistical mechanics that are commonly active in soft materials. Its structure and function largely depend on its building block(s). However, there is a huge amount of influences readily linked to it provided by the biological host or environment, that is, the dynamic lipid membrane. The host membrane and integral channels maintain a physical coexistence through mutual energetics agreements. That is to say, they follow a clear and simplified mutually inclusive energetics model, as shown in Figure 6.1.

Figure 6.1 demonstrates that the membrane ingredients, for example, lipids, cholesterol, hydrocarbons, etc., that are involved in channel formation exist in a membrane-specific energy state, $E_{Mem}$, similar to the channel-forming building blocks, for example, membrane proteins, peptides, drugs, etc., in channel building blocks (ChBB)-specific energy state $E_{ChBB}$. However, their overlapped area in Figure 6.1 represents a different or compromising energy state, $E_{Ch}$, in which the channel is formed, energetically involving both ChBBs and membrane ingredients. Therefore, the channel cannot exist independently in a membrane exercising its exclusive energetics. From a structural viewpoint, we may translate it as the channel is not a nonmembrane entity, instead the precondition to forming a channel is that a complex involving ChBBs and membrane building blocks (MBBs) must have already been constructed. We have demonstrated this matter in modeling, presented in my previous book *Nanoscale Biophysics of the Channel* (see Chapter 4) (Ashrafuzzaman, 2018).

Here, it is clear that both channels forming ChBBs and MBBs need to undergo energetic changes; hence, let's assume $\Delta E_{ChBB}$ and $\Delta E_{MBB}$, respectively, to fall into complex structures leading to the formation of the channel. As these physical systems exist and their structural transitions occur in a thermodynamic bath, they are expected to follow the principles of statistical thermodynamics.

DOI: 10.1201/9781003010654-6

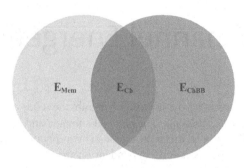

**FIGURE 6.1**  The overlapped region represents an energy state where the participating ingredients in a channel exist in an energy state $E_{ch}$. Channel building blocks (ChBB) and the membrane ingredients hosting ChBBs exist in exclusive/different energy states, $E_{ChBB}$ and $E_{Mem}$, respectively.

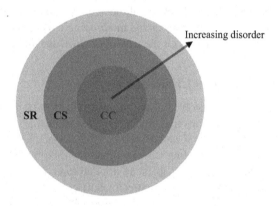

**FIGURE 6.2**  The central sphere (in the two-dimensional cross-sectional sketch, it looks like a circle) represents highly ordered CC, and the regions beyond it (radially outward) represent the CS and SR which are involved directly or through regulation in channel formation. Arrow direction represents progressing disorders, corresponding to moving to higher energy states.

Besides, the interactions among agents in the complex occur in bare or polarized electrostatic conditions following basic laws of statics, mechanics, and electrostatics. Using classical mechanics physics concepts, we may be able to theoretically calculate both of these energetic changes $\Delta E_{ChBB}$ and $\Delta E_{MBB}$ following our recently developed analytical treatments and numerical computations (NCs) (Ashrafuzzaman and Tuszynski, 2012a; 2012b).

From physical-order consideration, we may sketch the channel energetics, see Figure 6.2. Here, we consider that the ion channel interior core region (CC), channel (exterior or interior) surface region (CS), and channel surrounding region (SR) exist with different order parameters, namely, CC at highly ordered state, CS at generally ordered state, and SR at less-ordered state; beyond SR is the normal membrane state decoupling the channel state and existing at a less-ordered state experiencing thermodynamic fluctuations that are biologically relevant. From ion channel conduction perspectives, CC and (conditionally) CS represent channel conduction, while SR has nothing to do with channel conduction mechanism, although SR has a deterministic role in stabilizing the channels inside the membrane.

All these states (Figure 6.2) hold some sort of stochasticity, a natural criterion observed in biological systems. Modeling and simulating any biological structure and the possible transitions thereof require consideration of the principles defining stochastic systems. The CC may even be considered occasionally (if not always in all channel types) to host the quantum mechanical state (see a latter chapter dedicated to explaining the quantum mechanics of ion channels in this book). Therefore, we may have different energetics or even pattern of energetics (classical, semi-classical, and/or quantum mechanical) at different states, as shown in Figure 6.2. We envision ourselves doing some calculations in these comparable energetics areas. The physics may be different in channel energetics at different states among CC, CS, and SR. The electrical measurements involving electrophysiology (EP) records of channel currents can only deal with the phenomena of channels involving CS and SR states and the transitions between these states, and can never represent any phenomenon concerning CC state (though ions are expected to pass through CC state), if it is to be considered only a quantum energy state. EP records are always a time-dependent phenomena involving charge movement that produces currents, while in a quantum mechanical state, the charge movement happens without spending (almost) any time, raising uncertainty in position and momentum of the charges, which can only be explained using quantum wave functions. The channel conducting charges may then follow Heisenberg's uncertainly principle: if we know a particle's whereabouts, that is, the uncertainty of its position is small, we know almost nothing about its momentum, that is, the uncertainty of its momentum is large, and vice versa. On the other hand, charges, being conducted through channels (engaging mainly CS sate and conduction process being regulated due to SR structural state and any structural transitions thereof), can be located with momentum and at coordinates that are deterministic. Hence, the movement of not only the charges through channels but also the channel subunits that determine the channel structure is time-dependent. Additionally, the movement is also linked to classical mechanical and/or electrostatic energetics that is also measurable using conventional physics principles (Ashrafuzzaman and Tuszynski, 2012a; 2012b). In this chapter, we wish to cover the latter aspect by explaining the statistical nature of the current flow and the energetics of the channel structures. The quantum mechanical calculations are presented, based on available information, in another chapter in this book; here, a model where the quantum mechanical effects are to be fitted while completely addressing the channel function is presented.

## 6.2 CHANNEL-GATING MECHANISMS: AVAILABLE GENERALIZED METHODS

Environmental stimuli, such as voltage, ligand concentration, membrane tension, and temperature, cause the membrane-hosted ion channels to respond. Consequently, their catalytic sites, the ion-conducting pores, are found to open and close following specific gating mechanisms. Although several studies have addressed these ion channel responses including gating mechanisms, understanding is still considered incomplete due to unavoidable reasons. We have a huge amount of experimental data to address the phenomenology of channel systems, but we often find the theoretical parameters incorporated into cross-examining the data may not fit well. The reason behind this may be because we have not yet found the right approaches. An analysis has been discussed earlier in this chapter. A first-hand debate may emerge on an intriguing issue – whether the ion channel gating is a classical, semi-classical, or even quantum mechanical process, or a combination of all of them.

Three areas that have contributed to the understanding of ion channels, namely, traditional Eyring kinetic theory, molecular dynamics analysis, and statistical thermodynamics, have been analyzed quite rigorously in a recent review (Sigg, 2014). Although the primary emphasis is on voltage-dependent channels, the methods discussed here are generalizable to other stimuli and can be applied to any ion channel and indeed any macromolecule. Regarding the kinetic modeling of ion channels, we have to begin with the Hodgkin–Huxley (HH) scheme (Hodgkin and Huxley, 1952) and continue unabated today with the widespread use of multistate Markov models and single-channel analysis. The building block of the HH model is the gating particle, with transitions occurring between resting and activated states in a voltage-dependent manner. Ion channels are composed of various modular domains, each with its specialization and distinct evolutionary origin (Schreiber et al., 1999; Sasaki, 2006), so the gating particle concept remains a viable one, although gating schemes have become highly complex. The potassium ($K^+$) channel opening in the original HH scheme required four randomly fluctuating "n" particles to be simultaneously active, the sodium channel description was similar but used three activating "m" particles and one inactivating "h" particle. The conduction of K$^+$ in the HH model is, therefore, proportional to n$^4$, with particle kinetics following standard gating parameters (Sigg, 2014; Ashrafuzzaman, 2018). The Nobel Prize winning discovery of Nehar and Sakmann, single-channel ion current records using the patch clamp (Neher and Sakmann, 1976), confirmed the existence of stochastic transitions between discrete conductance levels, as predicted by the HH model. Since then, several studies have aimed to experimentally as well as theoretically understand various types of ion channel structural transition-based conductance processes using mostly classical mechanical approaches.

A recent study pointed out that although the active-state conformations of ion channels are known from typical X-ray structures, an atomic resolution structure of any voltage-dependent ion channel in the resting state is not currently available (Vargas et al., 2012). This is exactly the point I have raised earlier here and in my book (Ashrafuzzaman, 2018) while proposing a simplified model on the prepore structure of ion channels that might lead to creating the pore gating.

The voltage-sensor domains (VSDs) respond to changes in the cell membrane potential difference (see Figure 6.3). With membrane depolarization, the VSD in each subunit undergoes a conformational transition from resting to an activated state, and this information is predicted to be communicated to the ion-conducting pore to promote its opening (Bezanilla et al., 1994). The activation of the VSD and opening of the pore are associated with the transfer of a gating charge ΔQ across the membrane (Sigworth, 1994). The opening of the voltage-gated K$^+$ channel Shaker corresponds to the outward translocation of a large positive charge on the order of 12–14 elementary charges (Schoppa et al., 1992). Four highly conserved arginines along S4 (R1–R4) underlie the dominant contributions to the total gating charge of Shaker and appear to be mainly responsible for the coupling to the membrane voltage (Papazian et al., 1991). The overall structure of eukaryotic voltage-gated Na$^+$ channels, which are composed of four analogous subunits covalently linked in a single polypeptide, appears to be similar (Catterall, 2012).

The resting-state conformation of the VSD has been modeled using the Rosetta method (Yarov-Yarovoy et al., 2006), which was subsequently refined with all-atom molecular dynamics (MD) simulations (Khalili-Araghi et al., 2010) and with high-resolution Rosetta algorithms (Yarov-Yarovoy et al., 2011) predicting pairs of neighboring residues before they were identified experimentally (Campos et al., 2007). Subsequent refinement of structural models of the resting state helped demonstrate the existence of a consensus three-dimensional (3D) VSD conformation, consistent with experimental data (Vargas et al., 2011). Rosetta models for the bacterial sodium channel NaChBac are very similar to those of K$_V$ channels and have been extensively tested by disulfide cross-linking

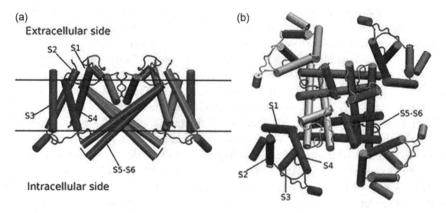

**FIGURE 6.3** Overall view of the voltage-activated Kv1.2 K$^+$ channel (Protein Data Bank accession no. 3LUT). (a) Two of the four channel subunits are displayed from a side view. The VSD comprises the transmembrane segments S1–S4, and the pore domain comprises the transmembrane segments S5–S6. (b) The tetramer is displayed from the extracellular side (each subunit is a different color). The two views are related by a 90° rotation (Vargas et al., 2012).

studies (Yarov-Yarovoy et al., 2011). Although much has been known on VSD, the ultimate dream would be to "visualize," atom-by-atom, how the channel and its subunits move, during especially prepore → pore formation phase, as a function of time in response to realistic membrane potential. This is yet to be achieved and our queries may demand the emergence of modeling(s) with realistic energetics (not based on computational guesses but high-resolution actual structural imaging) involving classical, semi-classical, or quantum mechanical treatment of the systems. Hopefully, in the near future, we will have answers.

## 6.3 DYNAMICS AND ENERGETICS OF ION CHANNEL SUBUNITS: MODELING

Ion channels are constructed using channel subunit proteins (or sections of protein structures), peptides, or drugs that have considerable hydrophobic properties relative to their hydrophilic counterparts (if any). Occasionally, a certain class of lipids is also found to participate in constructing channels (Colombini, 2010). Diversified channel structures involving different channel subunits have been discovered in various biological cell membranes. The dynamics and energetics of the channel subunits during prepore, pore, and postpore formation phases are distinguishable and need to be explored. While explaining EP-recorded currents of ion channels, we often focus on addressing the overall or resultant states of channels regarding their open and close states and the back-and-forth transitions between the states. Often the dynamics of channel subunits at these conduction states or transitions can be addressed using available molecular understandings based on various models. However, there are tricky molecular mechanisms active in the dynamics of channel subunits while opening the pore, accounting for the closed prepore → (open) pore transition, and while closing the pore, accounting for the (open) pore → closed postpore transition. The latter dynamics of the channel subunits are often unaddressed or poorly addressed. We have a plan to develop some calculations.

We may get a glimpse of how a pore finally gets formed starting with the random accumulation of drugs and lipids in a membrane, leading to complete cluster formation among drugs, A → B → C, as shown in Figure 6.4. I presented this as a model on essential steps leading to channel formation (Ashrafuzzaman, 2018).

We have assumed that drugs are in random motion in lipid membrane's hydrophilic (or occasionally hydrophobic) region, statistically penetrate the lipid cluster, or participate in creating drug-lipid cluster, which may slowly get enriched with more and more drugs and less and less lipids, and finally accomplish a conformation like in C. Here, the concept is that as a drug molecule penetrates the cluster (this happens concomitantly with the removal of certain number of lipids to compensate the geometrical penalty), the potential well (right panel, Figure 6.4) turns deeper and the drugs altogether feel stronger binding among them, making the cluster more stable. The process continues until reaching a point when the cluster collapses to its minimum energy conformation leading to the formation of a channel. In all three cases A, B, C, both per-drug and per-lipid energy in the lipid/drug cluster varies. An NC study is underway to address this energetics as the channel gets formed starting with the adsorption of drugs by lipid bilayer. However, we can predict that the binding energy per drug in the cluster in C is higher than that in B, which is higher than that in A. Or, we may say that the drugs in C exist at lower potential well than in B, and drugs in B exist at lower potential well than in A. We may configure the potential wells, as in Figure 6.4 (right panel), where the depth of the potential well corresponds to the strength of binding or simply the binding energy.

NCs on calculation of the per-drug binding energy in these three (A, B, C) or at least two (A or B and C) conformations have been under way for at least three different ion channels created by peptides gramicidin A (gA) and alamethicin (Alm), as well as chemotherapy drugs (CDs) to have an understanding that may eventually be generalized for all channels hosted in membranes.

Hypothetically, we may start with two conformations, see Figure 6.5 (prepore state) and Figure 6.6 (pore state) (Ashrafuzzaman, 2018). We then calculate the per-drug energy, $\Delta E_{ChBB}$, in

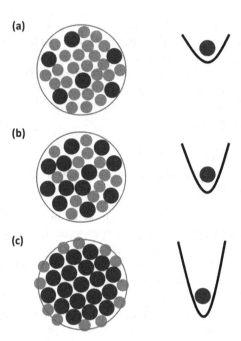

**FIGURE 6.4** Drugs penetrate into cell surface membrane that may initiate creation of clusters. (a) Drugs randomly penetrate into the lipid monolayer on cell surface. No considerable interdrug particle communication is yet established. (b) Substantial adsorption of drug particles by cell surface in lipid monolayer causes drugs and lipids initiating the creation of a so-called (I wish to name it) "drug-lipid cluster" where both drugs and lipids locate randomly. (c) Drugs stay together and make a cluster which is surrounded by lipids. I wish to call it a "drug-drug cluster." Hypothetically, we may consider the timing of these three clustering events (let's assume) $t_A$ for (a), $t_B$ for (b), and $t_C$ for (c). For clustering leading to the channel formation, $t_C > t_B > t_A$. Thus, the time-dependent dynamics in the cluster leads to channel formation. Right panel wells are three corresponding potential wells that host/bind the drug molecules inside the cluster. The varying depth (along negative vertical axis) suggests that for varying drug-drug interactions, where the deeper the well, higher is the binding energy accounting for higher stability of the drug cluster (a precondition to the formation of channels). The red sphere represents the drug particle and the brown sphere represents the lipid molecule. The circular boundary (virtual one) may be considered for a circle with radius $\xi_{radial}$ that determines the cluster size. The model is a two-dimensional representation (for simplicity), though the drug-lipid and drug-drug clusters actually appear in AFM phase imaging as three-dimensional as presented in a study showing chemotherapy drug adsorption on a cell surface (Ashrafuzzaman, 2018). (For color figure see eBook figure.)

both conformations for gramicidin A (gA), alamethicin (Alm), or CD as drug molecules and compare the energetics (depending on the free energy of transitions). Let's assume it to be $\Delta\Delta E_{ChBB}$, while transiting from the prepore to pore states (work underway) using theoretical method and NCs (Ashrafuzzaman and Tuszynski, 2012a; 2012b). These studies suggest that $\Delta\Delta E_{ChBB}$ actually depends on various scaling parameters, for example, geometry (drug molecule size and interdrug separation or drug density), charge density inside cluster, dielectric parameters qualifying the aqueous condition of the cluster, etc. Once $\Delta\Delta E_{ChBB}$ is calculated, we can easily apply this energetics in Boltzmann's statistics to calculate the probability of pore opening (p) using the following relation:

$$p \sim Exp\left\{-\Delta\Delta E_{ChBB}/k_B T\right\}$$

Where, T is the absolute temperature representing the thermodynamic condition (biological system temperature). The entire analysis can be repeated for channel closing using reverse calculations of energetics.

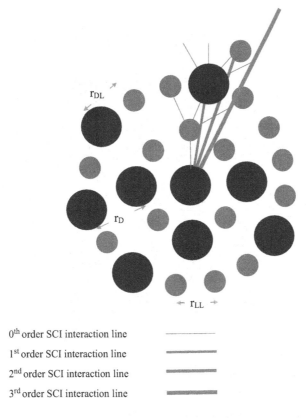

**FIGURE 6.5** Different screening orders in screened Coulomb interaction are shown (modeled here with interaction lines with varying thicknesses) in the interaction of a drug particle with surrounding lipids or drugs on cell surface for mixed drug cluster, as shown in Figure 6.4(b). As before, we have sketched an equivalent shell model presentation in a two-dimensional view. Here, average interparticle spaces can be three different distances, namely, lipid distance, $r_{LL}$, drug-lipid distance, $r_{DL}$, and drug-drug distance, $r_{DD}$ (Ashrafuzzaman, 2018).

## 6.4 THEORETICAL AND COMPUTATIONAL ENERGETICS AND DYNAMICS OF CHANNEL SUBUNITS: DETECTING ENERGIES AND THEIR UNIVERSAL PATTERNS

In this section, we wish to address a few of our own breakthroughs. We have been, during last decade, very successful in developing a novel theoretical technique to address molecular-level energetics, which were validated in dynamical calculations borne out of MD simulations.

### 6.4.1 Channel Energetics in a Lipid Membrane Hydrophobic Region

We have been successful in biophysical addressing, using NCs and MD simulations, of the channel energetics involving channel subunits and membrane lipids for small channels, such as gA, Alm, and CD channels in model membrane systems (Ashrafuzzaman and Tuszynski, 2012a; Ashrafuzzaman and Tuszynski, 2012b; Ashrafuzzaman et al., 2012; Ashrafuzzaman et al., 2014; Ashrafuzzaman et al., 2020a; Ashrafuzzaman et al., 2020b). In these publications, we established a single fundamental fact that the channel stability inside membrane is due to nothing but molecular mechanisms depending on charge-based screened Coulomb interaction (SCI) energetics among functional charge groups in the ion channel complex involving channel subunit peptides or drugs

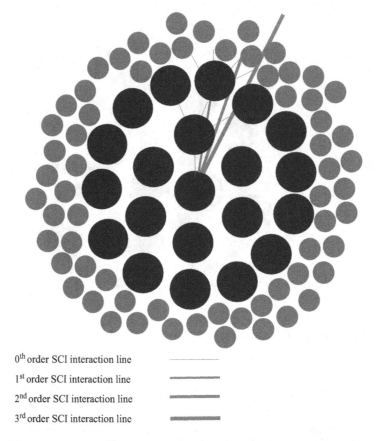

0th order SCI interaction line     ——————

1st order SCI interaction line     ——————

2nd order SCI interaction line     ——————

3rd order SCI interaction line     ——————

**FIGURE 6.6** Different screening orders in screened Coulomb interaction are shown (modeled here with interaction lines with varying thicknesses) in the interaction of a drug particle with the surrounding drugs for unitary drug cluster on the cell surface, as shown in Figure 6.4(c). As before, to explain the interaction with various screening orders, we have sketched an equivalent shell model presentation for the central unitary drug cluster in a two-dimensional view. Here, the central unitary drug cluster is modeled to be surrounded by the hydrophobic lipid environment (Ashrafuzzaman, 2018).

and membrane lipids. Our computational *in silico* assays (NCs and MD simulations) supported the experimental findings in distance and time-dependent channel subunit-lipid interaction energetics theoretically by evaluating their binding energy in the channel complex, and thus, helping us understand the statistical mechanical nature in the channel stability in a biological thermodynamic environment. Readers are invited to refer these articles for further details. Here, we shall brief a few of the vital results to familiarize ourselves with the crucial points.

NCs on gA channel energetics in model membrane system detected a universal trend in the free energy ($\Delta G_{\text{free, ch}}$) determining the channel stabilization, as follows (Ashrafuzzaman and Tuszynski, 2012a; Ashrafuzzaman and Tuszynski, 2012b):

$$\Delta G_{\text{free, ch}} \sim \text{Exp}\left\{-\left|(d_0 - 1)\right|\right\} \text{ and } \Delta G_{\text{free, ch}} \sim \left(q_L/q_g\right)^s$$

Here $\left|(d_0 - 1)\right|$ is the measure of the mismatch between channel-hosting lipid bilayer thickness (d) and channel length (l). $q_L$ and $q_g$ are effective charge on a lipid and gA channel subunit, respectively, that are involved in channel forming. The exponent $s = 1, 2$, etc. for first, second, etc. order SCI screening, respectively, suggesting that channel formation is harder in bilayers containing charged lipids. Thicker bilayer also destabilizes the channel due to higher mismatch. Due to the mismatch,

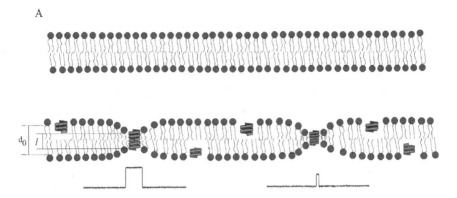

**FIGURE 6.7** A channel's deform lipid bilayer's resting thickness. With channel formation, bilayer conducts a current pulse with an average pulse width ($\tau$) (the channel lifetime) and height (the channel conductance). Two types of monomers have structured two different channels.

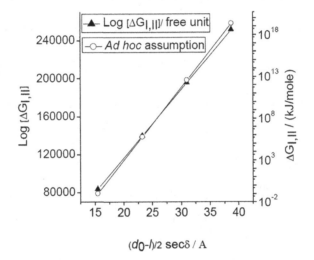

**FIGURE 6.8** Log[$\Delta G_{I, II}$] vs $d_0-1$ plot ($\Delta G_{I, II}= G_I-G_{II}$ ($\Delta G_{free, ch}$), free energy difference between free and dimer gA states). ($d_0-1$)sec$\delta$ is the distance covered by lipid head groups in the deformed bilayer region near channel. $\delta$ (~constant, within 0–90° in all screening orders) the angle at which lipids in the deformed portion of the bilayer couples with the extension of gA channel length. *Ad hoc* assumptions ($q_g$~electron charge and other relevant parameters) give an estimate of $\Delta G_{I, II}(O)/$ (kJ/mole) which depends on $q_L$ as $d_0$ increases. Results in this Fig. * perhaps falls within second-order screening ($d_0-1<40$ Å).

if the channel length is higher than the bilayer thickness, the channel is also expected to get destabilized. Figures 6.7 and 6.8 show the model composition and experimental energetics, along with the SCI interpretations (based on NCs on SCI energy expression) being matched with the experimental data, respectively. As we have detailed the theoretical backgrounds in Ashrafuzzaman and Tuszynski (2012a) and Ashrafuzzaman and Tuszynski (2012b), we do not repeat them here and ask the readers to go through our earlier book *Membrane Biophysics* (Ashrafuzzaman and Tuszynski, 2012a).

## 6.4.2 Molecular Dynamics Simulations to Determine the Time-Dependent Energetics of Channel Subunits

We performed *in silico* MD simulations to separately address lipid binding for three different classes of membrane active agents (MAAs), namely, antimicrobial peptides (AMPs), CDs, and aptamers (no study yet found regarding their channel formation potency, but they pose to have similar membrane-binding mechanisms as other channel-forming peptides, e.g., gA, Alm, CDs, etc.), and understand the molecular-level energetics and derive related probability functions.

**MD simulation helps detect the MAA-lipid interaction energies and discover the related probability functions.** Using MD simulations, we track the strength of interactions between any MAA and lipid (phosphatidylcholine (PC) or phosphatidylserine (PS)). We use AMPs gA and Alm, CDs thiocolchicoside (TCC) and taxol (TXL), and aptamers 5'-AAAAGA-3' (Aptamer I) and 5'-AAAGAC-3' (Aptamer II), which are the worst (SIAp1) and best (SIAp4) PS liposome-binding aptamers "PS aptamers" (Ashrafuzzaman et al., 2013; Ashrafuzzaman et al., 2020a). To specifically address the charge effects, we use zwitterionic charge-neutral PC and negatively charged PS.

*MD Methods and Analysis.* Using MD simulations, we provide computational support for agent-lipid physical interactions, which are postulated to be the mechanisms behind liposome binding or pore formation. This also sheds light on the complex interactions between stable structure (e.g., MAAs) and liquid crystal structure (e.g., membrane), a very important and still unresolved problem in biophysics and soft condensed matter physics. MD simulations for PS aptamer- and CD-lipid have been conducted earlier (Ashrafuzzaman et al., 2012). Here, similar MD simulations have been carried out for gA- and Alm-lipid pairs and three physical quantities: (a) the separation distance between the centers of mass of the agent and lipid, $d_{agent-lipid}$, (b) van der Waals (vdW), and (c) electrostatic (ES) energies were used to analyze simulation results (see explanations "MD simulation results of gA and Alm-lipid interactions" and Figures 1 and 3 in ref. Ashrafuzzaman et al., (2020b)). The solvent accessible surface area (SASA) was calculated using Amber Tools 11 (Case et al., 2010) when both drug and lipid molecules were completely separated. We can then expect them to be entirely exposed to the solvent, that is, the corresponding SASA values are at a maximum. The SASAs in all studies are roughly unchanged between the start and the investigated 20 Å length (Ashrafuzzaman et al., 2020a). So, within this range, the drug-lipid complexes stay at an equilibrium solvation condition. Therefore, we focus only on vdW and ES here. To investigate the features of physical interactions of all pairs of lipids and agents from MD results, a probabilistic description is proposed. We first evaluate the probability of observing a pair within $d_{agent-lipid}$ as $P(d_{agent-lipid}) = \Delta t(d_{agent-lipid})/T_{sim}$, where $\Delta t(d_{agent-lipid})$ is the duration during which the agent-lipid pair stays within $d_{agent-lipid}$ and $T_{sim}$ is the total simulation time. Second, the probability of having either vdW or ES energy of a lipid and an agent staying at distance $d_{agent-lipid}$ is given by Boltzmann distribution:

$$P\left(E\left(d_{agent-lipid}\right)\right) = \exp - \beta E\left(d_{agent-lipid}\right)\Big/Z,$$

where the partition function is $Z = \sum_{d_{agent-lipid}} \exp - \beta E\left(d_{agent-lipid}\right)$, $\beta = 1/k_B T$, T = 300 K.

*Universal footprint revealed by MD results.* Figure 6.9 shows the plots of $P(E(d_{agent-lipid}))$ against $P(d_{agent-lipid})$ and the corresponding $d_{agent-lipid}$ values are represented by symbol ze, which is illustrated in the bottom panel. Considering all three variables together, Figure 6.9 shows a corresponding 3D plot. The upper row shows the case with PC and lower row for PS and the bottom row shows the size of symbols, which denotes the distance between an agent and lipid. The left, middle, and right columns represent vdW, ES, and the sum of vdW and ES energies, respectively. This figure reveals several interesting features. First, it shows similar trends for all three categories of agents against either PC or PS from the vdW interactions point of view. Probabilities of having a pair within $d_{agent-lipid}$ and vdW energy $E_{vdW}(d_{agent-lipid})$ gradually decrease when a lipid and an agent separation distance $d_{agent-lipid}$ increases. Namely, the vdW force is likely to play a crucial role in all types of agents binding with lipids (short $d_{agent-lipid}$ range). Second, from the ES energy point of view, two CDs are the only type of agents to show similar trends, namely, the larger the separation distance $d_{agent-lipid}$ is, the lower the probabilities $P(d_{agent-lipid})$ and $P(E(d_{agent-lipid}))$ are for both PC and PS cases. It suggests that similar to the vdW force, the ES force is likely a mechanism for the binding process of CDs. Yet, the ES force is likely to play only a minor role in the binding of lipids and agents such

---

**TABLE 6.1**

**Total Charges of MAAs and Lipids Are Listed as Computed by AMBER Using the Antechamber AM1-bcc Method (Wang et al., 2005)**

| MAA | Total Charge (C) |
| --- | --- |
| TCC | −0.002 |
| TXL | −0.09 |
| gA | 36.448 |
| Alm | −18.208 |
| Aptamer I | −91.112 |
| Aptamer II | −91.112 |
| **Lipid** | |
| PC | −0.004 |
| PS | −18.224 |

*Source:* Details in Ashrafuzzaman et al. (2020a).

---

as peptides and aptamers. This is probably due to the polarities of charges on participating agents, and thus, the ES force can play a role either in favoring or disfavoring the binding (Table 6.1 shows that PC is nearly neutral, PS and all MAAs except gA are negatively charged). The negligible lipid binding for aptamer II is naturally valid as the sequence has been designed as a negative control not to bind with the target lipids. Despite having different membrane effects (see experimental results), all MAAs bind directly to lipids with considerable binding stability, and the vdW and ES interactions represent a universal molecular mechanism producing the necessary driving forces to bring charges together. We view this novel mechanism to be a very important finding in membrane science. The most important message here is that if we can determine all the important molecular-level interactions, we can correctly predict the binding phenomena. In this regard, detection of energies in a single MAA-lipid complex (Figure 6.9) provides better molecular-level understanding than in a usually expected MAA-lipid complex in a membrane. Consequently, we can detect the primary (not the collective) energy values. The overall energies in a MAA-lipid complex in a membrane are simply a combination of these individually detected contributing energies. Therefore, our MD simulation strategy and the detected energy-based discovery of the two associated probabilities appear as strong functions correlating the molecular-level information with the phenomenological observations of MAA-lipid complexes in lipid membranes, as predicted from various *in vitro* experiments. This new approach correlates information between *in silico* and *in vitro* experiments.

## 6.5 DYNAMICS AND ENERGETICS OF CONDUCTED CHARGES THROUGH ION CHANNELS

Ion conduction through ion channels rely on various crucial processes, but above all, the energetics of the ion channel system, its inside and outside, is highly important in maintaining the right dynamics. We will address this aspect briefly here using a few examples. The analysis will help us propose a generalized scheme to develop an ion channel conduction model that may be considered more complete than the available ones.

### 6.5.1 Statistical Equilibration of Channels in Membranes

Statistical mechanical formulation of the equilibrium properties of selective ion channels was developed by incorporating the influence of the membrane potential, multiple occupancy, and saturation

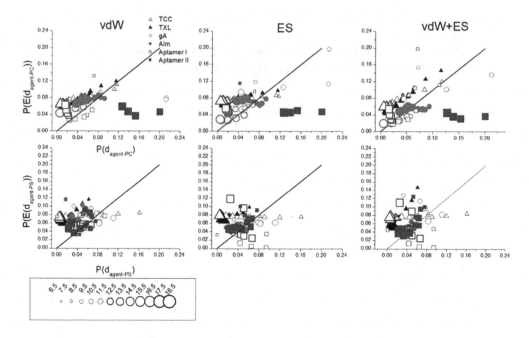

**FIGURE 6.9** Analysis of MD results of all lipid-agent pairs. Values of $d_{agent\text{-}lipid}$ are represented by symbol sizes (for simplicity only circle sizes are shown) (Ashrafuzzaman et al., 2020).

effects about two decades ago (Roux, 1999). The free energy profile along the channel axis, the cross-sectional area of the pore (see Figure 6.10), and probability of occupancy are provided using statistical mechanical concepts in detailed microscopic models. Here, in particular, the influence of the membrane voltage, the significance of the electric distance, and assumptions concerning the linearity of the membrane electric field along the channel axis have been examined. The following findings are important:

1. The equilibrium probabilities of occupancy of multiple occupied channels have the familiar algebraic form of saturation properties obtained from kinetic models with discrete states of the denumerable ion occupancy. This does not prove the existence of specific binding sites.
2. The total free energy profile of an ion along the channel axis can be separated into an intrinsic ion pore free energy potential of mean force, independent of the transmembrane potential, and other contributions that arise from the interfacial polarization.
3. As an example, the transmembrane potential calculated numerically for a detailed atomic configuration of the gramicidin A (gA) channel embedded in a bilayer membrane with explicit lipid molecules is shown to be closely linear over a distance of 25 Å along the channel axis.

As we have a lot of information on gA channel, let us stick to this example and try to understand further. A recent study modeled the local lipid redistribution, pulled by a gA channel inside a membrane, see Figure 6.11 (Beaven et al., 2017).

The lipid site 1 is in the first shell which shares at least a 4 Å tessellation border (the spatial boundaries between individual lipid areas) with the channel (in orange). Lipid site 2 is in the second shell because it shares a border with a lipid in the first shell (i.e., lipid site 1) and does not share a border with the channel. A lipid in the third shell would, therefore, be neighbor lipids in shell 2. This study has a strong preference for the better matching lipid near the channel hosted in a membrane

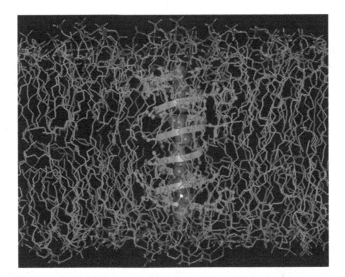

**FIGURE 6.10**   Atomic configuration of the gA in a DMPC bilayer used to calculate the transmembrane fieldfmp. The model incorporates one gA channel, a lipid bilayer with 100 DMPC molecules (50 molecules in the upper and lower leaflets, respectively), and 10 single-file water molecules. The model was assembled using an equilibrated configuration of a gA:DMPC system, and the final configuration was refined with energy minimization. Periodic boundary conditions were applied in the yz directions to simulate an infinite bilayer; the dimension of the central unit was chosen to be 58 Å to correspond roughly to the cross-sectional area of one gA and 50 DMPC molecules (Woolf and Roux, 1996; Roux, 1999).

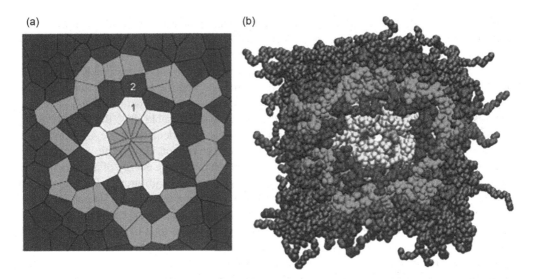

**FIGURE 6.11**   (a) Lipids shown with associated Voronoi area. The first four lipid shells surrounding the gA channel (orange) are defined (yellow, blue, green, and purple), with all lipids past these shells (gray). (b) Individual lipids around gA based on the same shell definition in (a) (Beaven et al., 2017). (For color figure see eBook figure.)

constructed using a combination of lipids. However, it is not actually surprising (Ashrafuzzaman and Tuszynski, 2012a; 2012b). Apart from the experimental facts, our findings differ strongly from the claims made by Beaven et al. (2017) that the membrane deformation energy, due to the pull of gA channels inside, changes quadratically with the hydrophobic mismatch between membrane thickness and channel length. We detected that the SCI among the charges on gA and surrounding lipids in the channel-membrane coupling regime are the determinants, and that the membrane deformation energetics follow an exponential change with the membrane deformation (hydrophobic mismatch) mentioned here. We also found that the elastic model produced quadratic changes with the hydrophobic mismatch in membrane deformation energetics automatically appear as a second-ary contributor into the membrane deformation energetics derived in our exponential energetics expression. Therefore, considering bilayer as an elastic entity only (Beaven et al., 2017; Nielsen et al., 1998) would not offer justice to the energetics, to be addressed requiring correct physical treatments using relevant principles active in this specialized biological problem. We believe that we have already resolved this dilemma in our SCI model calculations (Ashrafuzzaman and Tuszynski, 2012a; 2012b). However, this analysis clarifies one thing that the channel conduction length may be regulated at the channel's entry and exit locations due to membrane lipids.

### 6.5.2  Ion-Binding Sites to Determine the Statistical Ion Conduction Through Ion Channels

Ion conduction through the channel's conduction regions rely on various geometrical and physical parameters which create the conduction core of the channel. We wish to limit our discussion to only two channel classes, gA and KcsA channels, because these are well-explored channels providing us concrete evidence-based information that will help us address the subject here.

Regarding channel-binding sites specific for ions, we may consider the claim that the main bind-ing sites for sodium are near the gA channel's mouth, approximately $9.2\,\text{Å}$ from the center of the dimer channel, although the motion along the axis could be as large as $1–2\,\text{Å}$ (Woolf and Roux, 1997). In the binding site, the sodium ion lying off axis, contacting two carbonyl oxygens and two single-file water molecules.

The presence of an ion perturbs the periodic arrangement of the water molecules in the gA helix. Although both carry unit charges, the large difference between the water mean force on $Na^+$ and $K^+$ arises from their difference in size in relation to the helix periodicity. This is illustrated sche-matically in Figure 6.12 by the group of Nobel Laureate Martin Karplus (Roux and Karplus, 1991). During this simulation, the average distance between the $Na^+$ and the two nearest neighbor water oxygens is $2.32\,\text{Å}$, whereas it is $2.71\,\text{Å}$ for $K^+$, a $0.39\,\text{Å}$ difference in structure. The $2.23\,\text{Å}$ value for $Na^+$ is such that the waters ahead and behind the ion are in favorable sites. Because $Na^+$ is at a maximum of the water free energy, the free energy barriers are calculated to be $4.5$ kcal/mol for $Na^+$ and $1.0$ kcal/mol for $K^+$, making the channel conductance rely on ion-specific energetics. The water molecules are found to make a significant contribution to the free energy of activation. There is an increase in entropy at the transition state which is associated with greater fluctuations. The free energy profile of ions in the periodic channel is controlled not by the large interaction energy involving the ion but rather by the weaker water-water, water-peptide, and peptide-peptide hydrogen bond interactions. This complex energetics allows perhaps no more than two ions at a time inside the channel (Finkelstein and Andersen, 1981). Here, it is clear that the channel conduction length may have a fluctuation along its axis, and that hydration has a deterministic effect on the location, spatial distribution, and dynamics of ions inside channel interior regions.

We shall inspect the binding site of another well-explored channel, the potassium channel, whose X-ray structure was revealed by Nobel Prize winner MacKinnon and colleagues (Doyle et al., 1998). The selectivity filter was found to contain two $K^+$ ions approximately $7.5\,\text{Å}$ apart. This configuration promotes ion conduction by exploiting electrostatic repulsive forces to overcome attractive forces between $K^+$ ions and the selectivity filter. Recently, the crystal structures of $K^+$ channels displaying four $K^+$ ions bound to the selectivity filter has been investigated (Tilegenova et al., 2019). The study presented the atomic resolution crystallographic structure, the function, and the ion-binding prop-erties of the KcsA mutants, G77A and G77C, that stabilize the 2,4-ion-bound configuration (i.e., water, $K^+$, water, $K^+$-ion-bound configuration) of the $K^+$ channel's selectivity filter. Two mechanistic

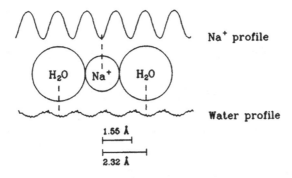

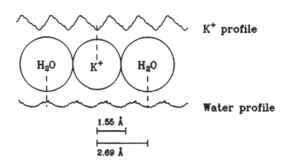

**FIGURE 6.12** Schematic representation of the periodic free energy profile of Na⁺ (upper panel) and K⁺ (lower panel) ions relative to the water oxygen profile. The water oxygen profile has been magnified by 10 relative to the ion profile for clarity. No translation or scaling were applied to any of the profiles in the x direction (Roux and Karplus, 1991).

models have been proposed to explain this experimental observation. The "canonical model" proposes two alternating ion-bound configurations (1,3 and 2,4) coexisting within a channel's filter. Alternatively, in the "direct knock-on model," all binding sites are occupied at any given time, and ions establish direct contact. Here, it is shown that the structure of a K⁺-channel selectivity filter stabilized in the 2,4-ion configuration, which provides a definitive experimental demonstration for the canonical model of ion permeation in K⁺ channels. In these structures, ions are separated by water molecules that seem to be cotransported during each ion transport cycle, which is in agreement with streaming potential measurements. The model is demonstrated in Figure 6.13.

### 6.5.3 SWITCHING OF THE ION CONDUCTION MECHANISMS

The mechanism of ion conduction through an ion channel may not follow a unique process, and may experience switching among different processes due to the effects of different ion concentrations (Kasahara et al., 2013). In EP records of ion channel currents, this kind of ion concentration-specific ion conduction mechanism cannot be identified. Kasahara and colleagues performed several micro-second MD simulations of the pore domain of the Kv1.2 potassium channel in KCl solution at four different ion concentrations. They observed that the conduction mechanism switched with different ion concentrations. At high ion concentrations, the potassium conduction occurred by Hodgkin and Keynes' knock-on mechanism (Hodgkin and Keynes, 1955), where the association of an incoming ion with the channel is tightly coupled with the dissociation of an outgoing ion in a one-step manner. On the other hand, at low ion concentrations, ions mainly permeated by a

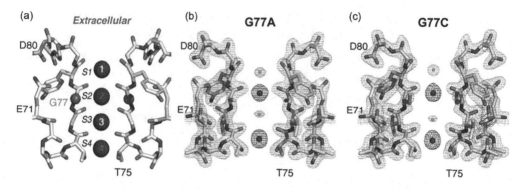

**FIGURE 6.13** Structure of the 2,4-ion-bound configuration of a $K^+$-channel selectivity filter (SF). (a) A cartoon representation of KcsA SF, underlining the position of the backbone carbonyl groups from two subunits relative to the four coordinated $K^+$ ions. (b) G77A structure (PDB is 6NFU) solved at 2.09-Å resolution. A 2Fo-Fc electron-density map (light blue, contoured at 2.3 σ) validating KcsA's SF structural model colored in yellow. Two $K^+$ ions are bound to the SF (shown as dark blue spheres), 1 at the S2 site, and the other at the S4 site (dark blue, contoured at 3σ). $K^+$ ions are interspaced with two water molecules, one coordinated by eight main-chain carbonyl groups at the S1 site, while the other is hydrogen-bonded to the amide nitrogen atoms of $Val^{76}$. (c) Crystal structure of the G77C mutant (PDB is 6NFV) at 2.13 resolution. A 2Fo-Fc electron-density map of the SF is shown in light blue (mesh, contoured at 2.5 σ). The SF is colored in white with the oxygen atoms in red. Two $K^+$ ions are modeled as blue spheres (dark blue mesh, contoured at 3.3 σ). As in (b), two water molecules were bound to the channel's SF and modeled as green spheres (at the S1 and S3 sites) (Tilegenova et al., 2019). (For color figure see eBook figure.)

two-step association/dissociation mechanism, where the association and dissociation of ions were not coupled and occurred in two distinct steps. This switch is triggered by the facilitated association of an ion from the intracellular side within the channel pore and by the delayed dissociation of the outermost ion, as the ion concentration increased.

In simulations, 30, 31, 27, and 71 $K^+$ ions were allowed to permeate through the pore, at the four ion concentrations in ascending order, which were equivalent to 4.8, 9.9, 8.7, and 11 pA, respectively (Kasahara et al., 2013). The order of magnitude and the ratio of these values roughly agree with the experimental results with a Shaker channel (Heginbotham and MacKinnon, 1993), and are comparable to the previous simulation (Jensen et al., 2010). The use of binding sites gradually changed with the increasing ion concentration of the solution, implying that changes in the ion concentration can cause differences in the ion conduction mechanisms. At higher ion concentrations, 600 mM over 150 mM, larger number of ions were kept in the pore. At 600 mM, ions were continuously supplied from the intracellular fluid, so the pore almost always retained three or four ions. Whereas, at 150 mM, the pore was found to have only two ions at about half of the simulation time. This hints that the release of the third ion from the pore in the three ion states occurred rapidly than the provision of a new ion at the lower ion concentrations. Figure 6.14 presents the ion-binding state graph for the simulation at 150 mM ion concentration.

### 6.5.4 Generalized Ion Conduction Model Involving Classical and Quantum Mechanical Approaches

Unlike the gA channel which has ion-binding site at the edges of the channel, the potassium channel's ion-binding sites exist inside. The mechanism of ion conduction through an ion channel may not follow a unique process, and may be influenced by ion concentrations (Kasahara et al., 2013). Besides, the channel conduction process largely relies on channel stabilization caused by the energetics of the channel coupled to the hosting bilayer. Therefore, the conducted ionic charges are considered not to conduct through channels freely as charges flowing through conductors experiencing

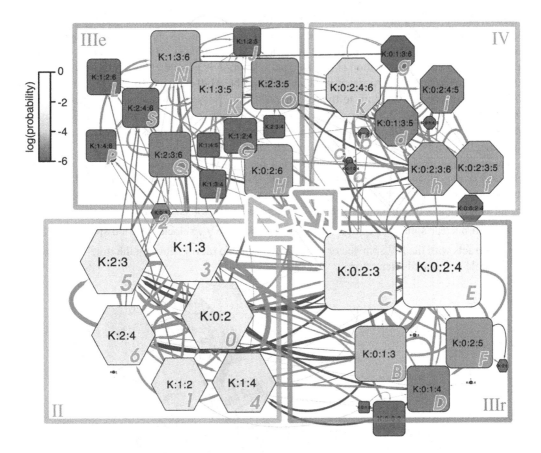

**FIGURE 6.14** The ion-binding state graph for the simulation at 150 mM ion concentration. A node and an edge indicate the state of the binding ions in the pore and the transition between two states, respectively. The size and color of each node mean log existence probability of the state during the simulation. The red edges mean that a new K⁺ ion is entering from the intracellular side, and the blue edges indicate that an ion in S0 is exiting to the extracellular side. The nodes are classified into the four groups: II, IIIr, IIIe, and IV. The two major ways to conduct K⁺ are shown as cyan and pink bold arrows (Kasahara et al., 2013). (For color figure see eBook figure.)

an overall resistance. Instead, these charges while passing through ion channels experience transient pinning effects due to their interactions in the binding sites, and the pinning effect may also get regulated due to certain types of causes, including the condition of the environment (Kasahara et al., 2013). In case where the charges see more than one binding sites, they may even undergo quantum tunneling between binding sites depending on the dielectric environment surrounding the sites. A generalized/universal ion current conduction model considers the following:

i. ions' interactions at the binding sites of channel (channel's mouth regions/exterior, e.g., gA channel and/or channel interior sites, e.g. KcsA channel), and

ii. conduction process to initiate considering the channel openning energetics (explained earlier in this chapter).

Both these aspects are important determinants related to the ion channel-gating mechanisms. They are ultimately recognized in detected ion conduction events using EP records of the channel currents, which is a time-dependent phenomenon. However, certainly if the charge movement has any reliance on quantum mechanical effects originating at the CC region in Figure 6.2 (expected to happen independent of consideration of time), EP records miss to detect them. There might be other methods capable of detecting the possible quantum phenomena involved alongside the classical ones in ion channel conduction process. Theoretical explorations have indicated that the real-time detection of ion channel operation at millisecond resolution is possible by directly monitoring the quantum decoherence of the nitrogen-vacancy (NV) probe (Hall et al., 2010). The quantum dynamics of an NV probe in proximity to the ion channel, lipid bilayer, and surrounding aqueous environment has been explored. Figure 6.15 presents the quantum decoherence imaging of ion channel operation. It estimates the sensitivity of the NV decoherence to various magnetic field sources indicating the ability to detect ion channel switch-on/off events.

The waiting time between ion ejections from the channel is assumed to follow a Poissonian distribution at a mean rate of 3 MHz (Hall et al., 2010). Each ion couples to the NV spin via a time-dependent interaction $H_{int}(t)$, presented in the following equation, with $\mathbf{r}_n(t)$ describing the spatial separation between the NV spin and the dipoles in the ion channel:

The nth nuclear spin with charge $q_n$, gyromagnetic ratio $\gamma_n$, velocity $\vec{v}_n$, and spin vector $\vec{S}_n$ interacts with the NV spin vector $\vec{P}$ and gyromagnetic ratio $\gamma_p$ through the time-dependent dipole dominated interaction

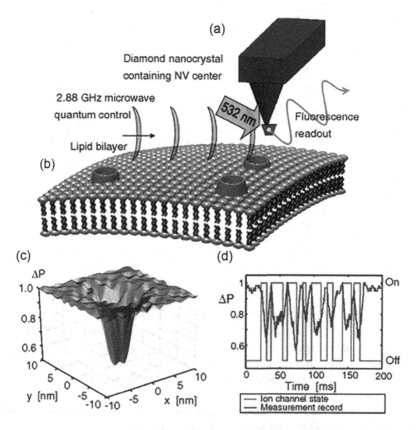

**FIGURE 6.15** Quantum decoherence imaging of ion channel operation (simulations). (a) A single NV defect in a diamond nanocrystal is placed on an atomic force microscope tip. The unique properties of the NV atomic-level scheme allows for optically induced readout and microwave control of magnetic (spin) sublevels. (b) The nearby cell membrane is host to channels permitting the flow of ions across the surface. The ion motion results in an effective fluctuating magnetic field at the NV position which decoheres the quantum state of the NV system. (c) This decoherence results in a decrease in fluorescence, which is most pronounced in regions close to the ion channel opening. (d) Changes in fluorescence also permit the temporal tracking of ion channel dynamics (Hall et al., 2010).

$$H_{int}(t) = \sum_{n=1}^{N} K_{dip}^{(n)} \left[ \frac{\vec{P} \cdot \vec{S}_n}{r_n^3(t)} - 3 \frac{\vec{P} \cdot \vec{r}_n(t) \vec{S}_n \cdot \vec{r}_n(t)}{r_n^5(t)} \right],$$

where $K_{dip}^{(n)} \equiv \frac{\mu_0}{4\pi} \hbar^2 \gamma_p \pi_n$ are the probe-ion coupling strengths, and $\vec{r}_n(t)$ is the time-dependent ion-probe separation. Upon exiting the channel, the timescales associated with the ion motion change to that described by the self-diffusion rate of nuclear spins in the electrolyte. As the addition of ions to the electrolyte at these frequencies is slow compared to the characteristic Brownian motion of the electrolyte, any ions exiting the channel rapidly diffuse. Thus, for simplicity, ions are terminated upon exiting the channel, making the nuclear dipole concentration near the channel opening equal to that of the bulk (Hall et al., 2010).

A quantum mechanical approach using the idea of quantum tunneling was used in a recent study to calculate the conductance of closed channels for different ions (Qaswal, 2019). The conductance due to quantum tunneling of ions through the closed channels was found not to affect the resting membrane potential. However, under different circumstances, including change in the mass or charge of the ion and the residues of the hydrophobic gate, the model of quantum tunneling would be useful to understand and explain several actions, processes, and phenomena in the biological systems. The quantum tunneling current $I_{tun}$ through the gate of closed channels and the quantum membrane conductance $\left(C_{QM} = I_{tun}/V\right)$ can be calculated using the following relations (Serway et al., 2005):

$$I_{tun} = \left( e^2 V / 4\pi^2 \hbar \right) T_{ion}$$

$$C_{QM} = \left( e^2 / 4\pi^2 \hbar \right) T_{ion}$$

Here, e is the electron charge, V is the voltage across the channel, $\hbar$ is reduced Planck constant, and $T_{ion}$ is the tunneling probability of the ion. The extracellular sodium and potassium concentrations are balanced by 67.7 and 96.8 mV, and the intracellular sodium and potassium concentrations are balanced by 6.7 and 2.8 mV, respectively. The quantum conductance (due to quantum tunneling current through the closed gate) of the channels and the tunneling probability have been calculated, as presented in Table 6.2.

The above explanations probably are enough to let us believe that the complete understanding of the mechanisms of ion channel conduction of ions is yet to be established due to additional classical and quantum mechanical approaches besides the existing ones. To develop a generalized ion current conduction model, we may need to consider the principles and techniques of either classical mechanics or quantum mechanics, or probably (for complete understanding) a combination of both. I wish to propose a model here. We may imagine modeling this conduction process as in Figure 6.16.

Here we consider that the resultant channel current $I_{ch}(t)$ is,

$$I_{ch}(t) = I_{ch, cl}(t) + I_{ch, qu},$$

## TABLE 6.2

### Single-Channel Quantum Conductance ($C_{Qion}$) and Tunneling Probability of Intracellular and Extracellular Ions (Qaswal, 2019)

| Ions | Intracellular Region $C_{Qion}$ (milli Siemens) | Extracellular Region $C_{Qion}$ (milli Siemens) | Intracellular Region Tunneling Probability | Extracellular Region Tunneling Probability |
|------|---------------------|---------------------|---------------------|---------------------|
| Na$^+$ | $4.2 \times 10^{-30}$ | $7.2 \times 10^{-22}$ | $6.77 \times 10^{-28}$ | $1.16 \times 10^{-19}$ |
| K$^+$ | $2.1 \times 10^{-24}$ | $1.1 \times 10^{-24}$ | $3.4 \times 10^{-22}$ | $1.7 \times 10^{-22}$ |

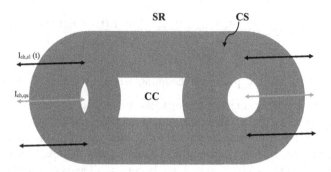

**FIGURE 6.16**   Instead of drawing a simple cylinder, a thick surface cylinder representing the ion channel has been drawn. The cylinder crosses through the thick bilayer (not shown here) to connect the cellular exterior and interior hydrophilic regions to bypass the membrane hydrophobic core. The reason is that the CS should be considered a region, not a demarkating line between channel interior and exterior regions, through which the charges may conduct following independent mechanics, if there is any physical demarkation with CC. CC most likely hosts all quantum mechanical states, as explained earlier, ion channel interior core region (CC), channel (exterior and/or interior) surface (CS), and channel surrounding region (SR). Birectional arrows (↔) represent the possible ion (entry/exit) movements through the channel under the influence of a membrane potential because a driving force acting on charges (current actually flows in one direction following (simple electrostatics meaning) the polarity of potentials and charges being conducted. Black arrow and $I_{ch, cl}(t)$ and green arrow and $I_{ch, qu}$ represent possible directions and current, respectively, conducted by channel due to time-dependent classical mechanical charge movement and time-independent quantum tunneling of charges, respectively.

If the two current pathways are independnet, shown in the model sketch presented here for simplicity. Or,

$$I_{ch}(t) = I_{ch, cl}(t)(1 + Q.F.),$$

Here quantum factor may be determined in special cases, for example, using the explanation and results of the effects of quantum coherence on ion currents (Hall et al., 2010).

Here $I_{ch, cl}(t)$ is current due to flow of charges following motion described by classical mechanics, while $I_{ch, qu}$ is current due to flow of charges following motion described by quantum mechanics (explained in a latter chapter). Although EP records can measure $I_{ch, cl}(t)$, the EP traces do not contain the time-independent fluctuations contributed to the current traces $I_{ch}(t)$, so $I_{ch, qu}$ most likely goes missing in EP recordings. Therefore, new experimental techniques are needed to measure the quantum currents, as a result of quantum tunneling.

## REFERENCES

Ashrafuzzaman, M. (2018). *Nanoscale Biophysics of the Cell.* Springer International Publishing Switzerland AG, Part of Springer Nature. doi:10.1007/978-3-319-77465-7

Ashrafuzzaman, M., Tseng, C. Y., Duszyk, M., & Tuszynski, J. A. (2012). Chemotherapy drugs form ion pores in membranes due to physical interactions with lipids. *Chemical Biology & Drug Design, 80*(6), 992–1002. doi:10.1111/cbdd.12060

Ashrafuzzaman, M., Tseng, C. Y., Kapty, J., Mercer, J. R., & Tuszynski, J. A. (2013). Computationally designed DNA aptamer template with specific binding to phosphatidylserine. *Nucleic Acid Therapeutics, 223,* 418–426.

Ashrafuzzaman, M., Tseng, C. Y., & Tuszynski, J. (2014). Regulation of channel function due to physical energetic coupling with a lipid bilayer. *Biochemical and Biophysical Research Communications, 445* (2), 463–468. doi:10.1016/j.bbrc.2014.02.012

Ashrafuzzaman, M., Tseng, C. Y., & Tuszynski, J. (2020a). Charge-based interactions of antimicrobial peptides and general drugs with lipid bilayers. *Journal of Molecular Graphics and Modelling, 95*, 107502. doi:10.1016/j.jmgm.2019.107502

Ashrafuzzaman, M., Tseng, C. Y., & Tuszynski, J. (2020b). Dataset on interactions of membrane active agents with lipid bilayers. *Data in Brief, 29*, 105138. doi:10.1016/j.dib.2020.105138

Ashrafuzzaman, M., & Tuszynski, J. A. (2012a). Regulation of channel function due to coupling with a lipid bilayer. *Journal of Computational and Theoretical Nanoscience, 9*(4), 564–570. doi:10.1166/jctn.2012.2062

Ashrafuzzaman, M., & Tuszynski, J. A. (2012b). *Membrane Biophysics*. Berlin, Heidelberg: Springer-Verlag. doi:10.1007/978-3-642-16105-6

Beaven, A. H., Maer, A. M., Sodt, A. J., Rui, H., Pastor, R. W., Andersen, O. S., & Im, W. (2017). Gramicidin A channel formation induces local lipid redistribution I: Experiment and simulation. *Biophysical Journal, 112*(6), 1185–1197. doi:10.1016/j.bpj.2017.01.028

Bezanilla, F., Perozo, E., & Stefani, E. (1994). Gating of shaker K+ channels: II. The components of gating currents and a model of channel activation. *Biophysical Journal, 66*(4), 1011–1021. doi:10.1016/s0006-3495(94)80882-3

Campos, F. V., Chanda, B., Roux, B., & Bezanilla, F. (2007). Two atomic constraints unambiguously position the S4 segment relative to S1 and S2 segments in the closed state of Shaker K channel. *Proceedings of the National Academy of Sciences, 104*(19), 7904–7909. doi:10.1073/pnas.0702638104

Case, D. A., Darden, T. A., Cheatham, T. E. et al. (2010). AMBER 11, University of California, San Francisco.

Catterall, W. A. (2012). Voltage-gated sodium channels at 60: Structure, function and pathophysiology. *The Journal of Physiology, 590*(11), 2577–2589. doi:10.1113/jphysiol.2011.224204

Colombini, M. (2010). Ceramide channels and their role in mitochondria-mediated apoptosis. *Biochimica Et Biophysica Acta (BBA) - Bioenergetics, 1797*(6–7), 1239–1244. doi:10.1016/j.bbabio.2010.01.021

Doyle, D. A., Cabral, J. M., Pfuetzner, R. A., Kuo, A., Gulbis, J. M., Cohen, S. L., . . . MacKinnon, R. (1998). The structure of the potassium channel: Molecular basis of K+ conduction and selectivity. *Science, 280* (5360), 69–77. doi:10.1126/science.280.5360.69

Finkelstein, A., & Andersen, O. S. (1981). The gramicidin a channel: A review of its permeability characteristics with special reference to the single-file aspect of transport. *The Journal of Membrane Biology, 59*(3), 155–171. doi:10.1007/bf01875422

Hall, L. T., Hill, C. D., Cole, J. H., Stadler, B., Caruso, F., Mulvaney, P., . . . Hollenberg, L. C. (2010). Monitoring ion-channel function in real time through quantum decoherence. *Proceedings of the National Academy of Sciences, 107*(44), 18777–18782. doi:10.1073/pnas.1002562107

Heginbotham, L., & Mackinnon, R. (1993). Conduction properties of the cloned Shaker K+ channel. *Biophysical Journal, 65*(5), 2089–2096. doi:10.1016/s0006-3495(93)81244-x

Hodgkin, A. L., & Huxley, A. F. (1952). A quantitative description of membrane current and its application to conduction and excitation in nerve. *The Journal of Physiology, 117*(4), 500–544. doi:10.1113/jphysiol.1952.sp004764

Hodgkin, A. L., & Keynes, R. D. (1955). The potassium permeability of a giant nerve fibre. *The Journal of Physiology, 128*(1), 61–88. doi:10.1113/jphysiol.1955.sp005291

Jensen, M. O., Borhani, D. W., Lindorff-Larsen, K., Maragakis, P., Jogini, V., Eastwood, M. P., . . . Shaw, D. E. (2010). Principles of conduction and hydrophobic gating in K+ channels. *Proceedings of the National Academy of Sciences, 107*(13), 5833–5838. doi:10.1073/pnas.0911691107

Kasahara, K., Shirota, M., & Kinoshita, K. (2013). Ion concentration-dependent ion conduction mechanism of a voltage-sensitive potassium channel. *PLoS ONE, 8*(2). doi:10.1371/journal.pone.0056342

Khalili-Araghi, F., Jogini, V., Yarov-Yarovoy, V., Tajkhorshid, E., Roux, B., & Schulten, K. (2010). Calculation of the gating charge for the Kv1.2 voltage-activated potassium channel. *Biophysical Journal, 98*(10), 2189–2198. doi:10.1016/j.bpj.2010.02.056

Neher, E., & Sakmann, B. (1976). Single-channel currents recorded from membrane of denervated frog muscle fibres. *Nature, 260*(5554), 799–802. doi:10.1038/260799a0

Nielsen, C., Goulian, M., & Andersen, O. S. (1998). Energetics of inclusion-induced bilayer deformations. *Biophysical Journal, 74*(4), 1966–1983. doi:10.1016/s0006-3495(98)77904-4

Papazian, D. M., Timpe, L. C., Jan, Y. N., & Jan, L. Y. (1991). Alteration of voltage-dependence of Shaker potassium channel by mutations in the S4 sequence. *Nature, 349*(6307), 305–310. doi:10.1038/349305a0

Qaswal, A. B. (2019). Quantum tunneling of ions through the closed voltage-gated channels of the biological membrane: A mathematical model and implications. *Quantum Reports, 1*(2), 219–225. doi:10.3390/quantum1020019

Roux, B. (1999). Statistical mechanical equilibrium theory of selective ion channels. *Biophysical Journal, 77*(1), 139–153. doi:10.1016/s0006-3495(99)76878-5

Roux, B., & Karplus, M. (1991). Ion transport in a model gramicidin channel. Structure and thermodynamics. *Biophysical Journal, 59*(5), 961–981. doi:10.1016/s0006-3495(91)82311-6

Sasaki, M. (2006). A voltage sensor-domain protein is a voltage-gated proton channel. *Science, 312*(5773), 589–592. doi:10.1126/science.1122352

Schoppa, N., Mccormack, K., Tanouye, M., & Sigworth, F. (1992). The size of gating charge in wild-type and mutant Shaker potassium channels. *Science, 255*(5052), 1712–1715. doi:10.1126/science.1553560

Schreiber, M., Yuan, A., & Salkoff, L. (1999). Transplantable sites confer calcium sensitivity to BK channels. *Nature Neuroscience, 2*(5), 416–421. doi:10.1038/8077

Serway, R. A., Moses, C. J., & Moyer, C. A. (2005). *Modern Physics*. Boston, MA: Thomson Learning.

Sigg, D. (2014). Modeling ion channels: Past, present, and future. *The Journal of General Physiology, 144*(1), 7–26. doi:10.1085/jgp.201311130

Sigworth, F. J. (1994). Voltage gating of ion channels. *Quarterly Reviews of Biophysics, 27*(1), 1–40. doi:10.1017/s0033583500002894

Tilegenova, C., Cortes, D. M., Jahovic, N., Hardy, E., Hariharan, P., Guan, L., & Cuello, L. G. (2019). Structure, function, and ion-binding properties of a $K^+$ channel stabilized in the 2,4-ion–bound configuration. *Proceedings of the National Academy of Sciences, 116*(34), 16829–16834. doi:10.1073/pnas.1901888116

Vargas, E., Bezanilla, F., & Roux, B. (2011). In search of a consensus model of the resting state of a voltage-sensing domain. *Neuron, 72*(5), 713–720. doi:10.1016/j.neuron.2011.09.024

Vargas, E., Yarov-Yarovoy, V., Khalili-Araghi, F., Catterall, W. A., Klein, M. L., Tarek, M., . . . Roux, B. (2012). An emerging consensus on voltage-dependent gating from computational modeling and molecular dynamics simulations. *Journal of General Physiology, 140*(6), 587–594. doi:10.1085/jgp.201210873

Wang, J., Wang, W., Kollman, P. A., & Case, D. A. (2005). Antechamber, an accessory software package for molecular mechanical calculation. *Journal of Computational Chemistry, 25*, 1157–1174.

Woolf, T. B., & Roux, B. (1996). Structure, energetics, and dynamics of lipid–protein interactions: A molecular dynamics study of the gramicidin A channel in a DMPC bilayer. *Proteins: Structure, Function, and Genetics, 24*(1), 92–114. doi:10.1002/(sici)1097-0134(199601)24:13.0.co;2-q

Woolf, T. B., & Roux, B. (1997). The binding site of sodium in the gramicidin A channel: Comparison of molecular dynamics with solid-state NMR data. *Biophysical Journal, 72*(5), 1930–1945. doi:10.1016/s0006-3495(97)78839-8

Yarov-Yarovoy, V., Baker, D., & Catterall, W. A. (2006). Voltage sensor conformations in the open and closed states in ROSETTA structural models of $K^+$ channels. *Proceedings of the National Academy of Sciences, 103*(19), 7292–7297. doi:10.1073/pnas.0602350103

Yarov-Yarovoy, V., Decaen, P. G., Westenbroek, R. E., Pan, C., Scheuer, T., Baker, D., & Catterall, W. A. (2011). Structural basis for gating charge movement in the voltage sensor of a sodium channel. *Proceedings of the National Academy of Sciences, 109*(2). doi:10.1073/pnas.1118434109

# 7 Nanotechnology of Ion Channels

Natural systems are run by their own processes and principles. In biological systems, meter-to-nanometer-long systems and subsystems exist. Several natural techniques make the biological processes continuously work and maintain the system's function. Several processes are active in the biological organ's building blocks, cellular systems. Among those active in cellular systems, those working at nanometer (nm) dimensions, which may be branded as nanosystems and nanoprocesses. As the soft condensed matter cellular system is dynamic, some of the timescales these nanosystems follow may also fall in the nanosecond order. Nanoprocesses working at the nanosecond-order timescale in cellular nanosystems may be considered as natural nanotechnology. The dimension of ion channel structures' functional sections is nothing but the order of cell membrane thicknesses, that is, 3–8 nm dimension. Therefore, the functional regions of the channels are also nanosystems. The active processes are nothing but nanoprocesses. The technologies that are naturally inbuilt are nanotechnologies. In this chapter, we aim to detect natural nanotechnologies that have deterministic roles in maintaining channel structures and functions. We often find some artificially developed nanotechnologies that are applicable in dysregulating the progression of ion channel disorders generally detected in disease states. That means, in channelopathies, nanotechnologies may also be utilized to find a cure.

## 7.1 NATURAL NANOTECHNOLOGY OF ION CHANNELS

The major structures and region of actions of ion channels extend to several nanometers. Within this small three-dimensional (3D) range, these complicated biological machineries, "bionanomachines" (Baumgaertner, 2008), maintain several nanotechnologies that are naturally inbuilt. These nanomachines including their technologies have matured over almost 4 billion years and can maintain tuned precisions that we need to know in detail if our goal is to understand the hidden mechanisms. My previous book *Nanoscale Biophysics of the Cell* addressed various aspects of cellular nanomachines including ion channels (Ashrafuzzaman, 2018). Here, we wish to inspect the channel structures where nanotechnologies are naturally inbuilt that help channels maintain their gating mechanisms. Hidden molecular mechanisms, mostly determined by principles of physics, may come to light if we can compartmentalize ion channels into nanomachines assembled and run by local nanotechnologies.

The term "machine" or "motor" in cellular system is used to describe some cell-based biomolecular complexes because they transduce one form of energy to another, for example, chemical binding to mechanical work. In case of ion channels, we have understood this in Chapter 6. The actual geometric range within which ion channels work falls at the dimension of cellular membranes, such as, plasma, mitochondrial, nuclear membranes, etc. These nanometer-scale compartments host the ion channels that work as not just machines but also nanomachines, and the technologies active there are nanotechnologies.

The concept of bionanotechnology (or being a synonym for nanobiotechnology) and the molecular machines of the biological systems evolved from initial thoughts of Richard Feynman about six decades ago. Most cellular functions are not carried out by a single protein enzyme colliding randomly within the cellular systems, but by macromolecular complexes acting as molecular machines. These complexes or machineries contain multiple subunits and are usually specific in their functions. Ion channels are similar complexes with specific roles, mostly related to localized

DOI: 10.1201/9781003010654-7

energetics which controls transport of materials and information through the channels. Channels are passive devices whose gating (opening $\leftrightarrow$ closing) is controlled externally by ligands, voltage, pH, or membrane-induced mechanical stresses. Similar to general machineries, ion channels also have moving parts that 'gate' the channel and directly control ion permeation, often in response to a specific external stimulus.

Ion channel gating mechanisms rely on two interrelated processes: first, the energy transduction machineries convert multiple types of physical stimuli, for example, voltage, ligand binding, force, etc., into protein motion; and second, the structural rearrangements defining the motion of the gating.

The gating process, related gating motion (movement or fluctuation of channel subunits), and the movement of materials across the gates are nanoprocesses occurring inside nanomachines at nanoscopic dimension. Timescale is below a millisecond, all the way down to the quantum tunneling timescale. We have explained most of these properties using appropriate examples in Chapters 1–3, though from mostly the physiological point of view. Ion channels as mechanical devices/machines require an analysis based on available information. However, the knowledge accumulated in physiological studies of ion channels is crucial for technologists to find means to manipulate or regulate these nanomachines' structures and functions. These bionanotechnolgical works have already started and slowly we may reach at certain points when better engineering techniques will be found to act upon natural ion channels, especially when their genes are mutated. Thus, we will be able to deal with channelopathies in finding cure for versatile cellular disorders linked to channel aberration.

## 7.2   APPLIED NANOTECHNOLOGY AND ION CHANNEL ENGINEERING

Natural nanotechnologies of ion channels are capable of self-running the nanomachines "ion channels." The required energies are instantaneously supplied locally through versatile molecular processes. The chapter on ion channel energetics has elaborated a few key molecular processes that actively work to maintain channel energetics. Adenosine triphosphate (ATP), the energy-carrying biomolecule, also participates in regulating ion channels (Mazzanti et al., 1994; Atkinson et al., 2002). An ion channel spanning the nuclear envelope, between the cytoplasm and the nucleus, was reported to be regulated by an ATP-binding receptor of the P2X7 subtype (Atkinson et al., 2002).

Despite the existence of ion channels' self-regulated biological processes, various genetic mutations in channel subunits may cause malfunctions in some of the ion channel nanotechnologies. Because the underlying molecular mechanisms and processes may not necessarily be active, the machineries may work poorly or even fail. This is most likely the technological perspective of a disease state concerning ion channels' structures and functions. The means to repair the machines using additional artificial techniques may be branded as applied nanotechnology. The applied nanotechnologies work to discover novel technologies or existing technology-modifying agents applicable in biological systems. Consequently, the relevant results may often get converted into marketable products. In case of channelopathies, these products are nothing but therapeutic drugs to target ion channel-contained nanomachines.

Various techniques have been adopted in addressing the biological machineries including ion channels. Two major aspects are of great interest, namely, the functions of the machineries and the source(s) of energies required to fuel the machines' operating requirements. Table 7.1 lists a few of them, especially those including ion channels related to the membrane transport phenomena (Baumgaertner, 2008).

The ion channel structural components often malfunction, especially in disease conditions. The channels then require artificial interventions to help them recover from some of their disorders. Whenever we deal with repairing or modulating machine components, we require appropriate engineering techniques. Ion-channel engineering (ICE) is applied to modify biological ion channels by chemical/biological synthetic means (Grosse et al., 2011; Subramanyam and Colecraft, 2015). The goal is to obtain synthetic ion channels (for artificial ion channel class) or obtain agents to repair,

**TABLE 7.1**

**A List of Biomolecular Machines Including Ion Channels That Have Been Studied Theoretically and by Simulations. ATP, Hydrolysis of Adenosine Triphosphate; *n*-Phos, Nucleotide Phosphorylation. $\Delta\psi$, Membrane Potential and Ion Gradient.**

| Biomolecular Devices | Function | Energy Source |
|---|---|---|
| Membrane transducers | | |
| GPCR | Signal relay | Ligands |
| Rhodopsin CCRS, CXCR4 | | |
| Catalytic receptors Insulin, EGF | Signal relay | Photons, ligands |
| Chemotactic sensors sR | Signal relay | Photons, ligands |
| Receptor-ligand Titian, GroEL, etc. | Signal relay | Photons, ligands |
| Membrane channels | | |
| Porins | Valve for ion, water | Gradients |
| OmpF, OmpT, OmpA, FhuA AQP Maltoporin | | |
| Ion channels | Selective ion valve | Gating by ligand |
| AChR | | |
| CIC | | |
| gA | | $\Delta\Psi$ |
| GluR | | Ligand |
| KcsA | | pH |
| KvAP | | Voltage |
| MscL, Mses | | Strain |
| Membrane transporters | | |
| Ion translocons | Ion pumps | Photons |
| bR | $H^+$ | |
| hR | $Cl^-$, anion | |
| Ion translocases | Ion pumps | |
| V-ATPase | $H^+$ | ATP |
| $Ca^{2+}$-ATPase | $Ca^{2+}$ | ATP |
| Na+-K+-ATPase | $Na^+, K^+$ | ATP |
| Solute translocases | Solute transport | ATP |
| HisP, MsbA, GLUT3 | | ATP |
| Protein translocases | Protein transport | ATP |
| Mitochondrial pore | | ATP, $\Delta\Psi$ |
| ER pore | | ATP, GTP |
| Nuclear pore | | |
| Membrane motors | | |
| ATP synthase | ATP synthesis | $\Delta\Psi$ |
| Flagellar motor | Bacterial motility | ATP, $\Delta\Psi$ |

alter, or regulate natural ion channel containing nanomachine components, with modified or novel functionality. The former are found best in molecular recognition and catalysis (Sakai and Matile, 2013). The latter are generally called "ion channel modulators" (Ashrafuzzaman et al., 2006), which may act directly upon ion channels or indirectly via their actions on ion channel-hosting bilayers. Cellular mechanisms and signaling pathways are linked to or involved in ion channel modulation, and that the changes to ion channel functions may affect key biological processes (Burke and Bender, 2019). Both the modulation of biological ion channel and/or the creation and insertion of synthetic ion channels are essential in certain applications to maintain correct membrane transport through channel routes.

Biological ion channels may be modified using ICE by chemical or biological synthesis. The goal is to obtain channels with modified or novel functionality. Three functional strategies have been summarized by Grosse et al. (2011):

- The engineering approach tries to manipulate the wider pores with robust β-barrel structures, such as those of α-hemolysin and porins.
- The engineering approach focuses on the modification of narrow (mostly α-helical) pores to understand selectivity and modes of action.
- The engineering approach addresses channel gating by (photo)triggering the biological receptor that controls the channel.

The first two functional strategies deal directly with the chemical modifications of the ion-transfer pathway, and employ either wide-pore channels such as porins or narrow-pore channels like potassium channels. The third functional strategy addresses the mechanism of ion channel activation and introduces chemical alterations in the channel-gating region.

Multiple synthetic strategies have been developed and utilized for the synthetic modification of biological ion channels (Grosse et al., 2011):

- The S-alkylation of specific cysteines,
- Protein semisynthesis by native chemical ligation,
- Protein semisynthesis by protein trans-splicing, and
- Nonsense-suppression methods.

Further reprogramming and reengineering of channels (Lutz and Bornscheuer, 2009) may be strategically used for sensing applications, such as in treatment of channelopathies and in chemical neurobiology.

In brief, ICE and subsequent reengineering can deal with a few key issues (Subramanyam and Colecraft, 2015):

- Engineering ion channels to elucidate their structure-function mechanisms,
- Engineering ion channels to probe and manipulate physiology, and
- Developing engineered ion channel modulators.

## 7.2.1  ICE ALTERING CHANNEL FUNCTION

The major functional characteristics of ion channels are mainly manifested by (i) ion-transfer pathway, (ii) ion channel selectivity, (iii) and surrounding environmental control over ion channel functions.

The simplest ICE is the manipulation of nonspecific ion-transfer pathways to introduce specificity, to make them switchable, or to use large ion channel currents to monitor the passage of analytes. ICE of wide-channel pores benefits from the rigid nature of the pores' ion conductance pathways and the lack of sterical restraints imposed on chemical modifications by the channel walls. The porins form wide, water-filled channels within their β-barrel structures. In addition to providing a pretty simple, regular architecture, these β-barrels often result in outstanding stability of these channels in a membrane (Pastoriza-Gallego et al., 2007). The porins are of great interest for ICE (Korkmaz-Özkan et al., 2010). There are a few requirements that must be satisfied by channels for use in ICE. Availability of their 3D structural information, simplicity of template structures, ease of recombinant overproduction and chemical modification, the stability of the refolded ion channel variants, and known conductance characteristics. Ion channels with wider pores meet these criteria, and their ion pathway modifications show great potential.

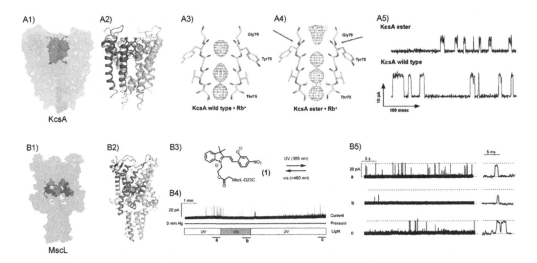

**FIGURE 7.1** Structural templates with narrow pore diameters used in ICE. (A) KscA and (B) MscL. (1) Sections through the corresponding pores: surface representation (gray), water-filled cavity (blue), and restricting amino acids (red). (2) Secondary structure representations with independently colored protein chains. (A3) Electron density for Rb$^+$ in the wild-type constriction site and (A4) in engineered KscA containing esters instead of peptide bonds (red arrows). (A5) Resulting differences in event duration of current recordings (Valiyaveetil et al., 2006). (B3) Switchable modulator 1 used in biohybrids. (B4) Fluctuations caused in current recording depending on UV irradiation (Kocer et al., 2005). (B5) Expanded traces for regions (a–c). For details see Grosse et al. (2011). (For color figure see eBook figure.)

α-helical ion channels, for example, the potassium channel KcsA, hugely surpass β-barrel-based porins in terms of selectivity, while maintaining high conductivities at ~108 ions/s. The selectivity filters (SFs) of these channels are extremely narrow, allowing the passage of partly dehydrated cations in a single file with water molecules. ICE of this ion channel was initially motivated to understand the basic mechanisms of cation discrimination, as well as control of the open and closed states. The role of the T$^{75}$VGYG$^{79}$ motif within the KcsA SF was investigated by Nobel Laureate MacKinnon and colleagues by engineering semisynthetic truncated KcsA channels, in which either the amide bond between Tyr78 and Gly79 was changed to an ester bond, or Gly77 was substituted with d-Ala, see Figure 7.1 (Grosse et al., 2011), and for details see Valiyaveetil et al. (2004; 2006). Both KcsA variants were functional, thus proving the structural necessity for Gly77, a local, left-handed α-helical backbone conformation is not favorable to other ʟ-amino acids, but can be adopted by Gly and ᴅ-Ala. The latter here demonstrated the tolerance to the single-atom replacement of the amide → ester substitution, and that atomic-scale interventions are capable of changing specificity, conductance, and distribution of alkaline ions within the SF of potassium channels.

A promising approach to control ion channel function is the use of tetherable and switchable ligand hybrids that are capable of subjecting ion channel activation to external control(s), see Figure 7.2 (Grosse et al., 2011). A tetraethylammonium (TEA) moiety was used as pore (the voltage-gated K$^+$-channels (Kv) of the Shaker family) blocker connected to a photo-switchable azobenzene group (Grosse et al., 2011).

## 7.2.2 ICE CONSIDERING SYNTHESIS OF CHANNEL SUBUNITS

Chemically modified channel protein synthesis is a typical example of protein engineering. In an earlier chapter on artificial/synthetic ion channels, we have addressed how we can achieve synthetic channel subunits using various strategic techniques. In ICE, we usually use four different synthesis strategies, as follows (Grosse et al., 2011):

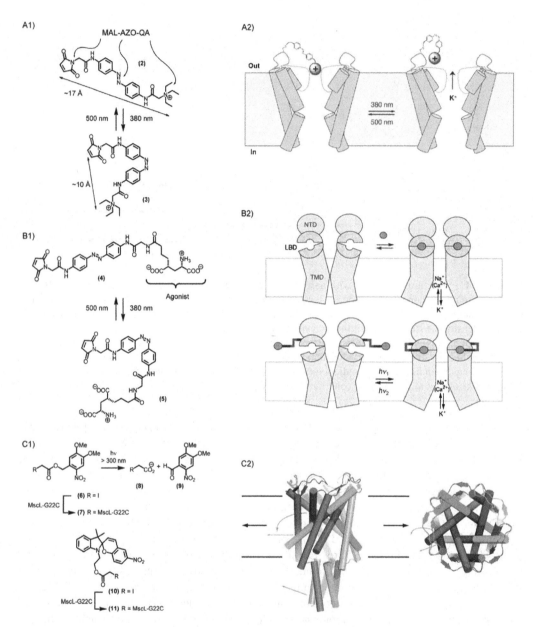

**FIGURE 7.2** Design of different photo-switchable biohybrid channels. (A1) Photo-switch bearing a maleimide for attachment to cysteines, a tethered tetraethylammonium (TEA) ligand (QA), and a switchable azobenzene moiety (AZO) applied to a K⁺(SPARK) channel in long (2) and short (3) configurations. (A2) Schematic representation of the photo-switch with the engineered channel (Banghart et al., 2004). (B1) synthetic photo-switch used to regulate a glutamate receptor bearing the agonist glutamate in long (4) and short (5) configurations. The glutamate receptor consists of an N-terminal domain, ligand binding domain, and transmembrane domain. (B2) Comparison of natural (top) and photoinduced channel opening (bottom) (Volgraf et al., 2005). (C1) Design of irreversible (upper) and reversible (lower) photo-switches used in the mechanosensitive channel MscL. Application to a central position within the constricted channel bears the possibility of pushing the channel by light into a permanently open position. (C2) Natural mode of action of MscL. Black arrows indicate changes in membrane pressure that result in a rearrangement (gray arrows) of helices, transmuting MscL to the open-channel state (Kocer et al., 2005). For details see Grosse et al. (2011).

- S-alkylation of specific cysteines,
- semisynthesis via native chemical ligation (NCL),
- semisynthesis by protein trans-splicing, and
- nonsense suppression.

All of these synthetic strategies have a common goal to introduce synthetic modulators for the channel function, such as (photo) switches and SFs. We avoid presenting other strategies here, which are used to chemically synthesize channel-forming peptides (Fields, 2001), for example, gramicidin and alamethicin types and their derivatives (Ashrafuzzaman et al., 2006; Ashrafuzzaman and Tuszynski, 2012).

In S-alkylation of specific cysteines, a specific Cys introduction by site-specific mutagenesis is an approach for the functionalization/conjugation of proteins. This ICE strategy has been popularly used by many groups. As an example, the linkage of a thiol function to an iodo acetamide was used to introduce a dibenzo crown ether into the K16C mutant of the OmpF porin [11], see Figure 7.3 (12 + 13 → 14), and for the linkage of the spiropyrane photo-switch to MscL-G22C, a mechanosensitive ion channel (Kocer et al., 2005). The thiol group addition to a maleimide led to the linkage of photo-switchable azobenzene units in the case of the K+(SPARK) channel (Glu422C; 15+16→17), see Figure 7.3 (Banghart et al., 2004).

Semisynthesis by NCL requires an N-peptide (thiol ester at the C-terminus) and a C-peptide (Cys at the N terminus). Both of these fragments can be prepared by recombinant expression and/or by chemical synthesis, despite the N-terminal peptide needing to be activated. The latter allows the synthetic building block incorporation. An early case of this protein engineering in the ion channel by NCL was with the KcsA channel; the first amide bond of the SF (between Tyr78 and Gly79)

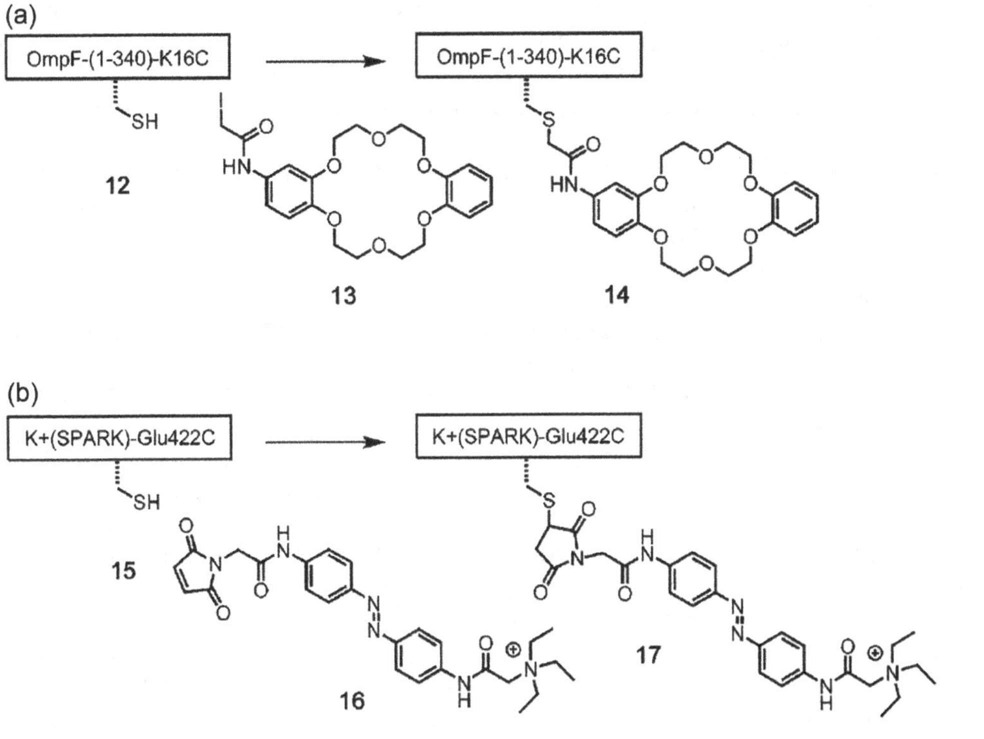

**FIGURE 7.3** Synthesis of modified ion channels by (a) S-alkylation of iodoacetamides, and (b) Michael addition to maleic imides (Grosse et al., 2011).

was changed to an ester bond (Banghart et al., 2004). The ligation product was obtained using a synthetic fragment and a recombinant fragment in its denatured form (Valiyaveetil et al., 2004).

In the nonsense suppression method, the synthetic approach requires two criteria to be fulfilled. One of the three stop codons must be introduced at the point of interest in the mRNA of the target protein, and then the presence of a stop codon-recognizing tRNA, to which the desired unnatural amino acid is attached. This method was used to engineer nAChR in Xenopus oocytes (Dougherty, 2008).

### 7.2.3  ICE Introducing Photosensitivity into Ion Channel Subunits

Application of artificial techniques to remotely control and monitor ion channel operations with light has become popular. Photosensitive unnatural amino acids (UAAs) have been utilized for directly introducing light sensitivity into proteins relying on the genetic code expansion technique. Introduction of UAAs results in a unique molecular-level control, and combined with the maximal spatiotemporal resolution and poor invasiveness of light, enables direct manipulation and interrogation of the ion channel functionality. Applications of light-sensitive UAAs in two ion channel superfamilies, voltage- and ligand-gated ion channels, and a summary of the existing UAA tools, their mode of action, potential, caveats, and technical considerations to illuminate ion channel structure and function have recently been detailed (Klippenstein et al., 2018).

In Chapter 1, we addressed the structures and gating mechanisms of both voltage-gated ion channels (VGICs) and ligand-gated ion channels (LGICs). VGICs represent a multiplicity of ion channel types, among other four-domain voltage-gated sodium and calcium channels, and single or multi-domain channels such as voltage-gated potassium channels and channels from the transient receptor potential (TRP) family. All these channels typically open their pores due to membrane voltage, which is mediated by a voltage-sensing domain (VSD) (Catterall, 2017). LGICs are activated by ligands binding extracellular domains. They can be trimeric, tetrameric, or pentameric (Lemoine et al., 2012). LGICs and VGICs are major potential drug targets involved in a wide variety of pathologies and channelopathies (Paoletti et al., 2013). Electrophysiology, molecular biology, biochemistry, pharmacology, and structural biology (X-ray crystallography and cryoelectron microscopy) have produced a large amount of information helping us understand the complex gating and regulation processes of these channels in native conditions, where their subunit compositions and microenvironments can influence their functions, which is key to drug discovery. These methods are not that competent and often miss detecting various crucial aspects of structural, spatial, or temporal resolution, limiting our understanding of ion channel structure and function. Methods with high spatial resolution and fast enough to synchronize with the channels' dynamic mode of action during normal and disease states are required. Light confers this high spatiotemporal resolution. Various techniques, such as fluorescence spectroscopy, optopharmacology (Kramer et al., 2013), and optogenetic pharmacology/synthetic optogenetics (Berlin and Isacoff, 2017), exploit of the power of light by engineering light-responsiveness into ion channels and receptors, see Table 7.2 (Klippenstein et al., 2018).

A method to introduce UAAs into proteins was recently developed that relies on the genetic encoding of UAAs (Liu and Schultz, 2010), see Figure 7.4. This technique utilizes bio-orthogonal tRNA/synthetase pairs, specifically engineered for the introduction of a particular amino acid. All components required for "nonsense" stop codon suppression, including amber version of the desired protein and specific tRNA/synthetase pair, are introduced into the cell in the form of genes. Once processed within the cell, the tRNA is charged with the UAA by its cognate synthetase to suppress the amber codon on the protein mRNA.

Ion channel function can be manipulated using light-sensitive UAAs. Ion channels decorated with light-responsive UAAs afford an artificial control mechanism that is orthogonal to the natural one provided by evolution (i.e., change in membrane potential and ligand binding). Various light-sensitive UAAs, carrying photo-cross-linking, photo-switchable, photo-cleavable, or

(*Continued*)

## TABLE 7.2
## Methodologies to Confer Light Sensitivity into Ion Channels (Klippenstein et al., 2018).

| Domain | Introduction of Light Sensitivity | Conditions |
|---|---|---|
| | | **Optical Manipulation of Ion Channel Activity** |
| Optopharmacology | Diffusible ligands; no genetic modification of target protein | Photosensitive ligands Caged compounds (irreversible); for example: MNI-glutamate Azobenzene-based photo-switches (reversible); example: 4-GluAzo |
| Optogenetic pharmacology | Tethered ligands; posttranslational labeling of genetically modified target protein (cysteine mutation) | Photo-sensitive tethered ligands (PTLs) Cysteine- and photo-reactive compounds (irreversible); for example: benzophenone-4-carboxamidocysteine methanethiosulfonate (BPMTS) |

**TABLE 7.2 (*Continued*)**

**Methodologies to Confer Light Sensitivity into Ion Channels (Klippenstein et al., 2018).**

| Domain | Introduction of Light Sensitivity | Conditions |
|---|---|---|
| | | Cysteine-reactive, azobenzene-based photo-switches (reversible); for example: maleimide-azobenzene-glutamate (MAG) 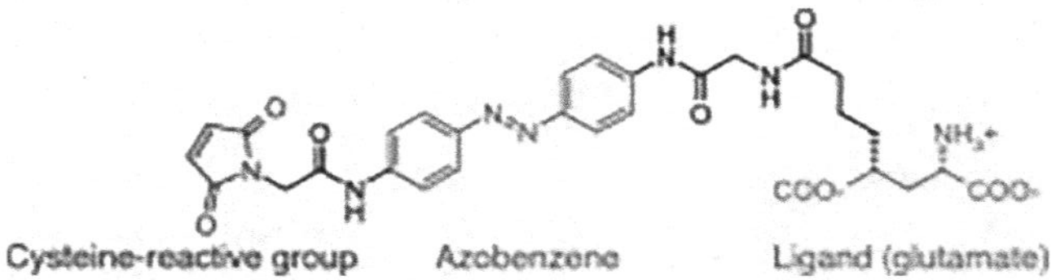 |

Cysteine-reactive group    Azobenzene    Ligand (glutamate)

| | UAAs; genetic incorporation | Light sensitivity is directly conferred by UAA |

**Optical Monitoring of Ion Channel Conformational Rearrangements**

| Optical monitoring (including VCF and/or PCF) | Tethered ligands as biophysical probes; posttranslational labeling of genetically modified target protein (cysteine mutation or small peptide tag insertion) | Tethered fluorophores Cysteine- and tag-reactive compounds; for example: tetramethylrhodamine maleimide (TMR) 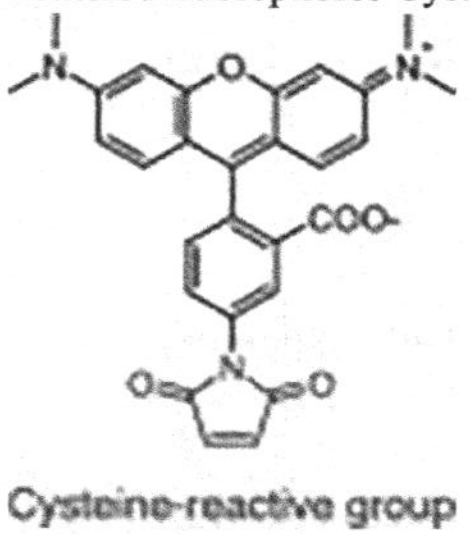 |

Cysteine-reactive group

| | UAAs; genetic incorporation | UAA is directly fluorescent and used as a genetically encoded biophysical probe |

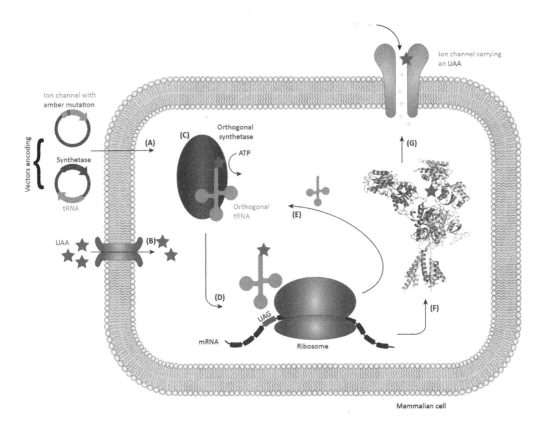

**FIGURE 7.4** Genetically encoding light-sensitive unnatural amino acids (UAAs) into ion channels. The UAA genetic encoding methodology, as performed for mammalian cell lines, is represented schematically. (a) Vectors carrying genes for an ion channel of interest (light blue) and an orthogonal suppressor tRNA (orange)/aminoacyl synthetase (dark blue) pair are introduced into the cell by transient transfection. The amber stop codon (TAG, red) replaces a native codon at a permissive site within the sequence of the ion channel gene. (b) The light-sensitive UAA (purple asterisks) is added to the cellular growth medium and spontaneously enters the cell through amino acid transporters (gray). (c) Within the cell, the orthogonal synthetase specifically aminoacylates the suppressor tRNA with the UAA, a catalytic reaction driven by ATP. (d) The UAA-carrying tRNA, which contains a CUA anticodon, enters the ribosomal machinery to incorporate the UAA in response to the complementary amber codon on the ion channel mRNA (black). (e) Once relieved from the charge at the ribosome, the tRNA can be reused for further UAA aminoacylation by the cognate synthetase. (f) The full-length polypeptide chain [shown here for two NMDA receptor (NMDAR) subunits, Protein Data Bank 4PE5[87]; light blue], site specifically carrying the UAA, undergoes folding and assembly into a functioning ion channel. (g) The newly formed membrane protein migrates to its assigned location (e.g., the cell surface) to selectively conduct ions (yellow), thus contributing to cellular functions (Klippenstein et al., 2018). (For color figure see eBook figure.)

photo-caged side-chains, have been successfully used to obtain remote and highly precise control of ion channel function. Examples of optical ion channel modulation by light-sensitive UAAs are presented in models, see Figure 7.5. For details, see Klippenstein et al. (2018).

## 7.3 ENGINEERED ION CHANNEL REGULATORS

Ion channel regulation may happen due to physical effects of agents on channel directly or channel-hosting bilayer directly that indirectly alters the channel function. However, there are a couple of general problems that create considerable issues while using small molecules as ion channel blockers or modulators. These include lack of targetability to specific populations of the cell and limited

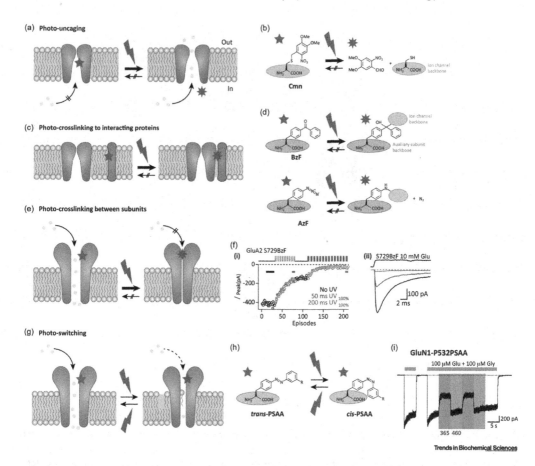

**FIGURE 7.5** Examples of optical ion channel modulation by light-sensitive UAAs. Ion channels are illustrated in blue. (a) Incorporation of a caged UAA (five-branch asterisk), here a caged cysteine, in the pore of a voltage-gated ion channel (VGIC; e.g., inwardly rectifying potassium channel hinders the conductance of ions (yellow). UAA photolysis by ultraviolet (UV) light (purple thunderbolt) releases the cage (eight-branch asterisk) and restores current flow. (b) Chemical reaction of 4,5-dimethoxy-2-nitrobenzyl-cysteine (Cmn) photolysis. (c) UV-induced photo-cross-linking of a VGIC to an auxiliary subunit (green). (d) Chemical reactions of the covalent photo-cross-linking by azido-phenylalanine (AzF) and benzoyl-phenylalanine (BzF) following UV illumination. (e) Photo-inactivation by intrareceptor UV-cross-linking of a ligand-gated ion channel (LGIC; e.g., an ionotropic glutamate receptor (iGluR)]. Photo-cross-linking UAAs, placed at key moving interface sites, physically link adjacent subunits. (f) (i) Representative time course of inactivation of the homomeric GluA2 AMPA receptor (with BzF at the S729site), induced by photo-cross-linking of subunits, following application of UV pulses (violet bars) of different durations (red and green circles).(ii) Example traces recorded before (black) and at the end (red and green) of the two UV protocols of panel (i), highlighting UV-mediated receptor silencing. (g) Reversible photo-modulation of ion channel activity. Photo-switchable UAAs, placed at key modulatory sites, can rapidly turn receptor activity on/off by toggling between a cis (purple asterisk) and a trans configuration (blue asterisk), induced by UV and blue/green range wavelengths (purple and blue thunderbolts), respectively. (h) Chemical structures of the cis and trans photoisomers of PSAA, a photo-switchable UAA. (i) Representative current trace of GluN1/GluN2A NMDA receptors carrying PSAA at GluN1-P532 under illumination by UV and blue light, demonstrating reversible photo-modulation (Reproduced from Klippenstein et al. (2014) (f) and (i). For details, see Klippenstein et al. (2018)). (For color figure see eBook figure.)

specificity/selectivity for particular ion channel isoforms. Often off-target effects of drugs are produced because of collaboration of these two issues, which limits their therapeutic applications or confounds interpretation of experimental results. There are various engineering nanotechniques that have been developed to address some of these limitations of small molecules as ion channel modulators, such as tethered toxins, engineering toxin molecules for improved channel selectivity, and genetically encoded intracellular channel inhibitors.

The use of toxins in living organisms has a drawback that they cannot be restricted to a particular type of cells and are soluble. A tethered toxin approach can help overcome this limitation (Ibañez-Tallon et al., 2004; Ibañez-Tallon and Nitabach, 2012). For example, a method was inspired by the prototoxin, lynx1, an endogenous nAChRs modulator in mammalian central nervous system (Miwa et al., 1999). The open reading frame of lynx1 contains a secretory signal sequence, a cysteine-rich region with homology tó secrete snake venom neurotoxins, and a hydrophobic C-terminus domain with a consensus site for GPI anchor addition. This fundamental design has been exploited to design tethered toxins with specificity for different ion channels. The general design principle includes fusing in frame a secretory signal, the toxin of interest, a linker sequence, and either a GPI anchor or single-pass transmembrane sequence. Additionally, variants may be included by incorporating fluorescent proteins or epitope tags that permit visual detection of the tethered toxin (Ibañez-Tallon and Nitabach, 2012). Several tethered bungarotoxins and conotoxins have been found to specifically and strongly inhibit particular nAChR isoforms in Xenopus oocytes and in zebrafish muscle in vivo (Ibañez-Tallon et al., 2004). The tethered conotoxins MrVIA and MVIIA selectively caused complete blocking of co-expressed recombinant $Na_V1.2$ and $Ca_V2.2$ currents, respectively (Ibañez-Tallon et al., 2004). Tethered conotoxin MVIIA and spider agatoxin IVA blocked $Ca_V2.2$ and $Ca_V2.1$ channels, respectively, and inhibited neurotransmission in cultured neurons and in vivo (Auer et al., 2010).

Often, the therapeutic potential of specific venom toxins is limited by their lack of target ion channel specificity. Therefore, toxins may be engineered to improve specificity and potency for the desired target. We may consider an example here to understand this clearly. Autoimmune diseases such as multiple sclerosis involve activated memory T cells exhibiting selective $K_V1.3$ channels upregulation controlling membrane potential and $Ca^{2+}$ signaling. Agents selectively blocking $K_V1.3$ channels on T lymphocytes are considered possible therapeutics for autoimmune diseases. A 35-amino-acid polypeptide from the sea anemone *Stichodactyla helianthus*, ShK, is found to potently block $K_V1.3$ and $K_V1.1$ with $IC_{50}$s in low picomolar range. Complementary mutagenesis of ShK and $K_V1.3$ combined with the mutant cycle analyses revealed two residues in ShK, lys22, and tyr23 to be essential for potassium channel block (Kalman et al., 1998). Using such structure-function information, several derivatives of ShK have been engineered with improved selectivity for $K_V1.3$ over $K_V1.1$, including ShK-Dap22, in which the unnatural amino acid diaminopropionic acid is substituted for the critical lys22, and ShK-170 featuring an L-phosphotyrosine attached to Arg1 of ShK via an aminoethyloxyethyloxy-acetyl linker (Kalman et al., 1998). Both these engineered toxins were found to block proliferation of memory T lymphocytes and suppressed hypersensitivity with limited off-target toxicity in animal model experiments.

A new engineering approach has been adopted to develop cytosolic genetically encoded ion channel blockers and modulators, demonstrated for $Ca_V1$ and $Ca_V2$ family channels. The method was inspired by the RGK (Rad, Rem, Rem2, and Gem/Kir) GTPases, a four-member family of Ras-like G-proteins that strongly inhibit $Ca_V1$ and $Ca_V2$ channels (Béguin et al., 2001). RGK protein inhibition of $Ca_V$ channels was found to show a dual requirement: direct anchoring of the G-protein to the plasma membrane, and binding of the G-protein to the cytosolic β subunit in the $Ca_V$ channel complex (Yang et al., 2012). This finding led to a hypothesis that membrane-targeted RGKs indirectly pull on the $\alpha_1$-subunit I-II loop via the associated $Ca_V\beta$ in a manner that is transmitted to the channel pore closure. Testing this hypothesis led to the discovery that diverse cytosolic proteins that bind $Ca_V\alpha_1$-subunits can be converted into inhibitors of $Ca_V$ channels with tunable selectivity, potency, and kinetics by anchoring them to the plasma membrane (Yang et al., 2013). The method

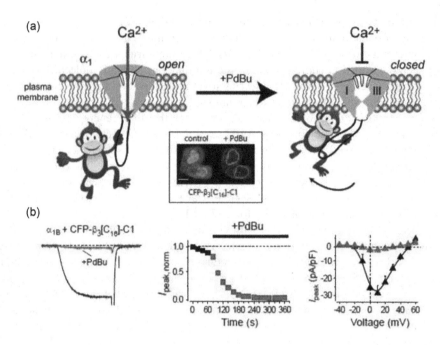

**FIGURE 7.6** ChIMP technique. (a) Cartoon showing the concept of ChIMP. A cytosolic protein associated with a cytoplasmic domain of a $Ca_V\alpha_1$ subunit is permissive for ionic conductance (*left*). Directly anchoring the cytosolic protein to the plasma membrane induces a conformational change that closes the channel (*right*). *Inset*, phorbol-12,13-dibutyrate (PdBu)-induced membrane translocation of a C-terminus truncated CFP-tagged $Ca_V\beta_3$ fused to the C1 domain of protein kinase C (CFP-$\beta_3$[C16]-C1). (b) Conversion of $Ca_V\beta_3$ into a PdBu-inducible $Ca_V2.2$ channel inhibitor using the ChIMP concept.

has been termed as the channel inactivation induced by membrane tethering an associated protein (ChIMP), see Figure 7.6 (Subramanyam and Colecraft, 2015).

## 7.4   DNA NANOTECHNOLOGY OF ION CHANNELS

DNA nanotechnology has revolutionized the capabilities to shape and control 3D structures at the nanometer scale. DNA-based designer sensors, nanopores, and ion channels have great potential for both cross-disciplinary research and technological applications. The concept of structural DNA nanotechnology, including DNA origami, and an overview of the work flow from design to assembly, characterization, and application of DNA-based functional systems have been detailed in a review article (Göpfrich and Keyser, 2019).

Recently, Göpfrich and colleagues reported ion channel formation from a single membrane-spanning DNA duplex (Göpfrich et al., 2016). They demonstrated ion conduction induced by a single DNA duplex that lacks a hollow central channel. Decorated with six porpyrin tags, this duplex is designed to span lipid membranes. Combining electrophysiology measurements with all-atom molecular dynamics simulations (see Figure 7.7), the microscopic conductance pathway was elucidated. Ions flow at the DNA-lipid interface as the lipid head groups tilt toward the amphiphilic duplex forming a toroidal pore filled with water and ions. Ionic current traces produced by the DNA-lipid channel show well-defined insertion steps, closures, and gating similar to those observed for traditional protein channels or synthetic pores.

The chemical functionalization of DNA has opened up pathways to transform static DNA structures into dynamic nanomechanical sensors. Exciting possibility is to create membrane-inserted DNA nanochannels that mimic their protein-based natural counterparts in form and function,

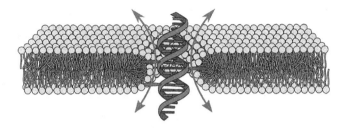

**FIGURE 7.7**   Conductance mechanism of DNA double helix.

detailed in an earlier chapter on artificial ion channels. Recently, a study reported that a water-in-oil microdroplet stabilized with amphiphilic DNA origami nanoplates (Ishikawa et al., 2019). DNA nanotechnology helped DNA nanoplates to be designed as a nanopore device for ion transportation and to stabilize the oil–water interface. Ion current measurements revealed the nanoplate pores functioning as a channel to transport ions. These nanoplates with pores might be tested after inserting into plasma membranes whether they would be stabilized there. If so, their interference with the natural channels could also be addressed. Future studies may conclude these. The creation of these DNA nanoplate ion pores may provide a general strategy for the programmable design of microcapsules to engineer artificial cells and molecular robots.

## 7.5   OPTICAL STIMULATION OF ION CHANNELS: A NEW TYPE OF THERMO-NANOTECHNOLOGY

An approach based on radiofrequency (RF) magnetic field heating of specific nanoparticles to remotely activate temperature-sensitive cation channels in cells was revealed about a decade ago (Huang et al., 2010). Superparamagnetic ferrite nanoparticles were targeted to specific proteins on the plasma membrane with expressed ion channel transient receptor potential vanilloid subtype 1 (TRPV1), and heated by an RF magnetic field. Molecular thermometers and fluorophores showed that the induced temperature increase was highly localized, within the not-so-larger nanometer-scale channel vicinity, making this magnetic heat induction as some sort of nanotechnology. The thermal activation of channels has been found to trigger important physiological states, action potentials, etc. in cultured neurons. Thus, it hints that this rather basic physics technique may eventually emerge as a remotely working manipulator of specific ion channel machineries and other general cellular machineries.

Manganese ferrite ($MnFe_2O_4$) nanoparticles (~6 nm dimension) were targeted to cells expressing the temperature-sensitive ion channel TRPV1, and heated with the use of an RF magnetic field. The local thermal stimulation due to the temperature increase opened the TRPV1 channels producing an influx of calcium ions, see Figure 7.8 (Huang et al., 2010). The TRPV1 protein activation temperature is 42°C (Caterina et al., 1997), close to the normal body temperature to permit quick stimulation while allowing the channels to be normally closed. TRPV1 has also been heterologous expressed in Drosophila neurons and stimulated with capsaicin to successfully evoke behavioral responses. This approach activates cells uniformly across a large volume, making it feasible for *in vivo* whole-body applications.

A high local nanoparticle density would be required to cause significant regional heating and to effectively heat TRPV1 channels in the membrane. This was achieved *in vitro* by targeting the streptavidin-conjugated nanoparticles to cells of interest, which have been genetically made to express the engineered membrane protein marker AP-CFP-TM (Huang et al., 2010). This protein marker contains a transmembrane domain (TM) of the platelet-derived growth factor, an extracellular fluorescent protein, and a biotin acceptor peptide (AP) that is enzymatically biotinylated to bind the streptavidin-conjugated nanoparticle (Howarth et al., 2005).

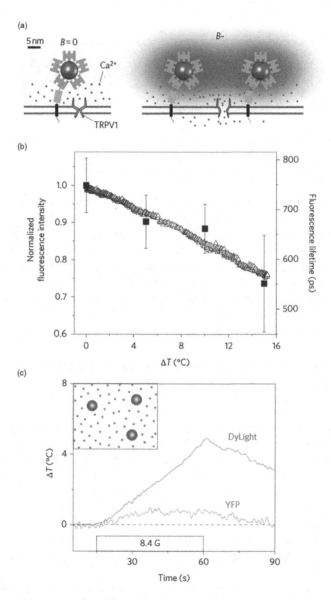

**FIGURE 7.8** Principles of ion channel stimulation using nanoparticle heating and local temperature sensing. (A) Schematic, drawn to scale, showing local heating of streptavidin–DyLight549 (orange)-coated superparamagnetic nanoparticles (gray) in a RF magnetic field (B~) and heat (red)-induced opening of TRPV1. The AP-CFP-TM protein binds the nanoparticles through the biotinylated AP domain (green box), which is anchored to the membrane by the TM (blue box) and CFP (cyan box) domains. (b) Temperature dependence of the fluorescence intensity and lifetime of streptavidin–DyLight549 ($\Delta F/F = -15\%°C^{-1}$, measured in an externally heated nanoparticle dispersion). (c) Applying an RF magnetic field to a nanoparticle dispersion increased the nanoparticle surface temperature (red trace, change in temperature measured by DyLight549 fluorescence) while only moderately changing the solution temperature (green trace, change intemperature measured by YFP fluorescence). Inset shows a schematic of the nanoparticle dispersion (green dots represent YFP; red rings indicate the streptavidin–DyLight549 coating around the nanoparticles (gray)). (For color figure see eBook figure.)

In another study, the whole-cell (mid-infrared (mid-IR) irradiated isolated retinal ganglion cells (RGCs) and vestibular ganglion cells (VGCs) from rodents) patch-clamp recordings revealed that both voltage-gated calcium and sodium channels contribute to the laser-evoked neuronal voltage variations (LEVV) (Albert et al., 2012). Albert and colleagues demonstrated that the mid-IR laser stimulations induced triggering of transient membrane potential variations in sensory neurons, as well as addressed how the transient membrane potential variations rely on voltage-gated sodium and calcium channels (Albert et al., 2012).

Selective blockade of the LEVV by ruthenium red and RN 1734 (at micromolar concentrations) identifies thermosensitive TRP vanilloid channels as the primary effectors of the chain reaction triggered by mid-IR laser irradiation.

This study inspected the association of TRPV channels with LEVV. It found that TRPV4 channels (investigated on TRPV1, 2, 3, 4 channels) mediated the IR laser-evoked response in sensory neurons. It is mentionable that the sequence of activation of the TRPV channels according to their heat threshold reported in expression systems, for example, HEK 293 cells and Xenopus oocytes is as follows: $TRPV4 > 27°C$, $TRPV3 > 32°C$, $TRPV1 > 43°C$, and $TRPV2 > 53°C$ (Caterina, 2007; Noël et al., 2009). In preparation of this study, the mean temperature reached during the laser irradiation ($37°C \pm 2°C$; $n = 10$) was above the TRPV4 activation threshold, therefore, supporting thermal activation of the TRPV4 channels. RN 1734 ($3.2\,\mu M$ in VGC and $5.2\,\mu M$ in RGCs), a TRPV4 channel blocker (Angelico and Testa, 2010), was found to prevent the induction of LEVV in both RGCs and VGCs. Altogether, these results indicated that the TRPV4 channels mediate the LEVV in the two types of sensory neurons (Albert et al., 2012).

The findings that TRPV4 channels are essential to LEVV's induction and that specific voltage-gated calcium and sodium channel blockers abolish the spike-like component of the response allow proposing a cellular cascade for the LEVV on activation, TRPV4 channels allow calcium influx to the cell cytosol. This cation entry appears to drive a membrane depolarization suitable to activate low-threshold voltage-gated calcium channels, leading to further depolarization and subsequent activation of voltage-gated sodium channels. This sequence of events allows the generation of sodium-based action potentials in the cells forming the optic and vestibular nerves following IR laser stimulation. Given that expression of ion channels in VGCs is heterogeneous, the small prespike hyperpolarization in a fraction of recorded VGCs likely reflects a different ion channel composition in this subpopulation. Because of minimal species differences in TRPV4 amino acid sequences and the similar TRPV4 protein expression in rodent and adult human RGCs, it is likely that the TRPV4-based process of laser-evoked cell response in these cells could apply to human tissue. This ion channel nanomachine involved thermo-nanotechnology may eventually help in creating future invasive therapeutics without needing any chemical drug intervention but laser therapy. However, several control experiments need to be done before achieving any such alternative form of therapy.

## 7.6 MAGNETOMECHANICAL AND MAGNETOTHERMAL REGULATION IN THE ION CHANNEL NANOTECHNOLOGY

Magnetogenetics, where ion channels are genetically engineered, is closely coupled to the iron-storage protein ferritin (Stanley et al., 2012; Wheeler et al., 2016). The plausibility of mechanisms of mechanical and thermal activation of ion channels has been challenged on physical grounds (Meister, 2016), and the current state of reported experimental observations in magnetogenetics and basic magnetic physics arguments that challenge those observations remain in conflict.

A magnetic field gradient is claimed to pull on ferritin, see Figure 7.9a (Meister, 2016). Paramagnetic particles experience a force proportional to the magnetic field gradient and the induced magnetic moment. In the experiments, the field strength was ~0.05 T and the field gradient was ~6.6 T/m (Wheeler et al., 2016). The question is, what is the resulting force on a ferritin particle? Meister

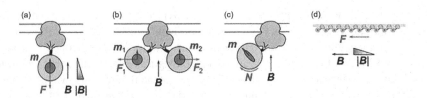

**FIGURE 7.9** A TRPV4 channel (pink) inserted in the membrane with a ferritin complex (green) attached on the cytoplasmic side, approximately to scale. The magnetic field B induces a moment m in the ferritin core, leading to a force F or a torque N on the ferritin particle, and resulting forces tugging on the channel. (For color figure see eBook figure.)

performed a simple physics calculations and found that $7 \times 10^{-23}$ N would be the force exerted by one ferritin complex on its linkage under the reported experimental conditions. This amount of force is about 10 orders of magnitude less than the force ($2 \times 10^{-13}$ N) needed to open an ion channel, measured directly for the force-sensitive channels in auditory hair cells (Howard and Hudspeth, 1988). This discrepancy is unacceptably large which requires a thorough reinvestigation.

Meister then measured the force ($3 \times 10^{-21}$ N) two ferritins use to pull on each other (Figure 7.9b). This force is also too tiny compared to the required range of force to induce ion channel opening. Meister also measured the magnetic field exerting a torque on the ferritin (Figure 7.9c). The induced magnetic moment may point at an angle relative to the field, resulting in a torque on the ferritin particle that could tug on the linkage with the channel protein. The magnitude of such effects is dwarfed by thermal fluctuations: The interaction energy between the moment and a magnetic field pointing along the easy axis is found to be $-3 \times 10^{-25}$ J and zero with the field orthogonal. This free energy difference is about four log units smaller than the thermal energy. The magnetic field can bias the alignment of the ferritins by only an amount of $10^{-4}$. Any torque exerted by the ferritin on its ion channel linkage will be 10,000 times smaller than the thermal fluctuations in that same degree of freedom.

Meister also deducted the stress many ferritins exert that gates mechanoreceptors in the membrane (Figure 7.9d). The membrane stress required to gate mechanoreceptors has been measured directly by producing a laminar water flow over the surface of a cell: For TRPV4 channels, it is ~20 dyne/cm²; and for Piezo1 channels, ~50 dyne/cm². If the membrane is decorated with ferritins attached by some linkage, and instead of viscous flow tugging on the surface one applies a magnetic field gradient to pull on those ferritins with a force. The density of ferritins one would need to generate the required membrane stress is $3 \times 10^{10}$ ferritins/µm². Unfortunately, considering the size effects, even if the membrane is close-packed with ferritin spheres, one could fit at most $10^4$ ferritins/µm². So, this hypothetical mechanism produces membrane stress at least six log units too weak to open any ion channel.

Magnetogenetics stands at the forefront of the tools for perturbation of neural activity, and this technique is applicable in regulating ion channel nanotechnology (Barbic, 2019). Here, the hypothesis is to regulate physical mechanisms of ion channel activation by external magnetic fields. Barbic has considered a wide range of possible spin configurations of iron in ferritin and analyzed the consequences these might have in magnetogenetics involving magnetomechanical and magnetothermal mechanisms of ion channel activation.

Any spin system consists of energy having terms related to the interaction of the individual spins with the external magnetic field, in addition to the interaction energy of the intraspin system. The later contribution can be quite large in ferromagnetism. This total energy must be compared with the thermal energy, and for interacting systems active in a thermodynamic bath. The thermal energy may be too small to dephase the spins. Thus, the experimental possibility to exploit the interaction of magnetic nanoparticles *in vivo* with external magnetic fields (~1 T) is reasonable. Barbic has presented the case that the physical viability of magnetogenetics depends critically on many physical parameters of the basic control construct-iron-loaded ferritin protein coupled to the

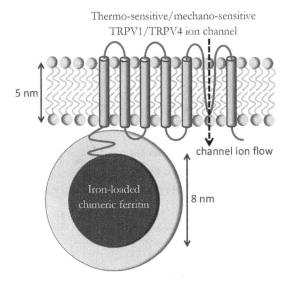

**FIGURE 7.10** Genetically engineered construct in magnetogenetics. Thermosensitive or mechanosensitive ion channel is closely coupled to the iron-loading protein ferritin. External DC or AC magnetic fields are applied to influence the ion channel function (Barbic, 2019).

thermosensitive or mechanosensitive ion channel in the cell membrane (Barbic, 2019). The critical parameters include the magnetism and magnetic spin configurations of iron atoms in the ferritin protein, as well as the realistic thermal, mechanical, and diamagnetic properties of ion channels and neural cell membranes coupled to the iron-loaded ferritin.

Figure 7.10 (Barbic, 2019) presents the fundamental genetically engineered construct in magnetogenetics, reported in other studies (Stanley et al., 2012; Wheeler et al., 2016). Thermosensitive or mechanosensitive TRP family channels (TRPV1 and TRPV4) were genetically engineered to be closely coupled to the novel chimeric iron-loading protein ferritin. External DC or AC magnetic fields were applied to this construct in *in vivo* experiments, and it was preliminarily concluded that thermal or mechanical magnetic effects were likely responsible for the observed biological responses, as was intended by the genetic engineering methods.

The commonly stated maximum possible number of iron atoms in ferritin are considered to be $N = 4,500$, and we may assume atomic moment of $m_{Fe} = 5 \mu_B$/iron atom (Coey, 2010). The actual number of iron atoms in ferritin in all the magnetogenetics reports is not presently known. Therefore, ferritin is undoubtedly an important candidate in creating substantial magnetic field in biological systems. Moreover, three distinct spin coupling configurations of iron atoms in ferritin are found.

They are as follows:

- paramagnetic state where N irons spins are magnetically independent from one another and noninteracting,
- clusterparamagnetic state where N iron spins are separated into n independent clusters of N/n exchange coupled spins, and
- superparamagnetic state where all N spins are strongly coupled by the magnetic exchange interaction and behave as a single macrospin.

While considering the force due to the magnetic fields and field gradients from the ferritin particle itself on the intrinsically weak diamagnetic mechanosensitive ion channel and neural cell membrane, it is suggestive that this diamagnetic repulsive force might be sufficient to mechanically deform the ion channel and affect its function, see Figure 7.11 (Barbic, 2019). This model clearly

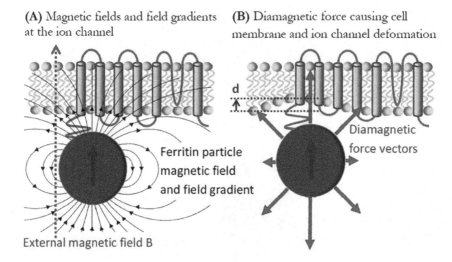

**(A)** Magnetic fields and field gradients at the ion channel

**(B)** Diamagnetic force causing cell membrane and ion channel deformation

Ferritin particle magnetic field and field gradient

Diamagnetic force vectors

External magnetic field B

**FIGURE 7.11**  Diamagnetic force deformation of ion channel and cell membrane. (a) Diamagnetic ion channel in the cell membrane experiences magnetic fields from the ferritin particle and the externally applied magnetic field B, as well as the large magnetic field gradient from the ferritin particle. This results in the repulsive diamagnetic force on the ion channel and the cell membrane in (b), which is sufficient to potentially mechanically deform them and affect the ion channel function (Barbic, 2019).

predicts a considerable technological alteration in the nanoscale structures of the ion channel and its hosting membrane. To prove the principle, appropriate confocal imaging of the system under the prescribed condition may be performed.

Another potential magnetomechanical mechanism, the Einstein-de Haas effect (Einstein, 1915), appears feasible in iron-loaded ferritin. It is a fundamental tenet of quantum mechanics that magnetic moment of a particle is proportional to mechanical angular momentum, and the proportionality parameter is the gyromagnetic ratio for spin angular momentum of iron. The Einstein-de Haas effect refers to the magnetomechanical effect, required by the conservation of angular momentum, where a reversal of a magnetic moment of a sample by an applied magnetic field has to be accompanied by a corresponding change in mechanical angular momentum of that sample, see Figure 7.12 (Barbic, 2019).

Ferritin particle with fluctuating magnetic moment may have potential effects on ion channels. For the potentially superparamagnetic spin arrangement of the ferritin particle, the magnetic field near the particle surface has a relatively large value of 0.65 T. This magnetic field from the particle fluctuates rapidly in time (Brown, 1963), see Figure 7.13 (Barbic, 2019). The frequency of this fluctuation at physiological temperature is significant, measured to be in the GHz frequency range through low-frequency magnetic susceptibility, Mossbauer spectroscopy, and neutron spin-echo spectroscopy. The ion channel in the vicinity of the potentially superparamagnetic iron-loaded ferritin experiences large magnetic field gradients and Tesla-scale magnetic fields at GHz frequencies, as well as the corresponding GHz frequency diamagnetic forces and torques. Upon application of the external DC magnetic field (~1 T), the magnetic moment saturates and the ion channel experiences DC magnetic field, field gradient, diamagnetic force, and torque, see Figure 7.13 (Barbic, 2019).

The above-explained various states of ion channels under the influence of magnetic fields may help us draw a few conclusions or at least make the following remarks:

- Ion channels under the influence of magnetic fields may experience considerable structural alterations. That may include considerable changes in its inherent nanotechnological moieties. Appropriate imaging experiments may be the start of inspecting the changes.

**(A)** Initial Magnetic Moment State
Positive magnetic field +B

**(B)** Final Magnetic Moment State
Negative magnetic field -B

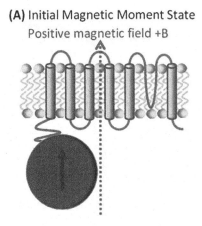
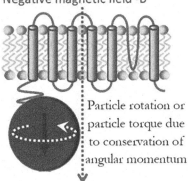

Particle rotation or
particle torque due
to conservation of
angular momentum

Initial angular momentum L=+m/γ

Final angular momentum L=-m/γ

**FIGURE 7.12** Einstein-de Haas effect in ferritin. (a) Ferritin magnetic moment +m is aligned with the field B and caries a mechanical angular momentum of L=+m/γ. (b) Magnetic moment reversal to −m results in the total change in angular momentum of ΔΔL=2m/γ that is compensated for by the mechanical rotation of the particle or by the mechanical torque on the particle.

**(A)** Fluctuating Magnetic Moment State

**(B)** Fixed Magnetic Moment State

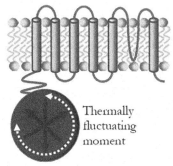
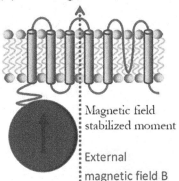

Thermally
fluctuating
moment

Magnetic field
stabilized moment

External
magnetic field B

Fluctuating 1-10 (GHz) Tesla-scale field
Fluctuating 1-10 (GHz) diamagnetic force
Fluctuating 1-10 (GHz) torque

DC Tesla-scale field
DC diamagnetic force
DC torque

**FIGURE 7.13** Magnetic moment fluctuations. (a) In zero external magnetic field, the ion channel experiences Tesla-scale magnetic fields and large field gradients from the fluctuating superparamagnetic particle moment at GHz-scale frequencies, as well as the corresponding AC diamagnetic forces and torques. (b) In the external field B, the ion channel experiences Tesla-scale DC magnetic fields and large field gradients from the stabilized ferritin magnetic moment, and the corresponding DC diamagnetic forces and torques.

• Ion channels experiencing both electric (due to membrane potential) and magnetic fields (using the techniques addressed here) together may show concomitant electromagnetic properties which are quite different from those observed at normal physiological condition where channels are active under the influence of membrane potential only. EP experiments may help address these concerns partially.

## REFERENCES

Albert, E. S., Bec, J. M., Desmadryl, G., Chekroud, K., Travo, C., Gaboyard, S., . . . Chabbert, C. (2012). TRPV4 channels mediate the infrared laser-evoked response in sensory neurons. *Journal of Neurophysiology, 107*(12), 3227–3234. doi:10.1152/jn.00424.2011

Angelico, P., & Testa, R. (2010). TRPV4 as a target for bladder overactivity. *F1000 Biology Reports.* doi:10.3410/b2-12

Ashrafuzzaman, M. (2018). *Nanoscale Biophysics of the Cell.* S.l.: SPRINGER. doi:10.1007/978-3-319-77465-7

Ashrafuzzaman, M., Lampson, M. A., Greathouse, D. V., Koeppe, R. E., & Andersen, O. S. (2006). Manipulating lipid bilayer material properties using biologically active amphipathic molecules. *Journal of Physics: Condensed Matter, 18*(28). doi:10.1088/0953-8984/18/28/s08

Ashrafuzzaman, M., & Tuszynski, J. (2012). Regulation of channel function due to coupling with a lipid bilayer. *Journal of Computational and Theoretical Nanoscience, 9*(4), 564–570. doi:10.1166/jctn.2012.2062

Atkinson, L., Milligan, C. J., Buckley, N. J., & Deuchars, J. (2002). An ATP-gated ion channel at the cell nucleus. *Nature, 420*(6911), 42–42. doi:10.1038/420042a

Auer, S., Stürzebecher, A. S., Jüttner, R., Santos-Torres, J., Hanack, C., Frahm, S., . . . Ibañez-Tallon, I. (2010). Silencing neurotransmission with membrane-tethered toxins. *Nature Methods, 7*(3), 229–236. doi:10.1038/nmeth.1425

Banghart, M., Borges, K., Isacoff, E., Trauner, D., & Kramer, R. H. (2004). Light-activated ion channels for remote control of neuronal firing. *Nature Neuroscience, 7*(12), 1381–1386. doi:10.1038/nn1356

Barbic, M. (2019). Possible magneto-mechanical and magneto-thermal mechanisms of ion channel activation in magnetogenetics. *ELife, 8.* doi:10.7554/elife.45807

Baumgaertner, A. (2008). Concepts in bionanomachines: Translocators. *Journal of Computational and Theoretical Nanoscience, 5*(9), 1852–1890. doi:10.1166/jctn.2008.903

Béguin, P., Nagashima, K., Gonoi, T., Shibasaki, T., Takahashi, K., Kashima, Y., . . . Seino, S. (2001). Regulation of $Ca_2$ channel expression at the cell surface by the small G-protein kir/Gem. *Nature, 411* (6838), 701–706. doi:10.1038/35079621

Berlin, S., & Isacoff, E. Y. (2017). Synapses in the spotlight with synthetic optogenetics. *EMBO Reports, 18* (5), 677–692. doi:10.15252/embr.201744010

Brown, W. F. (1963). Thermal fluctuations of a single-domain particle. *Physical Review, 130*(5), 1677–1686. doi:10.1103/physrev.130.1677

Burke, K. J., & Bender, K. J. (2019). Modulation of ion channels in the axon: Mechanisms and function. *Frontiers in Cellular Neuroscience, 13.* doi:10.3389/fncel.2019.00221

Caterina, M. J. (2007). Transient receptor potential ion channels as participants in thermosensation and thermoregulation. *American Journal of Physiology-Regulatory, Integrative and Comparative Physiology, 292*(1). doi:10.1152/ajpregu.00446.2006

Caterina, M. J., Schumacher, M. A., Tominaga, M., Rosen, T. A., Levine, J. D., & Julius, D. (1997). The capsaicin receptor: A heat-activated ion channel in the pain pathway. *Nature, 389*(6653), 816–824. doi:10.1038/39807

Catterall, W. A. (2017). Forty years of sodium channels: Structure, function, pharmacology, and epilepsy. *Neurochemical Research, 42*(9), 2495–2504. doi:10.1007/s11064-017-2314-9

Coey, J. M. (2010). *Magnetism and Magnetic Materials.* Cambridge University Press. doi:10.1017/cbo9780511845000

Dougherty, D. A. (2008). Cys-loop neuroreceptors: Structure to the rescue? *Chemical Reviews, 108*(5), 1642–1653. doi:10.1021/cr078207z

Einstein, A. (1915). Experimenteller Nachweis der Ampèreschen Molekularstrome. *Die Naturwissenschaften, 3*(19), 237–238. doi:10.1007/bf01546392

Fields, G. B. (2001). Introduction to peptide synthesis. *Current Protocols in Protein Science, 26*(1). doi:10.1002/0471140864.ps1801s26

Grosse, W., Essen, L., & Koert, U. (2011). Strategies and perspectives in ion-channel engineering. *ChemBioChem, 12*(6), 830–839. doi:10.1002/cbic.201000793

Göpfrich, K., & Keyser, U. F. (2019). DNA nanotechnology for building sensors, nanopores and ion-channels. *Advances in Experimental Medicine and Biology Biological and Bio-inspired Nanomaterials,* 331–370. doi:10.1007/978-981-13-9791-2_11

Göpfrich, K., Li, C. Y., Mames, I., Bhamidimarri, S. P., Ricci, M., Yoo, J., . . . Keyser, U. F. (2016). Ion channels made from a single membrane-spanning DNA duplex. *Nano Letters, 16*(7), 4665–4669. doi:10.1021/acs.nanolett.6b02039

Howard, J., & Hudspeth, A. (1988). Compliance of the hair bundle associated with gating of mechanoelectrical transduction channels in the Bullfrog's saccular hair cell. *Neuron, 1*(3), 189–199. doi:10.1016/0896-6273(88)90139-0

Howarth, M., Takao, K., Hayashi, Y., & Ting, A. Y. (2005). Targeting quantum dots to surface proteins in living cells with biotin ligase. *Proceedings of the National Academy of Sciences, 102*(21), 7583–7588. doi:10.1073/pnas.0503125102

Huang, H., Delikanli, S., Zeng, H., Ferkey, D. M., & Pralle, A. (2010). Remote control of ion channels and neurons through magnetic-field heating of nanoparticles. *Nature Nanotechnology, 5*(8), 602–606. doi:10.1038/nnano.2010.125

Ibañez-Tallon, I., & Nitabach, M. N. (2012). Tethering toxins and peptide ligands for modulation of neuronal function. *Current Opinion in Neurobiology, 22*(1), 72–78. doi:10.1016/j.conb.2011.11.003

Ibañez-Tallon, I., Wen, H., Miwa, J. M., Xing, J., Tekinay, A. B., Ono, F., . . . Heintz, N. (2004). Tethering naturally occurring peptide toxins for cell-autonomous modulation of ion channels and receptors in vivo. *Neuron, 43*(3), 305–311. doi:10.1016/j.neuron.2004.07.015

Ishikawa, D., Suzuki, Y., Kurokawa, C., Ohara, M., Tsuchiya, M., Morita, M., . . . Takinoue, M. (2019). DNA origami nanoplate-based emulsion with nanopore function. *Angewandte Chemie International Edition, 58*(43), 15299–15303. doi:10.1002/anie.201908392

Kalman, K., Pennington, M. W., Lanigan, M. D., Nguyen, A., Rauer, H., Mahnir, V., . . . Chandy, K. G. (1998). ShK-Dap22, a potent Kv1.3-specific immunosuppressive polypeptide. *Journal of Biological Chemistry, 273*(49), 32697–32707. doi:10.1074/jbc.273.49.32697

Klippenstein, V., Ghisi, V., Wietstruk, M., & Plested, A. J. (2014). Photoinactivation of glutamate receptors by genetically encoded unnatural amino acids. *The Journal of Neuroscience, 34*(3), 980–991. doi:10.1523/jneurosci.3725-13.2014

Klippenstein, V., Mony, L., & Paoletti, P. (2018). Probing ion channel structure and function using light-sensitive amino acids. *Trends in Biochemical Sciences, 43*(6), 436–451. doi:10.1016/j.tibs.2018.02.012

Kocer, A., Walko, M., Meijberg, W., & Feringa, B. (2005). A light-actuated nanovalve derived from a channel protein. *Science, 309*(5735), 755–758. doi:10.1126/science.1114760

Korkmaz-Özkan, F., Köster, S., Kühlbrandt, W., Mäntele, W., & Yildiz, Ö. (2010). Correlation between the OmpG secondary structure and its pH-dependent alterations monitored by FTIR. *Journal of Molecular Biology, 401*(1), 56–67. doi:10.1016/j.jmb.2010.06.015

Kramer, R. H., Mourot, A., & Adesnik, H. (2013). Optogenetic pharmacology for control of native neuronal signaling proteins. *Nature Neuroscience, 16*(7), 816–823. doi:10.1038/nn.3424

Lemoine, D., Jiang, R., Taly, A., Chataigneau, T., Specht, A., & Grutter, T. (2012). Ligand-gated ion channels: New insights into neurological disorders and ligand recognition. *Chemical Reviews, 112*(12), 6285–6318. doi:10.1021/cr3000829

Liu, C. C., & Schultz, P. G. (2010). Adding new chemistries to the genetic code. *Annual Review of Biochemistry, 79*(1), 413–444. doi:10.1146/annurev.biochem.052308.105824

Lutz, S., & Bornscheuer, U. T. (2009). *Protein Engineering Handbook*. Weinhein: Wiley-VCH.

Mazzanti, M., Innocenti, B., & Rigatelli, M. (1994). ATP-dependent ionic permeability on nuclear envelope in in situ nuclei of Xenopus oocytes. *The FASEB Journal, 8*(2), 231–237. doi:10.1096/fasebj.8.2.7509760

Meister, M. (2016). Physical limits to magnetogenetics. *ELife, 5*. doi:10.7554/elife.17210

Miwa, J. M., Ibañez-Tallon, I., Crabtree, G. W., Sánchez, R., Šali, A., Role, L. W., & Heintz, N. (1999). Lynx1, an endogenous toxin-like modulator of nicotinic acetylcholine receptors in the mammalian CNS. *Neuron, 23*(1), 105–114. doi:10.1016/s0896-6273(00)80757-6

Noël, J., Zimmermann, K., Busserolles, J., Deval, E., Alloui, A., Diochot, S., Guy, N., Borsotto, M., Reeh, P., Eschalier, A., & Lazdunski, M. (2009). The mechano-activated K+ channels TRAAK and TREK-1 control both warm and cold perception. *The EMBO Journal, 28*(9), 1308–1318. doi:10.1038/emboj.2009.57

Paoletti, P., Bellone, C., & Zhou, Q. (2013). NMDA receptor subunit diversity: Impact on receptor properties, synaptic plasticity and disease. *Nature Reviews Neuroscience, 14*(6), 383–400. doi:10.1038/nrn3504

Pastoriza-Gallego, M., Oukhaled, G., Mathé, J., Thiebot, B., Betton, J., Auvray, L., & Pelta, J. (2007). Urea denaturation of α-hemolysin pore inserted in planar lipid bilayer detected by single nanopore recording: Loss of structural asymmetry. *FEBS Letters, 581*(18), 3371–3376. doi:10.1016/j.febslet.2007.06.036

Sakai, N., & Matile, S. (2013). Synthetic ion channels. *Langmuir, 29*(29), 9031–9040. doi:10.1021/la400716c

Stanley, S. A., Gagner, J. E., Damanpour, S., Yoshida, M., Dordick, J. S., & Friedman, J. M. (2012). Radio-wave heating of iron oxide nanoparticles can regulate plasma glucose in mice. *Science, 336*(6081), 604–608. doi:10.1126/science.1216753

Subramanyam, P., & Colecraft, H. M. (2015). Ion channel engineering: Perspectives and strategies. *Journal of Molecular Biology, 427*(1), 190–204. doi:10.1016/j.jmb.2014.09.001

Valiyaveetil, F. I., Sekedat, M., Mackinnon, R., & Muir, T. W. (2004). Glycine as a D-amino acid surrogate in the K-selectivity filter. *Proceedings of the National Academy of Sciences, 101*(49), 17045–17049. doi:10.1073/pnas.0407820101

Valiyaveetil, F. I., Sekedat, M., Mackinnon, R., & Muir, T. W. (2006). Structural and functional consequences of an amide-to-ester substitution in the selectivity filter of a potassium channel. *Journal of the American Chemical Society, 128*(35), 11591–11599. doi:10.1021/ja0631955

Volgraf, M., Gorostiza, P., Numano, R., Kramer, R. H., Isacoff, E. Y., & Trauner, D. (2005). Allosteric control of an ionotropic glutamate receptor with an optical switch. *Nature Chemical Biology, 2*(1), 47–52. doi:10.1038/nchembio756

Wheeler, M. A., Smith, C. J., Ottolini, M., Barker, B. S., Purohit, A. M., Grippo, R. M., . . . Güler, A. D. (2016). Genetically targeted magnetic control of the nervous system. *Nature Neuroscience, 19*(5), 756–761. doi:10.1038/nn.4265

Yang, T., He, L., Chen, M., Fang, K., & Colecraft, H. M. (2013). Bio-inspired voltage-dependent calcium channel blockers. *Nature Communications, 4*(1). doi:10.1038/ncomms3540

Yang, T., Puckerin, A., & Colecraft, H. M. (2012). Distinct RGK GTPases differentially use α1- and auxiliary β-binding-dependent mechanisms to inhibit CaV1.2/CaV2.2 channels. *PLoS ONE, 7*(5). doi:10.1371/journal.pone.0037079

# 8 Channelopathies and Ion Channels as Therapeutic Targets

Ion channels and pumps are key transporters of materials and information across cellular boundaries, such as plasma, mitochondrial, and nuclear membranes. These transporters follow crucial biophysical, biochemical, and physiological processes linked to proper construction and functioning of various building blocks of biological cells. They not only regulate membrane potentials, ion homeostasis, and electric signaling in excitable cells but are also engaged in playing important roles in cell migration, proliferation, apoptosis, and differentiation. Disorders in their structures often lead to various temporary and permanent changes in their physicochemical state(s). Thus, they contribute occasionally in the rise of many cell-based diseases. The heterogeneous ion channel disorders resulting from several channel dysfunctions are branded as channelopathies. The malfunction of ion channel subunits, including various proteins and lipids involved directly in channel formation and/or indirectly in channel stabilization, leads to the rise in various channel-specific diseases, which are known as channelopathies. As channelopathies (or diseases thereof) originate at ion channels with mutations thereof, drug designing may be required to consider targeting the related ion channel(s) and/or its subunits to ensure discovery of drugs that are specific for binding with malfunctioning target structures. This way, achieving maximum therapeutic index of the drug can be ensured.

## 8.1 CHANNELOPATHIES

Channelopathies are the heterogeneous ion channel disorders resulting from channel dysfunctions. The abnormality of ion channels may occur due to various impaired physiological conditions of the channels' surrounding, alterations in channels' structural moieties, and modulation in the composition of the membranes hosting the channels. The disorders in ion channels may be both temporary and permanent. Temporary ion channel disorders refer to those that are occasionally repairable using modest external means (e.g., biophysical intervention) or local strength of channels' socio-chemical colocation with surrounding components. Mutation in ion channel genes may lead to vital channel function loss or gain causing often permanent channelopathy, while temporary channelopathy may get mediated by antibodies and toxins. Permanent ion channel disorders or permanent channelopathy may require therapeutic treatment to get repaired.

Channelopathies may be both acquired and inherited. Toxins accumulated or attached to the ion channel site and autoimmune phenomena are both recognized as premier causes of channelopathies. Genetic defects in both ligand and voltage-gated ion channels have long been found to cause some inherited neurological disorders (Ackerman and Clapham, 1997), which may lead to neurological channelopathies. A short list of both of these types of (the available genetic voltage and ligand gated) channelopathies were presented more than two decades ago (Hanna et al., 1998). Numerous developments have since been made on neurological channelopathies. To have a better understanding one may refer to Spillane et al. (2015). Considerable information and data with references are available in publications covering both channelopathies of various disease types and the disordered ion channel-targeted therapeutics.

DOI: 10.1201/9781003010654-8

Channelopathies are linked to diseases of various physiological systems as follows:

- Nervous system (e.g., epilepsy with febrile seizures, episodic ataxia, familial hemiplegic migraine, and hyperkalemic and hypokalemic periodic paralysis).
- Cardiovascular system (e.g., long QT syndrome, short QT syndrome, Brugada syndrome, and catecholaminergic polymorphic ventricular tachycardia)
- Respiratory system (e.g., cystic fibrosis).
- Endocrine system (e.g., neonatal diabetes mellitus, familial hyperinsulinemic hypoglycemia, thyrotoxic hypokalemic periodic paralysis, and familial hyperaldosteronism).
- Urinary system (e.g., Bartter syndrome, nephrogenic diabetes insipidus, autosomal-dominant polycystic kidney disease, and hypomagnesemia with secondary hypocalcemia).
- Immune system (e.g., myasthenia gravis, neuromyelitis optica, Isaac syndrome, and anti-NMDA [N-methyl-D-aspartate] receptor encephalitis).

Alongside channelopathies, progress has also been made in addressing the related pathophysiological mechanisms including finding novel molecules capable of targeting these ion channels, thus helping drug discovery scientists and clinicians to better understand, diagnose, and develop treatments for these diseases. We shall attempt to address a few example cases.

### 8.1.1 ONCOCHANNELOPATHY

Malfunctioning and/or overexpression of ion channels has been observed in several healthy and tumor cells (Lan et al., 2005; Han et al., 2007). Cancer is clearly linked to multiple cancer-specific channelopathies, therefore, ion channel aberrations in cancer cells are at the forefront of ongoing cancer research. Oncochannelopathy has recently been defined considering mainly the mechanisms in various stages involved, see Figure 8.1 (Prevarskaya et al., 2018). Multiple flowcharts in this figure are self-explanatory to define the oncochannelopathy as well as explain its rise in a few example cases. Table 8.1 provides a list of major oncochanelopathies that are currently known (Prevarskaya et al., 2018).

Oncochannelopathies are presented following the definition (Prevarskaya et al., 2018), and are grouped by the type of channels. Channels are presented in an alphabetic order. Only channels whose involvement has been demonstrated in primary human cancer tissue, in clinicopathological studies, or in the in vivo animal cancer xenograft models are presented. Question marks indicate lack or insufficient information.

### 8.1.2 IONIC CHANNEL DISORDERS CONCOMITANT TO AGING

Defects and/or mutations of voltage- and ligand-gated ion channels are physiologically associated with neurological disorders, for which the natural aging process (including the possibility of age-related diseases) progresses faster. Dysfunction of voltage-gated sodium, potassium, and calcium channel-subtype gene families have been linked to dyskinesia, seizure, epilepsy, and ataxia pathogenesis (Simms and Zamponi, 2014). Neurological disorders are caused due to permanent mutation or temporarily altered function in ion channels. These aberrant ion channels cause crucial negative effects by favoring the neurodegenerative disorders linked to the rise of several diseases, such as Alzheimer's disease, Parkinson's disease, Huntington's disease, and sclerosis (Kumar et al., 2016). Downregulation of the $Ca_V3.1$ T-type calcium channel in N2a cells was demonstrated using the $3 \times Tg$-AD mouse model for Alzheimer's disease (Rice et al., 2014).

Neurogenic inflammation and pain signaling transient receptor potential (TRP) ion channel family member(s), e.g., TRPA1, are activated in lung epithelial cells in response to cigarette smoke stress (Lin et al., 2015). In aged mice, increased expression of voltage-activated $K^+$ channels occurs, which enhances IL-6 production and neuroinflammation with age (Schilling and Eder, 2014).

(a) **Proposed definition of oncochannelopathy**

(b) **Examples**

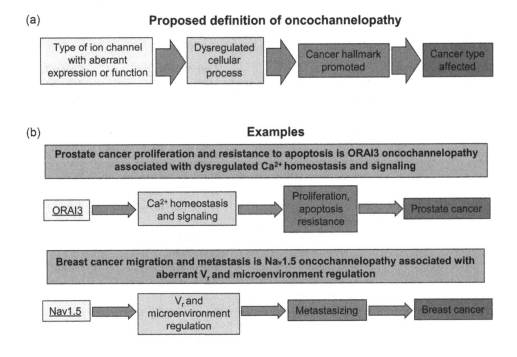

(c) **Differences between classical channelopathy and oncochannelopathy**

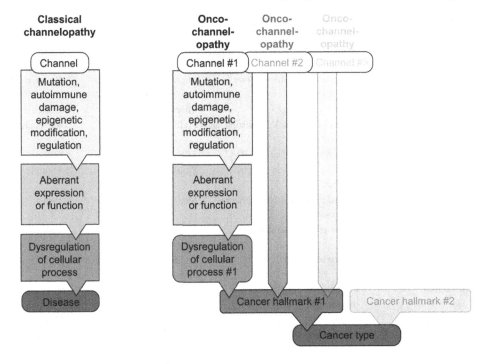

**FIGURE 8.1** Definition of oncochannelopathy has been proposed. Cancer hallmark for each particular cancer type is regarded as oncochannelopathy of specific class of channel(s). The aberrant expression/function of these channels contributes to this hallmark via dysregulation of certain cellular processes (a; i.e., $Ca^{2+}$ signaling, membrane potential regulation, volume regulation, mechanosensitivity, and microenvironment regulation). Examples of the application of such definition to prostate and breast cancers are presented in b, with the complete list being presented in the Table 8.1 (Prevarskaya et al., 2018). (For color figure see eBook figure.)

**TABLE 8.1**

**List of Major Oncochannelopathies (Prevarskaya et al., 2018).**

| Ion Channel with Aberrant Expression or Function | Cellular Process(es) Dysregulated | Cancer Hallmark(s) Promoted | Cancer Type Affected |
|---|---|---|---|
| ANO1 (TMEM16A) | Volume regulation | Proliferation (?), migration, invasion, and metastasis | Prostate, HNSCC, lung |
| AQP1 | Volume regulation | Apoptosis resistance, TEC migration, and angiogenesis | RCC, melanoma, breast |
| AQP8/9 | Volume regulation | Apoptosis resistance | HCC |
| ASIC1 | Microenvironment (pH) sensitivity | Migration and invasion | HCC |
| | | Proliferation and migration | GBM |
| | | Invasion and metastasis | Breast |
| ASIC2a/ASIC3 | Microenvironment (pH) sensitivity | ? | ACC |
| $Ca_v1.1/Ca_v1.3$ | $Ca^{2+}$ signaling: CCE | Migration, invasion, and metastasis | Breast |
| CLC-3 | Volume regulation | Migration and invasion | Glioma |
| $IP_3R$ | $Ca^{2+}$ signaling: ER filling, ER-mitochondria crosstalk | Proliferation and survival | All types of cancer |
| $K_{2p}2.1$ (TREK1) | $V_r$ regulation and driving force for $Ca^{2+}$ entry | Proliferation | Prostate |
| $K_{2p}3.1$ (TASK1) | Microenvironment ($Po_2$, pH) sensitivity | Proliferation, apoptosis resistance | NSCLC |
| $K_{2p}9.1$ (TASK3) | $V_r$ regulation and driving force for $Ca^{2+}$ entry | Proliferation | Breast, lung, colorectal |
| $K_{Ca}1.2$ (BK) | $V_r$ regulation and driving force for $Ca^{2+}$ entry | Migration | Breast |
| $K_{Ca}2.2$ (SK2) | Volume regulation, $V_r$ regulation, and $Ca^{2+}$ entry | Proliferation | Melanoma |
| $K_{Ca}2.3$ (SK3) | $V_r$ regulation and driving force for $Ca^{2+}$ entry | Migration and metastasis | Breast, colon |
| $K_{Ca}3.1$ (IK) | Volume regulation | Migration and invasion | Glioma |
| | Volume regulation, $V_r$ regulation, and $Ca^{2+}$ entry | Proliferation | Melanoma |
| | $V_r$ regulation, $Ca^{2+}$ entry in TEC | TEC proliferation | Colon |
| Kv10.1 (EAG1) | $V_r$ regulation, protein-protein interaction | Proliferation | Multiple cancers |
| | $V_r$ regulation and driving force for $Ca^{2+}$ entry | Migration and metastasis | Breast |
| | Protein-protein interaction | Angiogenesis via AF release by tumor cells | Cancer nonspecific |
| Kv10.2 (EAG2) | Volume regulation | Proliferation, migration, and metastasis | Medulloblastoma |
| Kv11.1 (HERG) | $V_r$ regulation, protein-protein interaction | Proliferation | Multiple cancers |
| | ? | Invasion and metastasis | Colorectal |
| | $V_r$ regulation and protein-protein interaction | Angiogenesis via AF release by tumor cells | Colorectal |
| $Na_v1.5$ | $V_r$ regulation, microenvironment (pH) regulation, protein-protein interaction | Invasion and metastasis | Breast, colon, ovarian |
| $Na_v1.6$ | ? | Invasion | Cervical cancer |

*(Continued)*

**TABLE 8.1 (*Continued*)**
**List of Major Oncochannelopathies (Prevarskaya et al., 2018).**

| Ion Channel with Aberrant Expression or Function | Cellular Process(es) Dysregulated | Cancer Hallmark(s) Promoted | Cancer Type Affected |
|---|---|---|---|
| Na$_v$1.7 | V$_r$ regulation, Na$^+$ entry | Invasion | Breast, prostate, cervical, NSCLC |
| ORAI1 | Ca$^{2+}$ signaling: SOCE | Migration and metastasis | Breast |
| ORAI1/TRPC1 | Ca$^{2+}$ signaling: SOCE | Migration and metastasis | Colon |
| ORAI3 | Ca$^{2+}$ signaling: SMOC | Proliferation, apoptosis resistance | Prostate |
| | Ca$^{2+}$ signaling: noncanonical SOC | Proliferation, apoptosis resistance, migration | Breast, SCLC |
| PIEZO1 | Ca$^{2+}$ signaling, mechanosensitivity | Migration | Breast |
| PIEZO2 | Ca$^{2+}$ signaling and mechanosensitivity in TEC | Angiogenesis via TEC proliferation, migration, and tube formation | Glioma |
| TRPC4 | Ca$^{2+}$ signaling in tumor cells: SMOC, CCE | Angiogenesis via AF release by tumor cells | RCC |
| TRPC6 | Ca$^{2+}$ signaling: SMOC, CCE | Migration and invasion, angiogenesis | Glioblastoma |
| | Ca$^{2+}$ signaling: SMOC, noncanonical SOC | Proliferation, apoptosis resistance | HCC |
| TRPM2 | Ca$^{2+}$ signaling | Apoptosis resistance, prosurvival autophagy | Neuroblastoma |
| TRPM3 | Ca$^{2+}$ signaling | Prosurvival autophagy | ccRCC |
| TRPM7 | Ca$^{2+}$ signaling | Proliferation | Breast |
| | Ca$^{2+}$ signaling | Proliferation, migration, and invasion | Ovarian |
| | Protein-protein interaction, mechanotransduction | Migration and metastasis | Breast |
| | [Mg$^{2+}$]$_i$ homeostasis and signaling | Migration and invasion | Pancreatic |
| TRPV2 | Ca$^{2+}$ signaling: CCE | Migration | Breast |
| | | Invasion and metastasis | Prostate |
| TRPV4 | Ca$^{2+}$ signaling in TEC | Angiogenesis | Breast, RCC |
| | Ca$^{2+}$ signaling and mechanosensitivity in TEC | Angiogenesis | Multiple cancers |
| TRPV6 | Ca$^{2+}$ signaling: CCE | Proliferation | Prostate |
| STIM1 | Ca$^{2+}$ signaling: SOCE | Proliferation, migration, angiogenesis | Cervical |
| STIM1/ORAI1 | Ca$^{2+}$ signaling: SOCE | Migration, invasion, and metastasis | Breast, melanoma, glioblastoma, ccRCC |
| STIM1/ORAI1/TRPC1 | Ca$^{2+}$ signaling in EPC: SOCE | Angiogenesis | RCC |
| VRAC | Volume regulation | Proliferation | All types of cancer |
| VRAC (LRRC8A/D) | Volume regulation, Pt drug uptake | Apoptosis resistance | All types of cancer, ovarian |

ACC, adenoid cystic carcinoma; AF, angiogenic factors; CCE, constitutive Ca$^{2+}$ entry; GBM, glioblastoma multiforme; EPC, endothelial progenitor cells; HCC, hepatocellular carcinoma; HNSCC, head and neck squamous cell carcinoma; NSCLC, non-small-cell lung cancer; RCC, renal cell carcinoma; SCLC, small cell lung cancer; SMOC, second messenger-operated channel; TEC, tumor endothelial cells.

For each row there are references, readily found from (Prevarskaya et al., 2018). We have avoided relisting them in the book to save space.

Dysfunctional calcium channel signaling is observed in cognitive and cardiac decline in model organisms, for example, Drosophila (Lam et al., 2018). Within the atrioventricular region of rats, sodium ($Na_v1.5$) is downregulated with age, while calcium ($Ca_v1.3$) channels are found upregulated, alongside augmented atrial-ventricular node functioning (Saeed et al., 2018). L-type and osmotically activated calcium channels were upregulated in human cardiomyopathy and aging, respectively (Jones et al., 2018).

Table 8.2 presents an overview of the dysfunction of ion channels and age-related diseases concerning the channelopathies (Strickland et al., 2019). Channel blockers corresponding to every channel disorder are listed. Ion channels are incriminated in many age-related dysfunctions (Santulli and Marks, 2015; Rao et al., 2016). Aging causes physiological alterations of ion channel function. It is known that abnormal changes in ionic gradients can underlie most of the age-dependent decline in physiological functions (Rao et al., 2016). As age progresses, functional changes in ion channels may lead to clinical phenotypes, referred to as channelopathies (Rao et al., 2016).

**TABLE 8.2**

**Ionic Channels and Their Age-Related Diseases by Blocker/Ligand Matches**

| Ionic Channel Subtypes | Age Related Diseases | Blockers/Ligands |
|---|---|---|
| Transient receptor potential channels (TRP1, TRPM3) | Polycystic kidney disease 2; polycystic kidney disease 2-like 1 protein; | No putative TRP1 blockers $Cd^{2+}$, $Ni^{2+}$ ligands |
| Voltage-gated potassium channels (Kv1.3) | Autoimmune diseases (i.e., diabetes, multiple sclerosis, and rheumatoid arthritis) | Noxiustoxin; |
|  |  | Charybdotoxin; |
|  |  | Margatoxin |
|  |  | Kaliotoxin |
|  |  | Maurotoxin |
| Voltage-gated calcium channels ($Ca_v1.4$) | Ocular albinism, ocular albinism type 2 | Dihydropyridine antagonists (verapamil, diltiazem) |
| Calcium- and sodium-activated potassium channels ($K_{Ca}2.3$) | Parkinson's disease | $K_{Ca}2.3$ blockers (Apamin, Leiurotoxin I) |
| TRPV1 Transient receptor potential channels | Inflammatory bowel disease, Crohn's disease; ulcerative colitis | Agatoxin |
| TRPA1 Transient receptor potential channels | Inflammation, inflammatory pain, and inflammatory diseases | Divalent cations modulators |
| $Ca_v2.2$ Voltage-gated calcium channels | Renal and cardiovascular diseases | Omega-conotoxins |
| $Na_v1.4$ Voltage-gated sodium channels | Susceptibility to periods of hyperactivity | Saxitoxin |
|  |  | Tetrodotoxin |
|  |  | Mu-conotoxins |
|  |  | Lidocaine |
| Kv8.1 Voltage-gated potassium channels | Epileptic disease | Kv8.1 is not functional on its own but modulates the properties of coexpressed Kv2.1 |
| $Ca_v1.3$ Voltage-gated calcium channels | Multiorgan disease | $Cd^{2+}$; Verapamil dihydropyridine antagonist |
| Kv4.3 Voltage gated potassium channels | Àtrial fibrillation, valvular heart disease | Phrixotoxin 1 |
| $K_{Ca}2.1$ Calcium- and sodium-activated potassium channels | Ataxia, epilepsy, memory disorders, pain and possibly schizophrenia and Parkinsons's disease | NS8593 gating inhibitor |
| $Na_v1.6$ Voltage-gated sodium channels | Motor end-plate disease | α scorpion toxins |

Guide to Immunopharmacology portal, http://www.guidetopharmacology.org/ (Strickland et al., 2019).

## 8.2 CHANNELOPATHIES OF THE PLASMA MEMBRANE AND THERAPEUTIC TARGETS OF ION CHANNELS

Disorders in plasma membrane ion channels and channel-constructing proteins are mainly responsible for plasma membrane channelopathies. Specific mutations in the membrane proteins constructing ion channels are enormous. We do not attempt to list all of them, but try more to address the plasma membrane channelopathies considering a few examples cases of channel mutations that concern the transport of materials and information across the membrane. This way, we will understand about the targets that might get considered while finding therapeutics for the channelopathies and related diseases.

### 8.2.1 Disordered Potassium Channels in the Plasma Membrane

Potassium channelopathies (or potassium channel disorders) in various disease conditions show specific potassium channel functions. We shall avoid going into any details, and will avoid covering huge amount of known potassium channelopathies. Using examples cases, we wish to explain the matter in enough details so that readers find necessary information to enrich their understanding on potassium channelopathies.

In hippocampal primary neurons, mutation of the C-terminal domain of voltage-gated ion channels (VGCs) and potassium channels Kv4.2LL/AV and Kv2.1S586A cause their mislocalization, with the Kv4.2 channel being expressed at the somatodendritic compartment, as well as in the axon, and the Kv2.1 channel being expressed all along the dendritic plasma membrane (Jensen et al., 2014). Jensen et al. (2014) also reported that the disruption of actin polymerization using latrunculin A altered the motility of Kv2.1-containing, but not Kv4.2-containing vesicles. Comparison of the trafficking mechanism of two mutant potassium channels, Kv4.2 and Kv2.1, have shown that channels are sorted at the Golgi apparatus into different vesicle pools, with each pool being transported in a compartment-specific manner (Jensen et al., 2014). Therefore, it may be concluded that the channelopathies may have channel specific, not perhaps universal-type, responses. Figure 8.2 demonstrates the molecular mechanisms behind segregated expressions of the dendritic Kv4.2 and Kv2.1 channels (Duménieu et al., 2017).

A novel epileptic encephalopathy mutation in KCNB1 was recently reported to disrupt Kv2.1 ion selectivity, expression, and localization (Thiffault et al., 2015). The de novo V378A variant in KCNB1 was found to fundamentally change the ion selectivity of Kv2.1 channels from potassium-selective to nonselective cation channels, as was also demonstrated for previously reported three de novo KCNB1 mutations (Torkamani et al., 2014). The loss of $K^+$ selectivity yields to a depolarizing inward cation conductance at negative voltages.

All of the other reported missense variants are located within the pore domain of Kv2.1 (Srivastava et al., 2014; Torkamani et al., 2014). The V378A channel was found gated by voltage (Thiffault et al., 2015). Whereas the effects as a result of various other disease-associated Kv2.1 mutations are yet unresolved (Torkamani et al., 2014).

The voltage-dependent V378A channels are sensitive to the Kv2-specific toxin GxTX, which provides compelling evidence that the nonselective cation currents recorded from cells expressing the mutant channels are from bona fide Kv2.1 channels.

Figure 8.3 demonstrates that the Kv2.1 channel's reversal potential was notably altered by the V378A variant (Thiffault et al., 2015). Compared with wild-type Kv2.1 channels, V378A channels were found to produce large inward tail currents at voltages that are much more positive than the calculated $K^+$ reversal potential, $-97\,mV$, see Figure 8.3a and b, suggesting that ions other than $K^+$ (such as $Na^+$) might permeate the mutant channels. This hypothesis was further tested by substituting extracellular sodium with $NMDG^+$, a large cation that cannot permeate most ion channels. Substitution of $NMDG^+$ reduced the large inward tail currents. This indicates that $Na^+$ was indeed permeating the V378A channel, see Figure 8.3 b, d, e. To further test if the V378A channel

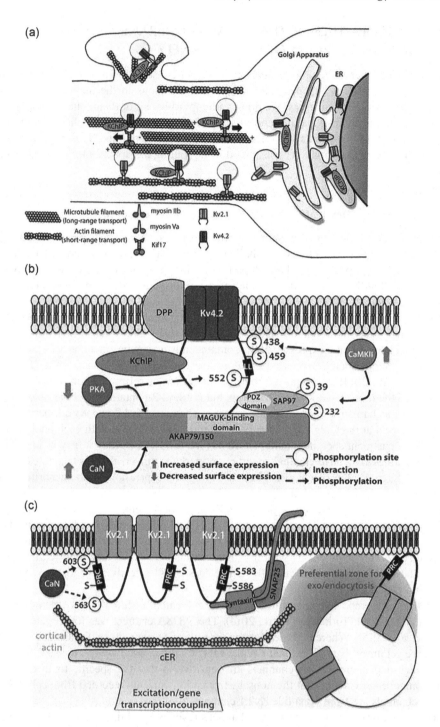

**FIGURE 8.2**  Demonstration of the molecular mechanisms behind segregated expressions of the dendritic Kv4.2 and Kv2.1 channels. (a) Sorting and trafficking potassium channels, Kv4.2 and Kv2.1. Association of the Kv4.2 channel with the auxiliary protein K⁺ channel interacting protein (KChIP) targets the channel to the surface of the cell. Kif17 ensures the microtubule-based transport, whereas myosin Va is used for actin-based transport of the Kv4.2 channel on proximal dendrites and spines. Myosin IIb transports the Kv2.1 to the somatodendritic compartment. (b) Intrinsic motifs, posttranslational regulations, and binding partners regulating Kv4.2 surface expression. (c) Intrinsic motifs, posttranslational regulations, and binding partners regulating Kv2.1 cluster formation and function (Duménieu et al., 2017).

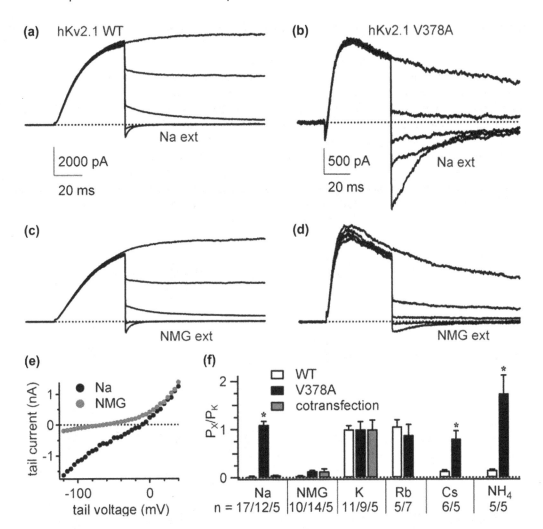

**FIGURE 8.3** The Kv2.1 V378A mutant is found to be a nonselective cation channel. Whole-cell patch-clamp recordings are presented here which were made from transiently transfected CHO-K1 cells. Currents are as usual responses to voltage steps; holding potential = $-100\,\mathrm{mV}$. (a) Kv2.1 wild-type channel currents. 50-ms steps to $40\,\mathrm{mV}$, and then to $-120$, $-80$, $-40$, $0$, or $40\,\mathrm{mV}$. (b) Kv2.1 V378A currents, measured using the same protocol as in (a, c, and d). Same cell and protocol as used in (a) and (b), but with external $Na^+$ replaced by $NMDG^+$. (e) Instantaneous current level from Kv2.1 V378A channel at the indicated voltage after a 50-ms step to $40\,\mathrm{mV}$ in control external (black) or NMDG external (gray). (f) Permeability ratios for the indicated ion over $K^+$, calculated from the reversal potential when external $Na^+$ is replaced with the indicated ion. Cotransfection was found to deliver an equal amount of wild-type and V378A DNA. *, $P < 0.01$, compared with wild-type by two-tailed Mann–Whitney $U$ test. Error bars represent means $\pm$ SEM. For details, see Thiffault et al. (2015).

completely lost the section ability between monovalent cations, the relative permeabilities of $K^+$, $Na^+$, $NMDG^+$, $Rb^+$, $Cs^+$, and $NH_4^+$ were measured.

Striking cell background-dependent differences in expression levels and subcellular localization of the V378A mutation were clearly observed in heterologous cells (in immunocytochemistry experiments). Furthermore, coexpression of V378A subunits and wild-type Kv2.1 subunits were found to reciprocally affect their respective trafficking characteristics (Thiffault et al., 2015).

Channelopathies and ion transporter defects are responsible for disorders of glucose homeostasis (hypoglycaemia and different forms of diabetes mellitus (DM)). Hyperinsulinemic hypoglycem

(HH) (a group of heterogeneous disorders characterized by the inappropriate insulin secretion from pancreatic β-cells) is linked to potassium channel subunit/component mutations (Demirbilek et al., 2019). Variants or polymorphisms in various plasma membrane-hosted ion channel genes and transporters are found in association with type 2 DM. For example, the nonsynonymous E23K variant in the *KCNJ11* was the first robustly replicating signal to emerge as a link to type 2 DM. Pancreatic $K_{ATP}$, Non-$K_{ATP}$, and a few calcium channelopathies and MCT1 transporter (such as GLUT1 and MCT1) defects (due to genetic mutations) can lead to various forms of HH. Mutations in the genes encoding the pancreatic $K_{ATP}$ channels can lead to different types of diabetes (including neonatal DM and maturity-onset diabetes of the young), and defects in the solute carrier family 2 member 2 (SLC2A2) leads to DM as part of the Fanconi–Bickel syndrome. Genome-wide association studies (GWAS) have identified variants or polymorphisms in channel and ion transporter genes associated with type 2 DM (Imamura et al., 2019). Figure 8.4 presents a flowchart summarizing the ion channel defects and related disorders leading to diabetes (Demirbilek et al., 2019).

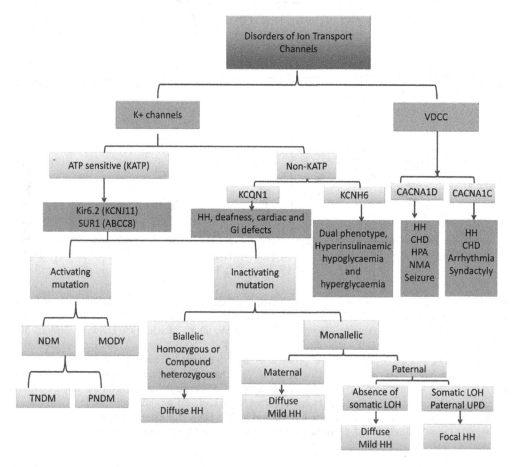

**FIGURE 8.4** A summary of ion transport channel defects and related disorders (K⁺, Potassium; VDCC, Voltage-dependent calcium channels; KATP, ATP-sensitive potassium channels ($K_{ATP}$); KIR6.2, Inward rectifier potassium channel subunit 2; KCNJ11, Potassium voltage-gated channel subfamily J member; SUR1, Sulphonylurea receptor 1; ABCC8, ATP-binding cassette subfamily C member 8; MODY, Maturity-onset diabetes of the young; CACNA1D, Calcium voltage-gated channel subunit Alpha1D; CACNA1C, Calcium channel; Voltage-dependent, L type; alpha 1C subunit; KCNH6, Potassium voltage-gated channel subfamily H member 6; NDM, Neonatal diabetes mellitus; TNDM, Transient NDM; PNDM, permanent NDM; HH, Hyperinsulinemic hypoglycemia; CHD, Congenital heart diseases; HPA, Hyperaldosteronism; NMA, Neuromuscular abnormalities; LOH, Loss of heterozygosity; UPD, Uniparental isodisomy) (Demirbilek et al., 2019).

## 8.2.2 Calcium, Sodium, and Potassium Channelopathies Cause Cardiac Arrhythmic Disorders

Cardiac action potentials are generated and function through a delicate balance of several ionic currents (Figure 8.5). An imbalance in this coordinated ion current flow system due to ion channel dysfunction leads to life-threatening cardiac arrhythmias or cardiac channelopathies. Mutations in calcium, sodium, potassium, and TRP channel genes cause cardiac arrhythmic disorders, see Table 8.3 (Kim, 2014).

**TABLE 8.3**

**Cardiac Channelopathies Involving Various Types of Channelopathies, Such as Calcium, Sodium, Potassium, etc. (Kim, 2014)**

| Disease | Channel Protein | Gene |
|---|---|---|
| Atrial standstill | Na$_v$1.5: sodium channel, voltage-gated, type V, α subunit | SCN5A |
| Brugada syndrome type 1 | Na$_v$1.5: sodium channel, voltage-gated, type V, α subunit | SCN5A |
| Brugada syndrome type 3 (short QT syndrome type 4) | Ca$_v$1.2: calcium channel, voltage-gated, L type, α1C subunit | CACNA1C |
| Brugada syndrome type 4 (short QT syndrome type 5) | Ca$_v$β2: calcium channel, voltage-gated, β2 subunit | CACNB2 |
| Brugada syndrome type 5 | Na$_v$β1: sodium channel, voltage-gated, type I, β subunit | SCN1B |
| Brugada syndrome type 6 | Potassium channel, voltage-gated, Isk-related subfamily, member 3 | KCNE3 |
| Brugada syndrome type 7 | Na$_v$β3: sodium channel, voltage-gated, type III, β subunit | SCN3B |
| Brugada syndrome type 8 | Hyperpolarization-activated cyclic nucleotide-gated potassium channel 4 | HCN4 |
| Catechoiaminergic polymorphic ventricular tachycardia type 1 | RyR2: ryanodine receptor 2 | RYR2 |
| Dilated cardiomyopathy type 1E | Na$_v$1.5: sodium channel, voltage-gated, type V, α subunit | SCN5A |
| Dilated cardiomyopathy type 10 | ATP-binding cassette, subfamily C, member 9 (sulfonylurea receptor 2) | ABCC9 |
| Familial arrhythmogenic right ventricular dysplasia type 2 | RyR2: ryanodine receptor 2 | RYR2 |
| Familial atrial fibrillation type 3 | Kv7.1: potassium channel, voltage-gated, KQT-like subfamily, member 1 | KCNQ1 |
| Familial atrial fibrillation type 4 | Potassium channel, voltage-gated, Isk-related subfamily, member 2 | KCNE2 |
| Familial atrial fibrillation type 7 | Kv1.5: potassium channel, voltage-gated, shaker-related subfamily, member 5 | KCNA5 |
| Familial atrial fibrillation type 9 | Kir2.1: potassium channel, inwardly-rectifying, subfamily J, member 5 | KCNJ2 |
| Familial atrial fibrillation type 10 | Na$_v$1.5: sodium channel, voltage-gated, type V, α subunit | SCN5A |
| Familial atrial fibrillation type 12 | ATP-binding cassette, subfamily C, member 9 | ABCC9 |
| Jervell and Lange-Nielsen syndrome type 1 | Kv7.1: potassium channel, voltage-gated, KQT-like subfamily, member 1 | KCNQ1 |
| Jervell and Lange-Nielsen syndrome type 2 | Potassium channel, voltage-gated, Isk-related subfamily, member 1 | KCNE1 |
| Long QT syndrome type 1 | Kv7.1: potassium channel, voltage-gated, KQT-like subfamily, member 1 | KCNQ1 |
| Long QT syndrome type 2 | Kv11.1: potassium channel, voltage-gated, subfamily H, member 2 | KCNH2 |

*(Continued)*

**TABLE 8.3 (*Continued*)**
**Cardiac Channelopathies Involving Various Types of Channelopathies, Such as Calcium, Sodium, Potassium, etc. (Kim, 2014)**

| Disease | Channel Protein | Gene |
|---|---|---|
| Long QT syndrome type 3 | Na$_v$1.5: sodium channel, voltage-gated, type V, α subunit | *SCN5A* |
| Long QT syndrome type 5 | Potassium channel, voltage-gated, Isk-related subfamily, member 1 | *KCNE1* |
| Long QT syndrome type 6 | Potassium channel, voltage-gated, Isk-related subfamily, member 2 | *KCNE2* |
| Long QT syndrome type 7 (Andersen-Tawil syndrome) | Kir2.1: potassium channel, inwardly-rectifying, subfamily J, member 2 | *KCNJ2* |
| Long QT syndrome type 8 (Timothy syndrome) | Ca$_v$1.2: calcium channel, voltage-gated, L type, α1C subunit | *CACNA1C* |
| Long QT syndrome type 10 | Na$_v$β$_4$: sodium channel, voltage-gated, type IV, β subunit | *SCN4B* |
| Long QT syndrome type 13 | Kir3.4: potassium channel, inwardly-rectifying, subfamily J, member 5 | *KCNJ5* |
| Nonprogressive familial heart block | Na$_v$1.5: sodium channel, voltage-gated, type V, α subunit | *SCN5A* |
| Paroxysmal familial ventricular fibrillation, type 1 | Na$_v$1.5: sodium channel, voltage-gated, type V, α subunit | *SCN5A* |
| Progressive familial heart block type IA (Lenegre-Lev syndrome) | Na$_v$1.5: sodium channel, voltage-gated, type V, α subunit | *SCN5A* |
| Pogressive familial heart block type IB | Transient receptor potential cation channel, subfamily M, member 4 | *TRPM4* |
| Short QT syndrome type 1 | Kv11.1: potassium channel, voltage-gated, subfamily H, member 2 | *KCNH2* |
| Short QT syndrome type 2 | Kv7.1: potassium channel, voltage-gated, KQT-like subfamily, member 1 | *KCNQ1* |
| Short QT syndrome type 3 | Kir2.1: potassium channel, inwardly-rectifying, subfamily J, member 2 | *KCNJ2* |
| Short QT syndrome type 4 (Brugada syndrome type 3) | Ca$_v$1.2: calcium channel, voltage-gated, L type, α1C subunit | *CACNA1C* |
| Short QT syndrome type 5 (Brugada syndrome type 4) | Ca$_v$β2: calcium channel, voltage-gated, β2 subunit | *CACNB2* |
| Short QT syndrome type 6 | Ca$_v$α2δ1: calcium channel, voltage-gated, α2/δ1 subunit | *CACNA2D1* |
| Sick sinus syndrome type 1, autosomal-recessive | Na$_v$1.5: sodium channel, voltage-gated, type V, α subunit | *SCN5A* |
| Sick sinus syndrome type 2, autosomal-dominant | Hyperpolarization-activated cyclic nucleotide-gated potassium channel 4 | *HCN4* |

Two familial disease groups are responsible for sudden cardiac death:

- cardiomyopathies; mainly hypertrophic cardiomyopathy, dilated cardiomyopathy, and arrhythmogenic cardiomyopathy; and
- channelopathies; mainly long QT syndrome, Brugada syndrome, short QT syndrome, and catecholaminergic polymorphic ventricular tachycardia.

Cardiac channelopathies are characterized by mainly lethal arrhythmias, resulting from pathogenic variants in genes encoding cardiac ion channels and/or associated proteins. The heart, an electromechanical pump, is electrically triggered by the generation and propagation of an action potential (AP) across myocytes. This is then followed by a period of muscle contraction and relaxation until the generation of the next impulse. Myocardial action potential is generated by

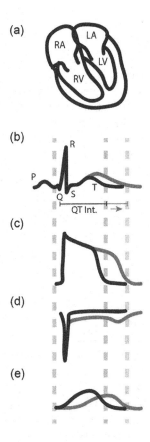

**FIGURE 8.5**  QT interval and action potential (AP) prolongation caused by genetic mutations. The ECG measures the electrical activity of the heart reflected on the body surface, and the waveform of ECG correlates with electrical depolarization and repolarization of the cardiac muscle in various chambers (a, b). The first upward deflection in panel B (P wave) corresponds to activation of the upper heart chambers, the left and right atria (LA and RA, respectively), which collect blood that is returning from the body (RA) and the lungs (LA). The prominent spike formed by the Q, R, and S points is linked to excitation of the massive lower chambers, the left and right ventricles (LV and RV, respectively). The magnitude of the QRS complex is a consequence of the larger ventricular muscle mass, which is needed to generate the force that pushes blood through the body (LV) and the lungs (RV). Finally, the T wave occurs when the ventricles return to an electrical resting state. Consequently, we would expect that a prolonged QT interval is caused by an increase in the time that the ventricles remain in an electrically excited state. The individual cells in the myocardium, myocytes, each generate an action potential (AP) (c) that is responsible for excitation. Therefore, the QT interval (b) corresponds to the duration of the ventricular AP (c), implying that any change in AP duration (APD) will affect the QT interval. APD is determined by a delicate balance of inward and outward ionic currents. The morphology of the AP (c) is the consequence of positively charged $Na^+$, $Ca^{2+}$, and $K^+$ ions entering and exiting the myocyte. For example, as the membrane potential ($V_m$) rises from its resting state, caused by the excitation of neighboring myocytes, $Na^+$ channels open and positively charged $Na^+$ ions enter the cell (d). This inward sodium current ($I_{Na}$) causes $V_m$ to rise quickly. Once $V_m$ is elevated by $I_{Na}$, voltage-gated L-type $Ca^{2+}$ channels open and bring in a sustained inward $Ca^{2+}$ flux, $I_{Ca, L}$ (not shown). It is this influx of $Ca^{2+}$ that signals contraction of the myocyte. The sustained ICa, L supports the AP plateau, which continues until the $K^+$ channels open to generate repolarizing outward current (e). Throughout the AP a small inward $Na^+$ current persists and is enhanced toward the end of the AP when channels begin to recover from inactivation, but are not deactivated yet (window current). In ventricular myocytes, there are two major repolarizing $K^+$ currents, one rapid component ($I_{Kr}$) and one slow ($I_{Ks}$). The $I_{Kr}$ α-subunit, Kv11.1, is encoded by the *KCNH2* (aka *HERG*) gene and the $I_{Ks}$ α-subunit, Kv7.1, is encoded by the *KCNQ1* (aka *KvLQT1*) gene. Once the outward $K^+$ flux overwhelms the inward $Ca^{2+}$ flux, the myocyte returns to its resting state. Therefore, APD is determined by the balance of inward and outward ion fluxes. Increasing inward currents (d) or reducing outward currents (e) will prolong APD (c) and therefore the QT interval (b). For further details, see Zaydman et al. (2012).

ionic changes across the membrane. Combined efforts of the sequential activation and inactivation of three ion channels that conduct depolarizing, inward currents (Na$^+$ and Ca$^{2+}$) and repolarizing, outward currents (K$^+$) are found to enable transmembrane ion currents and subsequent action potential creation. This mechanism is known since many decades. The ion current directs are determined by the electrochemical gradient of the corresponding ions. Changes in action potential, synchronization, and/or propagation of electrical impulse predispose to potentially malignant arrhythmias. These medically important modifications may be induced by pathogenic variants in genes that encode any of the three ion channels, or their associated proteins. For example, the main cardiac channelopathies found to be associated with sudden cardiac death are Brugada syndrome, long and short QT syndromes, and catecholaminergic polymorphic ventricular tachycardia (Kim, 2014).

The combined effects of three ion currents and their abnormalities (channelopathies) due to genetic mutations or any other defects in the channel or associated proteins may be explained (see the caption of the figure) easily by carefully inspecting Figure 8.5 (Zaydman et al., 2012).

### 8.2.3  Aquaporin Channelopathies

Water movement across biological cell membranes is enhanced or facilitated by water channel or aquaporins. There are currently 13 known aquaporin proteins in mammals and distributed in most tissues, but many more exist (have been identified) in lower organisms and in the plant kingdom. Defects in aquaporin function resulting from aquaporin mutations are known as "aquaporin channelopathies," which are related to various disease conditions and pathological states.

To demonstrate aquaporin channelopathies, as an example, Table 8.4 lists the mutations and related diseases linked to AQP2 (Bichet et al., 2012). AQP2 gene works in making an aquaporin 2 protein which forms an aquaporin channel that carries water molecules across cell membranes.

The assembly of the channel proteins (including the mutations shown in Table 8.4) in plasma membrane is modeled in Figure 8.6 (Bichet et al., 2012). The AQP2 mutants expression in Xenopus oocytes and in polarized renal tubular cells recapitulates the clinical phenotypes, and are found to reveal a continuum from severe loss of function with urinary osmolalities <150 mOsm/kg H$_2$O to milder defects with urine osmolalities >200 mOsm/kg H$_2$O (Bichet et al., 2012).

AQP2 mutations in nephrogenic diabetes insipidus are well addressed (Loonen et al., 2008). Water reabsorption in the renal collecting duct is regulated by the antidiuretic hormone vasopressin (AVP).

When the vasopressin V2 receptor, present on the basolateral site of the renal principal cell, becomes activated by antidiuretic hormone vasopressin (AVP), AQP2 water channels are inserted in the apical membrane. This way, water can be reabsorbed from the pro-urine into the interstitium. The physiological role of the vasopressin V2 receptor and AQP2 in body water homeostasis maintenance became clear when it was shown that mutations in their genes cause nephrogenic diabetes insipidus, which is a disorder in which the kidney is unable to concentrate urine in response to AVP. For details on aquaporin channelopathies in nephrogenic diabetes insipidus and in many other disorders, readers may consult published studies on PubMed.

### 8.2.4  Channelopathies Linked to Plasma Membrane Phospholipids

Plasma membrane channels are hosted in a lipid bilayer due to a physical energetic binding among channel subunit proteins and membrane subunit lipids (Ashrafuzzaman and Tuszynski, 2012a; 2012b; Ashrafuzzaman, 2018). Lipid cell-specific organization and composition in the vicinity of the channels are highly regulatory of the membrane-hosted channel function. Fatty acids of diets (that we take) can redistribute within membrane phospholipids in a very selective manner, with phosphatidylcholine being the preferred sink for this redistribution, found in mouse model experiments (Bacle et al., 2020). Consequently, in many cell types, perturbation occurred in the

**TABLE 8.4**
**Listing of 46 Putative Disease-Causing AQP2 Mutations**

| Count | No. of Families | Name of Mutation | Domain | Nucleotide Change | Predicted Consequence |
|---|---|---|---|---|---|
| | | Missense | | | |
| 1 | 1 | M1I | NH2 | ATG-to-ATT | Met-to-Ile |
| 2 | 1 | L22V | TMI | CTC-to-GTC | Leu-to-Val |
| 3 | 1 | V24A | TMI | | Val-to-Ala |
| 4 | 1 | L28P | TMI | CTC-to-CCC | Leu-to-Pro |
| 5 | 1 | G29S | TM1 | | Gly-to-Ser |
| 6 | 2 | A47V | TMII | GCG-to-GTG | Ala-to-Val |
| 7 | 2 | Q57P | TMII | CAG-to-CCG | Glu-to-Pro |
| 8 | 1 | G64R | CII | GGG-to-AGG | Gly-to-Arg |
| 9 | 1 | N68S | CII | AAC-to-AGC | Asn-to-Ser |
| 10 | 1 | A70D | CII | GCC-to-GAC | Ala-to-Asp |
| 11 | 2 | V71M | CII | GTG-to-ATG | Val-to-Met |
| 12 | 2 | G100V | TMIII | GGA-to-GTA | Gly-to-Val |
| 13 | 1 | G100R | TMIII | GGA-to-AGA | Gly-to-Arg |
| 14 | 1 | I107D | EII | ATC-to-AAC | Ile-to-Asp |
| 15 | 1 | T125M | EII | ACG-to-ATG | Thr-to-Met |
| 16 | 1 | T126M | EII | ACG-to-ATG | Thr-to-Met |
| 17 | 1 | A147T | TMIV | GCC-to-ACC | Ala-to-Thr |
| 18 | 2 | D150E | ICII | GAT-to-GAA | Asp-to-Glu |
| 19 | 1 | V168M | TMV | GTG-to-ATG | Val-to-Met |
| 20 | 1 | G175R | TMV | GGG-to-AGG | Gly-to-Arg |
| 21 | 1 | G180S | EIII | GGC-to-AGC | Gly-to-Ser |
| 22 | 1 | C181W | EIII | TGC-to-TGG | Cys-to-Trp |
| 23 | 1 | P185A | EIII | CCT-to-GCT | Pro-to-Ala |
| 24 | 3 | R187C | EIII | CGC-to-TGC | Arg-to-Cys |
| 25 | 1 | R187H | EIII | CGC-to-CAC | Arg-to-His |
| 26 | 1 | A190T | EIII | GCT-to-ACT | Ala-to-Thr |
| 27 | 1 | G196D | EIII | GGC-to-GAC | Gly-to-Asp |
| 28 | 1 | W202C | EIII | TGG-to-TGT | Trp-to-Cys |
| 29 | 1 | G215C | TMVI | GGC-to-TGC | Gly-to-Cys |
| 30 | 2 | S216P | TMVI | TCC-to-CCC | Ser-to-Pro |
| 31 | 1 | S216F | TMVI | TCC-to-TTC | Ser-to-Phe |
| 32 | 1 | K228E | CIV | | Lys-to-Glu |
| 33 | 1 | R254Q | CIV | CGG-to-CAG | Arg-to-Gln |
| 34 | 1 | R254L | CIV | CGG-to-CTG | Arg-to-Leu |
| 35 | 1 | E258K | CIV | GAG-to-AAG | Glu-to-Lys |
| 36 | 2 | P262L | CIV | CCG-to-CTG | Pro-to-Leu |
| | | Nonsense | | | |
| 1 | 2 | R85X | CII | CGA-to-TGA | Arg-to-stop |
| 2 | 1 | G100X | TMIII | GGA-to-TGA | Gly-to-stop |
| | | Frameshift | | | |
| 1 | 1 | 369delC | EII | 1bp deletion | Stop at Codon 131 |
| 2 | 1 | 721delG | CIV | 1 bp deletion | Post-elongation |
| 3 | 1 | 727delG | CIV | 1 bp deletion | Post-elongation |
| 4 | 1 | 763-772del | CIV | 10 bp deletion | Post-elongation |
| 5 | 1 | 779–780insA | CIV | 1 bp insertion | Post-elongation |
| 6. | 1 | 812-818del | CIV | 7 bp deletion | Post-elongation |
| | | Splice-site | | | |
| 1 | 1 | IVS2-1G>A | NA | G-to-A | NA |
| 2 | 1 | IVS3+1G>A | NA | G-to-A | NA |
| 46 | 53 | | | | |

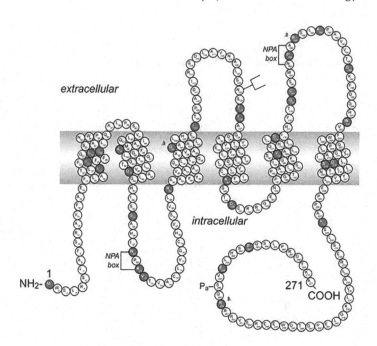

**FIGURE 8.6** A model representation of the AQP2 protein and the identification of 46 putative disease-causing mutations in AQP2. A monomer is represented by having six transmembrane helices. The location of the PKA phosphorylation site (Pa) is indicated. The extracellular, transmembrane, and cytoplasmic domains are defined following the literature (Deen et al., 1994). Solid symbols indicate the location of the mutations (in ref. (Bichet et al., 2012, see Table 1): M1I; L22V; V24A; L28P; G29S; A47V; Q57P; G64R; N68S; A70D; V71M; R85X; G100X; G100V; G100R; I107D; 369delC; T125M; T126M; A147T; D150E; V168M; G175R; G180S; C181W; P185A; R187C; R187H; A190T; G196D; W202C; G215C; S216P; S216F; K228E; R254Q; R254L; E258K and P262L. GenBank accession numbers—AQP2: AF147092, Exon 1; AF147093, Exons 2–4. NPA motifs and the N-glycosylation site are also indicated. Details in Bichet et al. (2012).

physiological balance within phospholipids between saturated fatty acids and monounsaturated or polyunsaturated fatty acids, which is known to regulate the biophysical properties of cellular membranes. In the high-fat and high-fructose (HFHF) rat model, the liver and skeletal muscles experience considerable fatty acid redistribution.

For example, in the liver, the fatty acid redistribution was paralleled by the deposition of a few neutral lipids, including diglycerides, triglycerides, cholesterol, and stearyl esters. The hepatic deposition of such neutral lipids is a hallmark of dyslipidemia in obesity and is thought to promote hepatic insulin resistance associated with nonalcoholic fatty liver disease, which is a major factor in the pathogenesis of type 2 diabetes and metabolic syndrome, where channelopathies are directly observed (explained in an earlier section). Interestingly, in obese individuals (mostly abnormal diet-induced), diglycerides is commonly known to inhibit insulin signaling by activation of protein kinase C isoforms.

Channelopathies have been found to be linked with the plasma membrane phosphoinositide phosphatidylinositol 4,5-bisphosphate ($PIP_2$) (Logothetis et al., 2010). For channels whose activities are $PIP_2$-dependent, and for which mutations can lead to channelopathies, the mutations may be found to alter channel–$PIP_2$ interactions. Similarly, diseases linked to disorders of the phosphoinositide pathway result in altered $PIP_2$ levels, which may also raise concomitant dysregulation of channel activity. Ion channels whose activity depends on interactions with $PIP_2$ (Tables 8.5 and 8.6) may be tested for crucial mechanisms that might play a role in creating defects on either the channel proteins or the phosphoinositide levels, leading to diseases.

## TABLE 8.5
## Phosphoinositide (PIP)-Sensitive Channels

| Channel | Tissue Distribution |
|---|---|
| Kir1.1 (ROMK1) | Kidney > skeletal muscle > pancreas > spleen > heart = brain > liver |
| Kir2.1 (IRK1) | Forebrain, skeletal muscle, heart, macrophage cells, aortic endothelial cells |
| Kir2.2 (IRK2) | Heart, forebrain, cerebellum, skeletal muscle, kidney |
| Kir2.3 (IRK3) | Heart, hippocampus, amygdala, caudate nucleus, thalamus, kidney |
| Kir2.4 (IRK4) | Retina, heart, striatum |
| Kir2.6 (IRK6) | Skeletal muscle |
| Kir3.2 (GIRK2) | Brain, pancreas, testis |
| Kir3.1/3.2 (GIRK1/2) | Brain |
| Kir3.1/3.4 (GIRK1/4) | Heart atria |
| Kir3.4-S143T (GIRK4*) | Experimental construct |
| Kir4.1 | Forebrain, cerebellum, striatum, kidney, retina |
| Kir4.2 | Kidney, pancreas > lung > prostate, testes, leukocytes |
| Kir4.1/Kir5.1 | Brainstem nuclei, kidney |
| Kir6.1 | Heart, ovary, adrenal gland > skeletal muscle, lung, brain, stomach, colon, testis, thyroid, pancreatic islet cells > kidney, liver, small intestine, pituitary gland |
| Kir6.2-$\Delta$36 | Experimental construct |
| Kir6.2/SUR1 | Brain, pancreas |
| Kir6.2/SUR2A | Heart |
| Kir6.2/SUR2B | Vascular smooth muscle cells |
| Kir7.1 | (Human) small intestine > stomach, kidney, brain, thyroid, choroid plexus, retinal pigment epithelium (rat) lung, testis |
| KirBac1.1 | Bacteria |
| $K_{2P}2.1$ (TREK1) | Brain, heart |
| $K_{2P}3.1$ (TASK1) | Brain, heart, lung, kidney, pancreas, others |
| $K_{2P}4.1$ (TRAAK) | Brain, kidney, small intestine, placenta, prostate |
| $K_{2P}9.1$ (TASK3) | Brain |
| ENaC ($\alpha/\beta/\gamma$) | Kidney, lung, gastrointestinal tract and skin |
| CFTR | Heart, exocrine glands (pancreas, airways, gastrointestinal tract and sweat glands) |
| P2X1 | Smooth muscle, brain, cerebellum, dorsal horn spinal neurons and platelets |
| P2X2 | Nervous system (widespread) |
| P2X3 | Sensory neurons |
| P2X2/3 | Sensory neurons |
| P2X4 | CNS synapses |
| P2X5 | Brain, heart, spinal cord, adrenal medulla, thymus and lymphocytes |
| P2X7 | Macrophages, lymphocytes, microglia, osteoclasts, osteoblasts |
| Native NMDA receptors | Rat cortical pyramidal neurons |
| NR1/NR2A | Mainly CNS |
| NR1/NR2B | Mainly CNS |
| NR1/NR2C | Mainly CNS |
| Native AMPA receptors | Rat CA1 pyramidal neurons |
| Ins(1,4,5)$P_3$ receptor | Ubiquitous |
| Ryanodine receptor (RyR1) | Skeletal muscle |
| Connexin Cx43 | Widespread, dermal fibroblasts, glial cells, heart |
| Hair cell mechanotransduction channels | Frog saccular hair cells |
| Volume-regulated anion channel (VRAC) | Widespread, epithelial cells |

(Continued)

## TABLE 8.5 (*Continued*)
## Phosphoinositide (PIP)-Sensitive Channels

| Channel | Tissue Distribution |
|---|---|
| Plasma membrane $Ca^{2+}$-ATPase (PMCA1) | Ubiquitous |
| $Na^+/Ca^+$ exchanger (NCX1) | Cardiac myocytes, neurons, skeletal muscle, smooth muscle, kidney |
| $Na^+/H^+$ exchanger (NHE1) | Ubiquitous |
| $Na^+/HCO_3$ cotransporter (NBCe1-A) | Renal proximal tubule, eye |

This table summarizes ion channels that have been shown to be sensitive to phosphoinositide modulation and can be categorized in the following groups: inwardly rectifying potassium (Kir) channels, two-pore potassium (K2P) channels, sodium ($Na^+$) channels, chloride ($Cl^-$) channels, P2X receptors, ionotropic glutamate receptors (iGluRs), calcium ($Ca^{2+}$)-release channels, and other channels. Transporters and exchangers that have been shown to be sensitive to phosphoinositides also appear in the corresponding group. The tissue distribution of each protein is provided in the second column, largely drawing from the IUPHAR database (Harmar et al., 2008) but also from specific references that complement the database, as indicated. Specific references for the regulation of each channel/transporter by phosphoinositides are cited in the third column.

## TABLE 8.6
## PIP-Sensitive Channels

| Channel | Tissue Distribution |
|---|---|
| Kv1.1/Kvβ1.1 | Brain, heart, retina, skeletal muscle |
| Kv1.3 | (Human) T- and B-lymphocytes, alveolar macrophages, monocyte-derived macrophages, prostate epithelium, platelets, cerebral cortical grey matter, (rat) testis |
| Kv1.4 | (Human) brain, heart, pancreatic islet, (rat) skeletal muscle, arterial smooth muscle, retina |
| Kv1.5/Kvβ1.3 | (Rat) pulmonary arterial smooth muscle, spinal cord, brain, heart, (mouse) skeletal muscle, (human) pancreatic islet, atrial myocytes, atrium, ventricle |
| Kv3.4 | (Rat) parathyroid, prostate, brain, skeletal muscle, (mouse) pancreatic acinar cells |
| Kv7.1 (KCNQ1) | (Human) heart=pancreas>kidney>lung=placenta, (mouse) intestine, stomach, liver, thymus |
| Kv7.1 (KCNQ1)/KCNE1 | Heart, inner ear |
| Kv7.2 (KCNQ2) | (Human) brain (cortex and hippocampus), (rat) sympathetic ganglia, high expression levels in the cerebellum, cortex, and hippocampus |
| Kv7.3 (KCNQ3) | (Human) brain (cortex and hippocampus), (rat) sympathetic ganglia, lower expression levels in the cerebellum than in cortex and hippocampus |
| Kv7.2/7.3 (KCNQ2/3) | Brain |
| Kv7.4 (KCNQ4) | Cochlea (outer hair cells), brainstem auditory nuclei |
| Kv7.5 (KCNQ5) | Brain, skeletal muscle |
| Kv11.1 (HERG) | Heart, brain, gut, pancreatic beta cells, kidney, others |
| Native rod CNG channel | Bovine rod outer segment |
| Native olfactory CNG channel | Rat olfactory receptor neurons |
| CNGA1 | Retina, pineal gland, some neurons |
| CNGA1/B1 | Retina |
| CNGA2 | Olfactory neurons, hippocampus |
| CNGA2/CNGA4 | Olfactory neurons |
| CNGA3/CNGB3 | Retina |
| HCN1 | Brain, retina, sinoatrial node cells |
| HCN2 | Brain, retina, heart |
| HCN4 | Brain, retina, heart, testis |

(*Continued*)

**TABLE 8.6 (*Continued*)**
**PIP-Sensitive Channels**

| Channel | Tissue Distribution |
|---|---|
| $K_{Ca}1.1$ (SLO1) | Ubiquitous, brain, skeletal muscle, smooth muscle, adrenal cortex, cochlear hair cells, pancreas, colon, kidney |
| $Ca_v1.2$ (L-type) | Heart, brain, prostate, bladder, uterus, stomach, colon, small intestine, placenta, adrenal gland, spinal cord |
| $Ca_v2.1$ (P/Q-type) | Brain, heart, pancreas, pituitary |
| $Ca_v2.2$ (N-type) | Brain, spinal cord |
| TRPV1 | Dorsal root ganglia, brain, kidney, pancreas, testes, uterus, spleen, stomach, small intestine, lung, liver |
| TRPV1t (taste receptor variant) | Rat chorda tympani taste nerve |
| TRPV5 | Kidney, prostate, testes, placenta, pancreas, brain |
| TRPV6 | Intestines, stomach, placenta, salivary glands, liver, prostate, pancreas, kidney, testes, mammary glands |
| TRPM4 | Small intestine, prostate, colon, kidney, testes, heart, lymphocytes, spleen, lung, brain, pituitary, skeletal muscle, stomach, adipose tissue, bone |
| TRPM5 | Taste tissue, stomach, intestines, uterus, testis |
| TRPM7 | Ubiquitous, heart, pituitary, bone, adipose tissue |
| TRPM8 | (Rat) dorsal root ganglia, trigeminal ganglia, (human) prostate, testis, bladder>breast, thymus |
| TRPA1 | Brain, heart, small intestine, lung, skeletal muscle, and pancreas (mouse) dorsal root ganglia, trigeminal ganglia, nodose ganglia, nociceptive neurons, inner ear (organ of corti) |
| TRPC1 | Heart, brain, lung, liver>spleen, kidney, testis |
| TRPC3 | Pituitary gland>brain, heart, lung, dorsal root ganglia |
| TRPC4 | Heart, brain, pancreas, placenta, kidney |
| TRPC5 | (Mouse) brain, testis, kidney, uterus |
| TRPC6 | Heart, lung, kidney, muscle, intestine, stomach, pancreas, prostate, bone, brain |
| TRPC7 | Kidney, intestine, pituitary gland, brain |
| TRPC1/C5/C6 | Rabbit coronary artery myocytes |

This table summarizes ion channels that have been shown to be sensitive to phosphoinositide modulation and can be categorized in the following groups: voltage-gated potassium (Kv) channels, cyclic nucleotide-gated (CNG) channels, hyperpolarization-activated (HCN) channels, calcium-activated potassium (KCa) channels, voltage-gated calcium ($Ca_v$) channels, and transient receptor potential (TRP) channels. The tissue distribution of each channel is provided in the second column, drawing from the IUPHAR database (Harmar et al., 2008) and from specific references complementing the database, as indicated. The tissue distribution shown for Kv1.1/Kvβ1.1 and Kv1.5/Kvβ1.3 heteromers pertains to Kv1.1 and Kv1.5, respectively. Specific references for the regulation of each channel by phosphoinositides are cited in the third column.

## 8.3  CHANNELOPATHIES OF THE MITOCHONDRIAL MEMBRANE AND THERAPEUTIC TARGETS OF ION CHANNELS

A large proportion of cellular diseases originate primarily due to malfunction in mitochondrial ion channels hosted in both mitochondrial outer membrane (MOM) and mitochondrial inner membrane (MIM). Defects, disorders, or even mutations in ion channels, commonly known as channelopathies, are increasingly found in a large spectrum of pathologies. Mutations in genes encoding ion channel proteins, which disrupt channel functions, are the most commonly identified cause(s) of channelopathies.

Ion channel function is largely controlled or regulated by the state of the hosting membrane potential. For details on this issue, readers may consult published articles including our books *Membrane Biophysics* (Ashrafuzzaman and Tuszynski, 2012b) and *Nanoscale Biophysics of the Cell* (Ashrafuzzaman, 2018). The state of the membrane potential of mitochondria is slightly differently inbuilt than that of the plasma membrane, as explained in these two books in detail. Lemeshko (2002) proposed a model in which the membrane potentials of MOM are linked to that of MIM. In this model, the MIM potential is considered to be divided into two resistances in series, the resistance of the contact sites of MIM and MOM and the resistance of the voltage-dependent anion channels (VDACs) localized beyond the contacts in the MOM. The MOM potential is generated and the VDAC channels in the MOM are strongly regulated by electrical potential. The main principle of the proposed mechanism is illustrated by kinetic and electric models, see Figures 8.7 and 8.8 (Lemeshko, 2002).

The mitochondrial membrane potential ($\Delta\Psi_m$) is generated by proton pumps. The proton gradient ($\Delta pH$) across the membrane and $\Delta\Psi_m$ together form the transmembrane potential of hydrogen ions, harnessed to synthesize ATP. $\Delta\Psi_m$ results in a driving force for transporting ions (other than $H^+$) and proteins, necessary for healthy mitochondrial functioning. For details, readers may consult a recent report by Zorova et al. (2018). Approximate physiological values of $\Delta\psi_m = 150\,mV$ and the proton gradient force $\Delta p = 180\,mV$ (Perry et al., 2011; Scaduto and Grotyohann, 1999). In disease condition, the membrane potentials are changed. For example, cancer cells have a more hyperpolarized MIM potential of approximately $-220\,mV$ than in normal cells of approximately $-140\,mV$

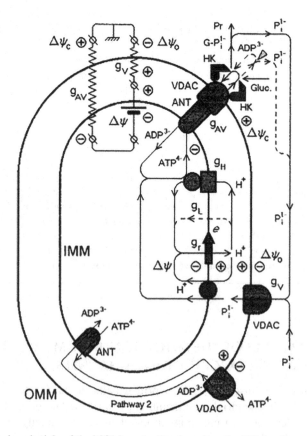

**FIGURE 8.7**  The main principle of the MOM generation, based on the MIM voltage ($\Delta\psi$) division between the intermembrane contact sites ($\Delta\psi_c$) and the outer membrane, beyond the contacts ($\Delta\psi_o$). For details on other parameters presented in the model, see Lemeshko (2002).

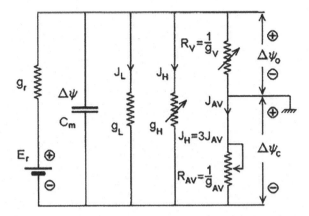

**FIGURE 8.8** Electric model of the MOM potential generation mechanism due to the MIM voltage division between the intermembrane contact sites and the outer membrane in mitochondria. Er, The redox potentials difference in the coupling sites of the respiratory chain; $1/g_r$, the inner resistance of the battery $E_r$; $C_m$, electric capacity of the MIM; $g_{AV}$, $g_V$, $g_L$, and $g_H$. For details, see Lemeshko (2002).

(Forrest, 2015). The altered physiological conditions including the membrane potentials that control the vital functions of the membrane channels are expected to regulate the mitochondrial channel transport phenomena to help raise the pathological condition of the channels in mitochondria called mitochondrial channelopathies. The versatile roles these membrane channels may play in various disease cells (compared to their conditions in healthy cells) is briefly addressed here. The channels may also experience considerable mutations or alterations in structures of the channel-forming membrane proteins (MPs) and other agents in the membrane, which may induce altered physiological roles in channels' energetics, thus resulting in transport mechanisms. We shall also inspect them carefully here for mitochondrial channels in cells experiencing a few vital diseases that humankind is dealing with to find a cure. The knowledge helps address drug discovery targeting mitochondrial membrane channels in the altered conditions due to diseases.

### 8.3.1 MITOCHONDRIAL ION CHANNELS LINKED TO CANCER AND RELATED ONCOLOGICAL TARGETS

Ion channels hosted in MOM and MIM are found linked to several cancer hallmarks. Therefore, these are branded as cancer channelopathies. Mitochondria generally play multiple central roles in the development of cancer. The classical cancer hallmarks such as metabolic reprogramming, invasion and induction of angiogenesis, apoptosis resistance, and sustained proliferation are all linked to mitochondria (Hanahan and Weinberg, 2011). The main role of mitochondrial ion channels is to actively participate in ion or metabolite transport and in fine-tuning of mitochondrial membrane potential, as well as in reactive oxygen species (ROS) release. Abnormal or altered functioning in these vital molecular processes, originating in mitochondrial ion channels, are somehow linked to cancer development, which works against the natural molecular process apoptosis. A recent review has elaborated on the participation and role of mitochondria-hosted ion channels leading to the acquisition of cancer hallmarks, and thus to cancer progression (Bachmann et al., 2019). Contribution of various mitochondrial ion channels to cancer hallmarks has been classified in Figure 8.9 (Bachmann et al., 2019).

The possible mechanisms that might link ion channel functions to the signaling events affecting cancer progression are shown in Figure 8.10 (Bachmann et al., 2019). Key kinases, for example, the protein kinase A (PKA), PTEN-induced kinase-1 (PINK1), c-Jun N-terminal kinase (JNK), and AMPK are found to readily translocate to MOMs. However, the role of the interface of

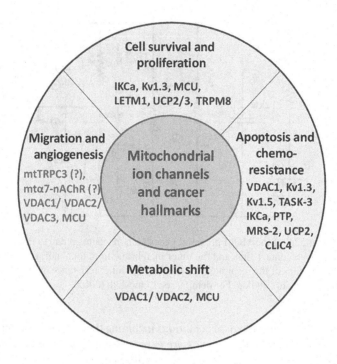

**FIGURE 8.9**  Mitochondrial ion channel contributions to cancer hallmarks. The major hallmarks, where ion channels' participation has been documented, are listed. The crucial role of uniporter and porins (VDAC) in the progression of cancer by acting on different hallmark processes is now evident. A putative role of some ion channels in angiogenesis is marked with questions (Bachmann et al., 2019).

mitochondria–cell communication remains to be yet found, concerning the actual mechanisms that ion channels might use to modulate the actions of these kinases. One possibility is the signaling due to ATP concentration (e.g., in the case of AMPK). Another possibility might be due to ROS. For details, readers should consult Bachmann et al. (2019).

As MOM and MIM ion channels are linked to cancer hallmarks, they naturally appear as oncological targets; research is currently ongoing on this matter. Drugs targeting these channels may perturb cancer progression via altering crucial molecular mechanisms that are responsible for the disorders. Figure 8.11 demonstrates ion channels with documented roles in the regulation of cell death (Leanza et al., 2014). For uncoupling proteins (UCPs), and acid-sensing ion channel 1a (ASIC1a) channels of mitochondria, the nature of the ions transported remains unknown (hence question marks are added!).

Table 8.7 lists numerous agents acting as pharmacological modulators of mitochondrial ion channels linked to various types of cancer cells (copied and modified from Leanza et al. (2014)).

### 8.3.2  MITOCHONDRIAL CHANNELOPATHY IN CARDIAC DISEASES

Mitochondria play crucial roles in ensuring essential physiological activity of the cardiovascular system. This mainly occurs by maintaining correct bioenergetic and anabolic metabolism, and most importantly by ensuring their phenomenological roles in intracellular $Ca^{2+}$ fluxes. The transport of materials, ions, and information in mitochondria occurs through versatile channels hosted in the MOM and MIM. A recent review presented the detail of the association of mitochondrial function with cardiovascular orders, and considered mitochondria as targets for repairing certain aspects of cardiovascular disorders (Bonora et al., 2018).

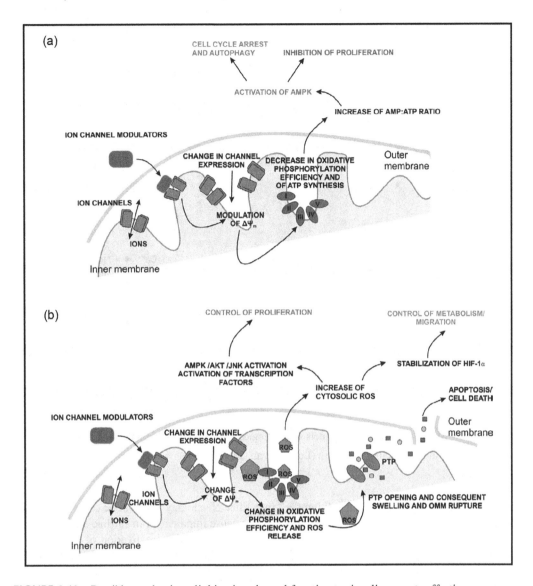

**FIGURE 8.10** Possible mechanisms linking ion channel function to signaling events affecting cancer progression. (a) A change in ion channel expression or function might affect the efficiency of respiration by modulating the transmembrane potential of the mitochondrial membrane. If the efficiency decreases, less ATP is expected to be produced, and so the activated protein kinase (AMP): ATP ratio increases leading to AMPK activation. The phosphorylation of downstream targets may be found to modulate different signaling pathways inducing the cell cycle arrest and inhibition of the proliferation. (b) Ion channel function, following a similar mechanism shown in (a), might be found to affect ROS release, prevalently at respiratory chain complexes I and III. However, ROS production is known to occur at the level of the complex II to a lesser extent.

Sustained release of the ROS in the matrix might induce the opening of the permeability transition (MPTP) and lead to depolarization, rupture, swelling, etc. of the MOM membrane and the mitochondrial release of multiple pro-apoptotic factors. ROS, produced at sub-lethal concentrations and then released to the cytosol, might also activate various kinases and transcription factors, may also help stabilize hypoxia-inducible factor 1-alpha (HIF1-$\alpha$). Thereby, ROS may control various processes, such as proliferation, migration, metabolic shift, and possibly the angiogenesis. For further details, see Bachmann et al. (2019).

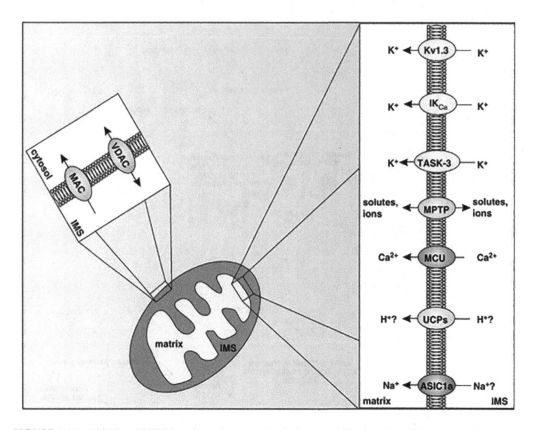

**FIGURE 8.11**  MOM and MIM ion channels as oncological targets. The ion channels cartooned here have documented roles in regulation of cell death (Leanza et al., 2014).

The role for homeostasis of intracellular $Ca^{2+}$ in cardiac physiology and genetic defects in L-type $Ca^{2+}$ channels of plasma membranes are well known to cause impairment in cardiac signal conduction, favoring arrhythmia development (Venetucci et al., 2012). Hyperactivation of the cytosolic $Ca^{2+}$-responsive enzyme, calcium/calmodulin-dependent protein kinase II (CaMKII), is known to be etiologically linked to various cardiovascular disorders, which often reflect the mitochondrial functions' regulation potential of CaMKII. Due to important physiological roles, mitochondrial $Ca^{2+}$ fluxes have been considered elusive drug targets. The $Na^+/Ca^{2+}$ exchanger NCLX inhibitors, such as CGP-37157 (CGP), KB-R7943, and SEA0400, are found to mediate promising cardioprotective effects in animal model studies of HF1 (Liu et al., 2014; 2018).

Liu et al. (2014) found in cardiomyocytes from failing hearts, insufficient amount of mitochondrial $Ca^{2+}$ accumulation, secondary to cytoplasmic $Na^+$ overload decreases the $NAD(P)H/NAD(P)^+$ redox potential and increases the oxidative stress when workload increases. These effects are abolished by enhancing the mitochondrial $Ca^{2+}$ with the acute treatment using CGP, an inhibitor of the mitochondrial $Na^+/Ca^{2+}$ exchanger (Liu et al., 2014).

Substantial evidence can be found O'Rourke et al. (2008) supporting the hypothesis that heart cell mitochondria act as a network of coupled oscillators. The network is collectively capable to produce frequency-and/or amplitude-encoded ROS signals under physiological conditions, see Figure 8.12. This intrinsic mitochondrial property may lead to a mitochondrial "critical" state. An emergent macroscopic response is predicted to manifest as complete collapse or synchronized oscillation in the network of mitochondria (Zamponi et al., 2018) under conditions of stress. The large amplitude depolarizations of the mitochondrial membrane potential $\Delta\Psi_m$ and ROS bursts

**TABLE 8.7**

**Modulating the Mitochondrial Ion Channels and Benefitting in Anticancer Treatment by Targeting the Cancer Cells**

| Channel | Cancer Cell Studies | Agents | Development Stage | Observations |
|---|---|---|---|---|
| VDAC | Adenocarcinoma (Simamura et al., 2008) | Furanonaphthoquinone (FNQs) (Simamura et al., 2008) | Preclinical development | FNQ acts on VDAC and causes alteration of mitochondrial membrane potential, NADH-dependent ROS production and cytochrome *c* release (Simamura et al., 2008) |
| | Non-small-cell lung cancer (Grills et al., 2011) | | | |
| | Glioblastoma (Wolf et al., 2011) | | | |
| | HELA cervical cancer cells (Shoshan-Barmatz and Mizrachi, 2012) | | | |
| | Melanoma (Fingrut and Flescher, 2002) | | | |
| | Prostate cancer cell line (Fingrut and Flescher, 2002) | | | |
| | Breast cancer cell line (Fingrut and Flescher, 2002) | | | |
| | Lymphoblastic leukemia (Fingrut and Flescher, 2002) | | | |
| | Multiple myeloma (Klippel et al., 2012) | | | |
| | El-4 lymphoma (Fingrut and Flescher, 2002) | | | |
| | | Clotrimazole (Quast et al., 2012) | Clinical trial for Sickle cell disease (SCD) (see ref. 52 in ref. 74) | Commonly used as fungicide (Quast et al., 2012) |
| | | | | Inhibits KCa3.1 calcium-dependent potassium channel (Marchi et al., 2009) |
| | | Erastin (Yagoda et al., 2007) | In vitro studies (Yagoda et al., 2007) PRLX 93936, erastin homolog in clinical trial for multiple myeloma (NCT01695590) | RAS/RAF/MEK-dependent action (Yagoda et al., 2007) |
| | | | | Oxidative cell death (not apoptosis) was induced via VDAC2 or VDAC3 (Simamura et al., 2008) |
| | | Methyl jasmonates (Fingrut and Flescher, 2002) | Preclinical development (Fingrut and Flescher, 2002; Klippel et al., 2012) | Plant hormones induce MOMP in isolated mitochondria. |
| | | | Topical application in cancerous skin lesions in humans (Palmieri et al., 2011) | Disruption of the HK–VDAC interaction (Galluzzi et al., 2008); activation of BAX and Caspase-3 via ROS production (Kim, 2004; Klippel, 2012) |
| | | | | Indirect activation of MPTP |

*(Continued)*

**TABLE 8.7 (*Continued*)**

**Modulating the Mitochondrial Ion Channels and Benefitting in Anticancer Treatment by Targeting the Cancer Cells**

| Channel | Cancer Cell Studies | Agents | Development Stage | Observations |
|---|---|---|---|---|
| | | 3-Bromopyruvate (Pedersen, 2012) | Clinical trial in hepatocellular carcinoma (Ko et al., 2012) | A pyruvate mimetic Acts as an inhibitor of glycolysis (Shoshan, 2012) but also as a covalently alkylating agent Pyruvylates certain proteins at cysteine residues (Ko et al., 2012) |
| MPTP | Melanoma (Pereira et al., 2007) Prostate Breast cancer Lymphoblastic leukemia (Flescher, 2005) Chronic lymphocytic leukemia (Flescher, 2005) Acute promyelocytic leukemia (Elliott et al., 2012) Multiple myeloma (Flescher, 2005) | CSA (Leanza et al., 2012) | Clinical use for different diseases | CSA inhibits MPTP (Szabó and Zoratti, 1991; Bernardi, 2013) |
| | | 4-(*N*-(*S*-glutathionylacetyl)-amino) phenylarsenoxide (GSAO) (Elliott et al., 2012) | Development for phase I clinical trial (Elliott et al., 2012) | Novel antineovascular agent that interacts with redox active, mitochondrial protein dithiols in endothelial cells (Elliott et al., 2012) |
| | | Mastoparan-like sequences (Jones et al., 2008) | Preclinical development (Jones et al., 2008) (in clinical trial but not for cancer) | Cell penetrant, mitochondriotoxic and apoptogenic (Jones et al., 2008) |
| | | Betulinic acid (Lena et al., 2009) | Phase I/II clinical trials NCT00346502 | Steroid-like structure. Induces MOMP in isolated mitochondria (Galluzzi et al., 2008) |
| | | Gold complex AUL12 (Chiara and Rasola, 2013) | Preclinical development | Inhibits complex I Increases ROS levels leading to activation of GSK3α/β, favoring MPTP opening (Chiara and Rasola, 2013) |
| | | CD437 (Lena et al., 2009) | Preclinical development Retinoid analog NRX195183 is in clinical trial (NCT00675870) | Synthetic retinoid (triterpenoid) able to promote MOMP independently of nuclear receptors (Galluzzi et al., 2008) |

*(Continued)*

**TABLE 8.7 (*Continued*)**
**Modulating the Mitochondrial Ion Channels and Benefitting in Anticancer Treatment by Targeting the Cancer Cells**

| Channel | Cancer Cell Studies | Agents | Development Stage | Observations |
|---|---|---|---|---|
| | | Berberine (Pereira et al., 2007) | In vitro studies (Pereira et al., 2007) | Induces mitochondrial fragmentation and depolarization, oxidative stress and decreased ATP levels (Gulbins et al., 2010) |
| | | Honokiol (Li et al., 2007) | In vitro studies on different cancer cell lines and in clinical trial for microbial infection, anxiety, oxidative stress and platelet aggregation (Li et al., 2007) | Induces MPTP-dependent cell death (Li et al., 2007) |
| | | α-Bisabolol (Cavalieri et al., 2009) | In vitro studies on glioma cell lines; (Cavalieri et al., 2009); Preclinical, ex–in vivo studies on human acute leukemia (Cavalieri et al., 2011) | Induces intrinsic apoptotic pathway through the loss of mitochondrial inner transmembrane potential and release of cytochrome *c* (Cavalieri et al., 2009) |
| | | Shikonin (Han et al., 2007) | In vitro studies (Han et al., 2007) Clinical trial on human lung cancer (Guo et al., 1991) | Induces necroptosis (Han et al., 2007) |
| Kv1.3 | T-cell leukemia cell line (Leanza et al., 2012) Chronic B lymphocytic leukemia (CLL) (Leanza et al., 2013) Osteosarcoma cell line Melanoma (Leanza et al., 2012) Colon rectum cancer (Schonherr, 2005) Breast cancer (Schonherr, 2005) Prostate cancer (Schonherr, 2005) | Psora-4 (Leanza et al., 2012) | Preclinical development. Methoxypsoralen is in clinical trial for cutaneous T-cell lymphoma NCT00056056 | Blocks inward K$^+$ flux inducing ROS, mitochondrial depolarization, cytochrome *c* release and cell death (Leanza et al., 2012) |
| | | PAP-1 (Leanza et al., 2012) | Preclinical development | Blocks inward K$^+$ flux inducing ROS, MT depolarization, cyt *c* release and cell death (Leanza et al., 2012) |

*(Continued)*

**TABLE 8.7 (*Continued*)**

**Modulating the Mitochondrial Ion Channels and Benefitting in Anticancer Treatment by Targeting the Cancer Cells**

| Channel | Cancer Cell Studies | Agents | Development Stage | Observations |
|---|---|---|---|---|
| | | Clofazimine (Leanza et al., 2012) | Preclinical development, ex–in vivo studies on chronic lymphocytic leukemia (Leanza et al., 2013). Used clinically to treat leprosy and autoimmune diseases (Ren et al., 2008) | Induces mtROS production and triggers cell death (Leanza et al., 2012) |
| $IK_{Ca}$ | Colon cancer cells (Marchi et al., 2009) Melanoma (Quast et al., 2012) Breast cancer (Schonherr, 2005) Prostate cancer (Schonherr, 2005) Pancreas cancer (Schonherr, 2005) | Clotrimazole (Marchi et al., 2009) | Clinical trial for Sickle cell disease | Systemic application is not possible because of hepatotoxicity induced by nonspecific effects on cytochrome *P450* (Quast et al., 2012) |
| | | TRAM-34 (Marchi et al., 2009) | | Clotrimazole analog, lacking cytochrome *P450* inhibitory effects, able to enhance TRAIL–induced apoptosis via the mitochondrial pathway (Quast et al., 2012) |
| UCP2 | Breast cancer (Ayyasamy et al., 2011) leukemia, ovarian, bladder, esophagus, testicular, colorectal, kidney, pancreatic, lung and prostate tumors (Baffy et al., 2011) | Genipin (Ayyasamy et al., 2011) | In vitro and in vivo studies (Ayyasamy et al., 2011) | Abolishes UCP2-mediated proton leak (Ayyasamy et al., 2011) |
| TASK-3 | Ovarian cancer (Innamaa et al., 2013) | Zinc and methanandamide | | Nonspecific inhibitors of plasma membrane TASK-3 |
| Ceramide | Normal and cancer cells (general) | Aptamers | Theoretical design, *in silico and in vitro* target binding studies, ongoing research (Ashrafuzzaman, 2020) | Direct ceramide lipid binding assays |

Abbreviations: ATP, adenosine triphosphate; CSA, cyclosporin A; GSK3α/β, glycogen synthase kinase-3α/β; MOMP, mitochondrial outer membrane permeabilization; MPTP, mitochondrial permeability transition pore; ROS, reactive oxygen species; TASK-3, TWIK-related acid-sensitive $K^+$ channel-3; TRAIL, tumor necrosis factor-related apoptosis-inducing ligand; UCP2, uncoupling protein-2; VDAC, voltage-dependent anion channel. The table summarizes the pharmacological tools used to elucidate the role of the indicated ion channels in several cancer cell types, with either in vitro or in vivo studies. SK2, MCU and ASIC1a are not mentioned in this table because pharmacological evidence for their involvement in cancer cell death regulation is not available.

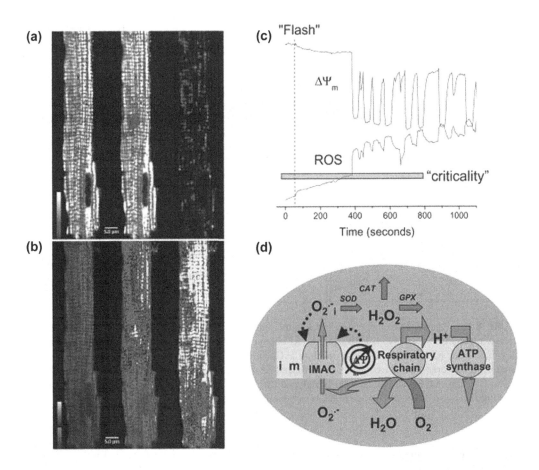

**FIGURE 8.12** Mitochondrial network depolarization following a local laser flash. (a) Upper panels. $\Delta\Psi_m$ signal before the flash, close to criticality, and after global depolarization. Lower panels: ROS signal before the flash, close to criticality, and after the global depolarization. (b) The mitochondrial number with ROS above threshold increases to ~60% at criticality, just prior to the global depolarization and limit cycle oscillation of the network (Aon et al., 2004). (c) Model of mechanism of the ROS-dependent mitochondrial oscillator (Cortassa et al., 2004). For details, see O'Rourke et al. (2008).

have widespread effects on all cellular subsystems, including energy-sensitive plasma membrane ion channels (Chen and Zweier, 2014). This affects electrical and contractile dysfunction at an organ level. Mitochondrial ion channels appear to play key roles in the mechanisms of this nonlinear network phenomenon, and are important targets for therapeutic intervention.

### 8.3.3 MITOCHONDRIAL CHANNELOPATHY AND AGING

Mitochondrial channelopathies are found in aging. This affects the $K^+$, $Ca^{2+}$, VDAC, and permeability transition pore ($Ca^{2+}$; PTP) channels. A recent review addressed this important issue (Strickland et al., 2019).

Mitochondrial $Ca^{2+}$ cycling is impaired with aging in neurons, resulting from reduced $Ca^{2+}$ channel activity and reduced recovery after synaptosomal stimulation (Figure 8.13) (Satrustegui et al., 1996). The reduced recovery rate of calcium results in reduced mitochondrial membrane potential and delayed repolarization, thus causing mitochondrial dysfunction with aging. This effect was reported in the hearts of 2-year-old senescent rats (Jahangir et al., 2001).

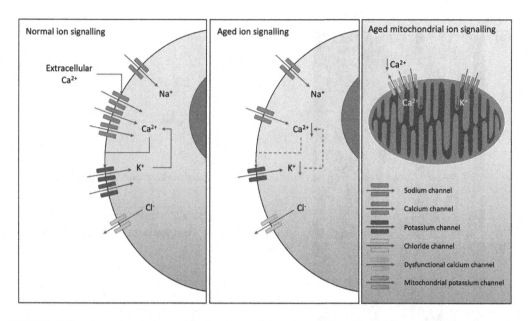

**FIGURE 8.13** Age-related changes in ion channel function. $Ca^{2+}$ release-activated $Ca^{2+}$ channel helps increase intracellular $Ca^{2+}$ levels, activating the $K^+$ channel opening and sustained $Ca^{2+}$ signaling, while the efflux of chloride ($Cl^-$) ions are found to inhibit the $Ca^{2+}$ influx. Downregulation of $Ca^{2+}$ channels is demonstrated in AD. Decreased expression of $Ca^{2+}$-activated $K^+$ channels is noted with aging. Within mitochondria, reduced $Ca^{2+}$ ion channel activities result in reduced $Ca^{2+}$ cycling. The expression of potassium channels on mitochondria is also reduced with age in heart sarcolemma (Strickland et al., 2019).

Regarding potassium channels, their density on mitochondrial surface is found to significantly decline with age and with metabolic syndromes in the heart sarcolemma, see Figure 8.13 (Strickland et al., 2019; Ranki et al., 2002; Truong et al., 2016). This altered physiological condition has been shown to reduce the tolerance to ischemia reperfusion, as well as increased injury in aged guinea pig, rat hearts, and humans (Roscoe et al., 2000; Kamada et al., 2008). These effects have repercussions in causing increased susceptibility to myocardial infarction and reduced neuronal activity in elderly subjects as mitochondrial $K^+$ channels are known to play a neuroprotective role in the neurological reperfusion injury in postnatal mouse pups (Connors et al., 2016). Amyloid-β (A β) plaques in Alzheimer's disease are found to increase intracellular calcium levels (Demuro et al., 2011), and $Ca^{2+}$ uptake into the mitochondria through VDAC and calcium uniporter has been shown to increase substantial mitochondrial stress responses; the process is found to initiate apoptosis in rat cortical neurons in vitro and hippocampal slices ex vivo (Alberdi et al., 2010). Studies on Parkinson's disease have revealed that α-synuclein acts via VDACs and promote the mitochondrial toxicity of respiratory chain components, yeast model of Parkinson's disease (Rostovtseva et al., 2015).

With aging, the mitochondrial electron transport chain efficiency weakens. Consequently, the cellular ATP production decreases, and electron leakage and ROS production increase, observed in model organisms, such as *Caenorhabditis elegans* and *Drosophila* (Ferguson et al., 2005; Rea et al., 2007; Chistiakov et al., 2014). While the relationship of the electron transport chain activity with aging appears with conflicting trends (Doria et al., 2012), mitochondria form the theoretical basis examining the phenomenon of aging: the accepted one is the free radical theory of aging (Harman, 1982; 1992). The theory deals with the increased ROS production linked to aging, which leads to progressive cellular damage, see Figure 8.14 (Strickland et al., 2019). ROS is known to favor mitochondrial DNA mutations with aging.

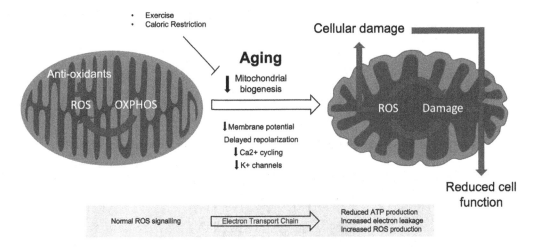

**FIGURE 8.14** Mitochondrial dysfunction during aging. Healthy mitochondria produce ROS through regular oxidative (OXPHOS) activity which aid in normal cell processes; this ROS production is kept in check by various antioxidant systems to prevent oxidative damage. During aging, dysfunctional mitochondria accumulate due to reduced biogenesis and ROS control. This increased ROS production induces both further mitochondrial and cellular damages, resulting in reduced cell function, and eventual apoptosis (Strickland et al., 2019).

### 8.3.3.1  Mitochondrial Channelopathy and Alzheimer's Disease

Amyloid-β (Aβ) plaques in Alzheimer's disease are found to increase the level of calcium ions $Ca^{2+}$ intracellularly. This intracellular increase of calcium and the mitochondria uptake through VDAC and calcium uniporter are found to increase multiple mitochondrial stress-related responses, which eventually work towards initiation of apoptosis. Mitochondrial channels may have specific and nonspecific roles in aging (briefly explained earlier) and Alzheimer's disease (a complex neurodegenerative disorder briefed here) (Strickland et al., 2019).

An article recently summarized mitochondrial membrane channels, which are potential candidates as drug targets during the treatment of neurodegenerative diseases (Peixoto et al., 2015). These channels include VDAC, protein-import channels, $Ca^{2+}$ channels, protein-import channels, and the Mrs2 – $Mg^{2+}$ channel. The mitochondrial channels known with established roles in neuronal death include the mitochondrial apoptosis-induced channel (MAC) and the VDAC of the MOM and the MIM-hosted mitochondrial permeability transition pore (mPTP). The mitochondrial ATP-dependent potassium channels (KATP) are also predicted to be linked in neuronal degeneration.

Brain atrophy that is caused by neuronal loss is a prominent pathological feature of Alzheimer's disease. Mitochondrial ASIC1a may serve as an important regulator of mitochondrial permeability transition (MPT) pores, which contributes to oxidative neuronal cell death (Wang et al., 2013). Figure 8.15 presents the schematics of possible mechanisms of action of ASIC1a in mitochondria (Wang et al., 2013).

### 8.3.3.2  Mitochondrial Channelopathy and Parkinson's Disease

The mitochondrial accumulation of $Ca^{2+}$ is known to lead to oxidative phosphorylation activation and subsequent increase in ATP production (Gleichmann and Mattson, 2011). The mitochondrial $Ca^{2+}$ channels of the MIM are responsible for the mitochondrial uptake of the cytosolic $Ca^{2+}$ (Williams et al., 2013). These chain molecular processes help meet the metabolic demands that are associated with the neuronal electrical activity, whose permanent-type disruption may be linked to the rise of Parkinson's disease. Figure 8.16 presents a network of mitochondrial dysfunctions, including in mitochondrial membrane transport mechanisms and in membrane-based channel functions that lead to Parkinson's disease (Perier and Vila, 2011).

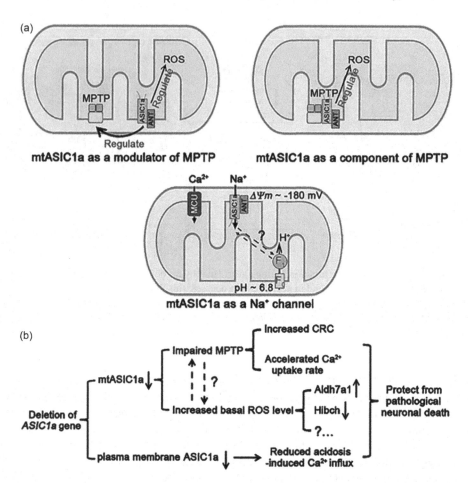

**FIGURE 8.15** Schematics of possible mechanisms of ASIC1a actions in mitochondria. (a) (1) Nonconducting role(s) of mtASIC1a include possible functioning as a modulator of mitochondrial permeability transition (MPT) pores (upper left panel), possible functioning as one of MPT pore (MPTP) components (upper right panel). (2) mtASIC1a may mediate the Na⁺ influx, reducing the driving force and counteracting MCU. (b) The protective effects of ASIC1a gene deletion in the pathological neuronal death. Here, we have used ANT for adenine nucleotide translocase; MCU for mitochondrial Ca²⁺ uniporter; MPTP for mitochondrial permeability transition pore, $\Delta\Psi_m$ for mitochondrial membrane potential (may be altered due to disease condition). For further details, see Wang et al. (2013).

Aggregation of the α-synuclein protein (encoded by the SNCA gene), in the form of Lewy bodies and Lewy neurites, is one of a few of the known neuropathological hallmarks of Parkinson's disease. However, the mechanisms that might specifically work for α-synuclein in Parkinson's disease remain obscure (Rostovtseva et al., 2015; Xu and Pu, 2016). Recent studies addressing Parkinson's disease have revealed that α-synuclein acts via the VDAC to promote mitochondrial toxicity of respiratory chain components in a yeast model of Parkinson's disease. Figure 8.17 shows the blockage of VDAC by α-syn in nanomolar concentrations (Rostovtseva et al., 2015). α-Syn mutations A53T and A30P have also been found not to affect VDAC blockage, see Figure 8.18 (Rostovtseva et al., 2015). Mitochondrial channel VDAC involvement in Parkinson's disease hints that mitochondrial channelopathy is a direct case in regard to Parkinson's disease.

Voltage dependences of on-rates and residence times for the α-syn wild-type and two mutants, A53T and A30P have been studied. Synucleins, added to both sides of the membrane, were at 50 nm concentrations (Rostovtseva et al., 2015). Another article has also addressed additional in-depth

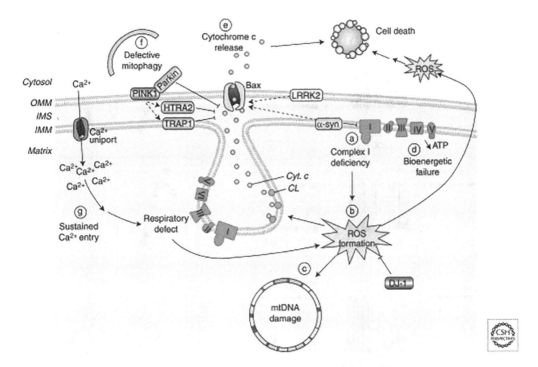

**FIGURE 8.16** Mitochondrial dysfunction in Parkinson's disease (PD). Alterations in several aspects of mitochondria biology have been linked to the pathogenesis of PD. (a) Reduced activity of complex I, (b) increase in production of the mitochondria-derived ROS, (c) ROS-mediated damage of mtDNA, (d) failure in bioenergetics, (e) Bax-mediated release of the cytochrome $c$ and activation of mitochondria-dependent apoptotic pathways, (f) defective mitophagy, or (g) increased mitochondrial $Ca^{2+}$-buffering burden. Many of the mutated nuclear genes linked to familial forms of PD, for example, PINK1, Parkin, α-synuclein, DJ-1, or LRRK2, etc., have been shown to affect many features of mitochondria. CL: cardiolipin, Cyt. C: cytochrome $c$, HTRA2: high temperature requirement A2, IMM or MIM: inner mitochondrial membrane, IMS: intermembrane space, LRRK2: leucine-rich-repeat kinase 2, OMM or MOM: outer mitochondrial membrane, PINK1: phosphatase and tensin homolog-induced kinase 1, ROS: reactive oxygen species, TRAP1: tumor necrosis factor receptor-associated protein 1, α-syn for alpha-synuclein. For details, see Perier and Vila (2011).

information on the involvement of impaired mitochondrial structural components including channels involved in the pathogenesis of PD (Chen et al., 2019). A model on this issue has been presented in Figure 8.19 (Chen et al., 2019).

## 8.4 CHANNELOPATHIES OF THE NUCLEAR MEMBRANE AND THERAPEUTIC TARGETS OF ION CHANNELS

Genetic mutations encoding proteins of the nuclear envelope (NE) cause an array of general disorders or specific diseases, known as "nuclear envelopathies" or "laminopathies." The mutations in NE proteins first start causing disturbances in the various structural and functional moieties of NE responsible for maintaining the localized transport processes across the two nuclear bilayers NOM and NIM. Figure 8.20 presents a schematic diagram of the NE-based localizations and interactions of different proteins, including the highlighted diseases caused due to genetic mutations encoding the indicated proteins (Worman and Bonne, 2007).

The disorders and mutations in structural components of NE may ultimately affect tissues and even damage vital organ systems. Table 8.8 lists an array of genetic disorders and mutations encoding NE proteins (Dauer and Worman, 2009).

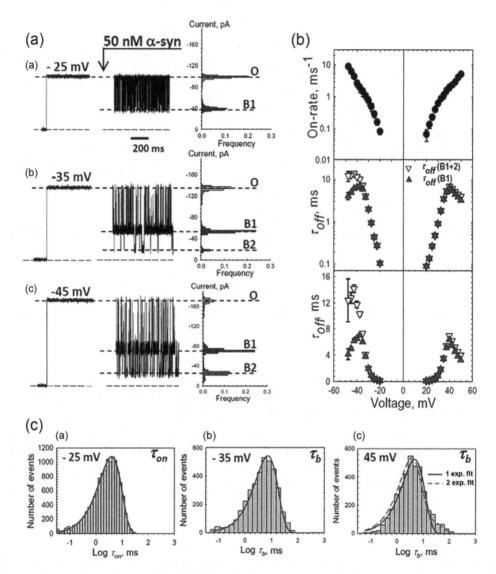

**FIGURE 8.17** VDAC is reversibly blocked by α-syn in nanomolar concentrations. (a) representative traces of currents representing ion flow through the single VDAC channel before (left panel) and after (right panel) adding 50 nm α-syn to both sides of membrane at indicated voltages. The blockage events are characterized by two well-conducting states, "blocked state 1" (B1) and "blocked state 2" (B2). B2 is seen at |V|≥30mV. The event amplitude histograms show the relative probability of B2 that increases with voltage. The dashed lines indicate open (O) and blocked (B1, B2) states of conductance and zero current. The solution bathing the membrane contained 1 m KCl buffered using 5 mm HEPES, pH 7.4. Current records were then filtered using a 5-kHz 8-pole digital Bessel filter. (b) voltage dependences of on-rate of α-syn blockage, ⟨τon⟩ − 1, in the presence of 50 nm α-syn (top panel) and of the residence time, τoff = ⟨τb⟩, where τb is the time of the blockage event. The residence time is presented in both logarithmic (middle panel) and linear (bottom panel) scales. τoff(B1) was calculated as the average time at the first blocked conductance level; τoff(B1 + B2) gives the average time in the blocked states without discrimination between the two states. Error bar, S.E. C, corresponding log-binned distributions (33) of the open time, τon, at −25 mV (a) and of the time of the blockage events, τb, calculated for both closed states (B1 + B2) at −35 mV (b) and 45 mV (c) from statistical analysis of the current records at 50 nm α-syn such as those shown in A. Solid lines, logarithmic single exponential fittings with characteristic times ⟨τon⟩ equal to 3±0.1 ms (a) and ⟨τb⟩ equal to 6.5±0.1 ms (b) and to 4.7±0.1 ms (c). A two-exponential fit of the blockage time histogram (dashed line) at 45 mV (c) with characteristic times of 3.5 and 22.1 ms fits the long-time events satisfactorily but not the short-time blockages. For details, see Rostovtseva et al. (2015).

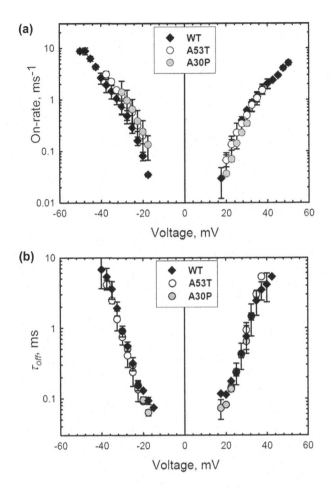

**FIGURE 8.18** α-Syn mutations A53T and A30P have not been found to affect VDAC blockage.

Numerous diseases caused by mutations in genes ($N_{mut.gene}$) encoding nuclear lamins and/or associated proteins grow following a power law (with an exponent higher than 1) with time in year (t), as follows:

$$N_{mut.gene} \sim t^s \text{ with } 1 < s < 2.$$

The value of s is deduced from the data presented in Dauer and Worman (2009). This trend is due to more and more investigations that reveal new mutations over time. This trend may continue until a threshold leading to a plateau is achieved only after the application of super high-resolution inspecting techniques in biological investigations (yet to be achieved) when most of the genetic mutations will be clarified. A technological revolution is needed to achieve this.

Tissue-specific diseases mostly result from a multitude of mutations in genes encoding NE proteins, and these diseases are also cell-type-specific, resulting in varied NE functions including altered actions and molecular mechanisms. The tissue-specific diseases result notably from mutations in the gene encoding the widely expressed A-type lamins, LMNA. LMNA mutations are found to cause varieties of clinical disorders that generally affect tissues, and some even the aging process. These mutations are expected to cause alterations in NE-localized function including nucleocytoplasmic transport (mediated by nuclear pore complexes) and specific functions of nuclear pore proteins (nucleoporins). We wish to also brief on the biophysical aspects of various sectional disorders, such as those found in NPC, NOM, and NIM, where the nuclear transports mainly

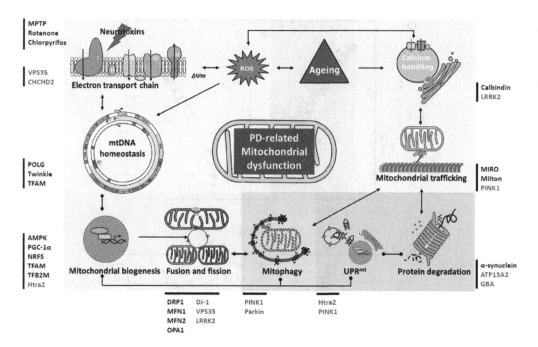

**FIGURE 8.19** Schematic presentation of mitochondrial involvement in the pathogenesis of Parkinson's disease (PD). This diagram serves to highlight the complex links between the changes in mitochondrial homeostasis, turnover, quality control, and trafficking in cases of PD. These mitochondrial alternations are also intricately associated with the aging process and impairments of the ubiquitin protease system that are attributed to Lewy body pathology. Lines with dots represent interactive effects, lines with arrows represent regulatory effects. Genes associated with familial PD are shown in blue, while we have also highlighted other proteins and toxins, recently been associated with PD, which impact mitochondrial function. MPTP: 1-methyl-4-phenyl-1,2,3,6-tetrahydropyridine; VPS35: vacuolar protein sorting 35; CHCHD2: coiled-coil-helix-coiled-coil-helix domain containing 2; TFAM: mitochondrial transcription factor A; AMPK: AMP-activated protein kinase; PGC-1α: peroxisome proliferator-activated receptor γ (PPARγ) coactivator 1α; NRF: nuclear respiratory factors; TFB2M: dimethyladenosine transferase 2; DRP1: dynamin-1-like protein; MFN: mitofusin; LRRK2: Leucine-rich repeat kinase 2; PINK1: phosphatase and tensin homolog (PTEN)-induced putative kinase 1; GBA: lysosomal enzyme glucocerebrosidase; MIRO: mitochondrial Rho GTPase1; ΔΨm: mitochondrial membrane potential; ROS: reactive oxygen species; mtDNA: mitochondrial DNA; UPRmt: the mitochondrial unfolded protein response. For details, see Chen et al. (2019).

happen through or in connection with various pores and ion channels. The disorders caused due to mutations in ion channels and proteins thereof are known as channelopathies associated with the nuclear ion channels. Various diseases may be caused due to ion channel mutations and vice versa.

## 8.4.1 NUCLEAR PORE COMPLEX TARGETS

The nuclear pore complexes (NPCs) are composed of several nucleoporin (Nup) units, which are found to arrange structurally and form the main channels. These channels facilitate the transport between the nucleoplasm and cytoplasm and allow small molecules (e.g., small proteins, metabolites, ions, etc.) to pass through the gates by passive diffusion, whereas large ones are found to require particular transport receptors (such as importins and exportins). Phe-Gly (FG) Nups are known to form the central channel of the NPC. Figure 8.21 demonstrates the structural components of NPC including a list of the Nups (Khan et al., 2020).

Mutations in Nups may cause defects in the central channels of NPCs. Mutations in a few of the Nups have already been observed. For example, Nup93, observed in all kidney cells, experiences mutation

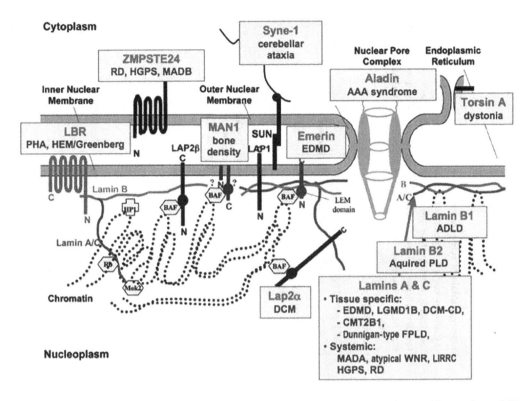

**FIGURE 8.20** Schematic diagram of the nuclear envelope (NE)-based localizations and interactions of different proteins including the highlighted (in boxes) diseases caused due to genetic mutations encoding the indicated proteins. Restrictive dermopathy (RD), Hutchinson-Gilford progeria syndrome (HGPS), Pelger-Huet anomaly (PHA), Emery–Dreifuss muscular dystrophy (EDMD), Adult onset autosomal dominant leukodystrophy (ADLD), Barraquer–Simons syndrome (BSS), limb-girdle muscular dystrophy (LGMD); dilated cardiomyopathy (DCM); Charcot-Marie-Tooth (CMT), familial partial lipodystrophy (FPLD), mandibuloacral dysplasia (MADA/MADB), Werner syndrome (WNR). For details including other relevant information, see Worman and Bonne (2007).

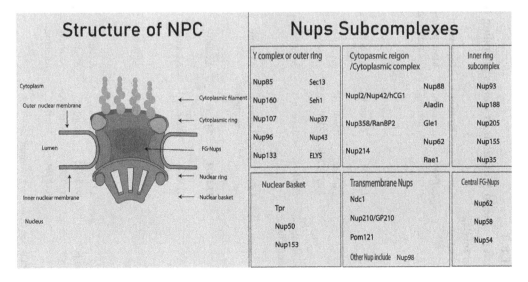

**FIGURE 8.21** Structure of the nuclear pore complex and Nups subcomplexes (Khan et al., 2020).

**TABLE 8.8**

**Disorders Due to Mutations in Genes Encoding NE Proteins**

| Nuclear Envelope Component | Gene | Protein(s) | Expression | Major Protein Function | Phenotype (Main Affected Tissues) |
|---|---|---|---|---|---|
| Lamina | LMNA | Lamin A, Lamin C | Most differentiated somatic cells | Structural support of nucleus; implicated in DNA synthesis and transcription | (1) Cardiomyopathy sometimes with muscular dystrophy (cardiac and skeletal muscle); (2) partial lipodystrophy (adipose); (3) peripheral neuropathy (peripheral nerve); (4) progeroid features (early aging affecting several tissues including skin, adipose, arteries); and (5) mandibuloacryl dysplasia (bone, adipose, some progeroid features) |
| | LMNB1 | Lamin B1 | All or most somatic cells | As above | Leukodystophy (CNS glial cell) |
| | LMNB2 | Lamin B2 | As above | As above | Possible partial lipodystrophy (adipose) |
| | ZMPSTE24 | ZMPSTE24 | As above | Processing of prelamin A | Progeroid features and restrictive dermopathy (skin and bone) |
| Nuclear membranes and perinuclear space | EDM | Emerin | As above | Binds lamins and other proteins | Cardiomyopathy with muscular dystrophy (cardiac and skeletal muscle) |
| | LBR | Lamin B receptor | As above | Binds B-type lamins, DNA, and heterochromatin protein 1; sterol reductase | (1) Pelger–Huët anomaly (heterozygous; neutrophils) and (2) Greenberg skeletal dysplasia (homozygous; multi-systemic developmental abnormalities most prominently affecting bone that causes neonatal lethality) |
| | LEMD3 | MAN1 | As above | Binds rSmads, antagonizing TGF-β signaling | Sclerosing bone dysplasia (bone and sometimes also skin) |
| | SYNE1 | Nesprin-1 | As above | Component of LINC complex that connects nucleus to cytoplasmic cytoskeleton | Cerebellar ataxia (CNS neurons) |
| | TOR1A | TorsinA | As above | AAA+ ATPase of ER and perinuclear space; binds lamina-associated polypeptide 1 | DYT1 dystonia (CNS neurons) |
| Nuclear pore complex | AAAS | Aladin | As above | Nuclear pore-associated | Triple-A syndrome (lower esophageal sphincter, adrenal gland, lacrimal gland) |
| | NUP155 | Nup155 | As above | Nucleocytoplasmic transport | Atrial fibrillation (heart) |
| | NUP62 | Nup62 | As above | Nucleocytoplasmic transport | Infantile striatal necrosis (CNS) |
| | RANBP2 | RanBP2 | As above | Nucleocytoplasmic transport | Acute necrotizing encephalopathy (CNS) |

For details, see (Dauer and Worman, 2009).

**TABLE 8.9**

**Phenotypes of Mice Possessing Nucleoporin Mutations (Sakuma and D'Angelo, 2017)**

| Nucleoporin | Mutation | Phenotype |
|---|---|---|
| Aladin/AAAS | Homozygous null | Sterility |
| ELYS | Homozygous null | Embryonic lethality |
| ELYS | Intestinal epithelium null | Juvenile growth delay |
| NDC1 | Homozygous null | Sterility and developmental defects |
| Nup35 | F192L Mutant | Degenerative colonic smooth muscle myopathy |
| Nup50 | Homozygous null | Embryonic lethality |
| Nup88 | Overexpression | Increased genetic tumor model tumorigenesis |
| Nup88 | Homozygous null | Not viable |
| Nup96 | Homozygous null | Embryonic lethality |
| Nup96 | Heterozygous null | Reduced antigen presentation and immunodeficiency |
| Nup98 | Homozygous null | Embryonic lethality |
| Nup98 | HoxA9/HoxD13 fusion | Spontaneous leukemia |
| Nup133 | Homozygous null | Embryonic lethality |
| Nup155 | Heterozygous null | Atrial fibrillation and early sudden cardiac death |
| Nup155 | Homozygous null | Embryonic lethality |
| Nup214 | Homozygous null | Embryonic lethality |
| Nup358/RanBP2 | Homozygous null | Embryonic lethality |
| Nup358/RanBP2 | Heterozygous null | Reduced size and impaired glucose homeostasis |
| Nup358/RanBP2 | Hypomorph | Spontaneous and carcinogen-induced tumorigenesis |
| Rae1 | Homozygous null | Embryonic lethality |
| Nup98 + Rae1 | Double Heterozygous Null | Increased carcinogen-induced tumorigenesis |
| Sec13 | Homozygous null | Embryonic lethality |
| Sec13 | Hypomorph | Reduced antigen presentation |

which leads to the pathogenesis of focal segmental glomerulosclerosis (Hashimoto et al., 2019). A reduction in human podocyte proliferation was achieved by knocking Nup93 down. Nup93 interacts with SMAD4 signaling protein. Mutation in Nup93 is found to lead to the abrogation of SMAD activity and causes steroid-resistant nephrotic syndrome (Braun et al., 2016). Mutation in Nup155 leads to atrial fibrillation and early sudden cardiac death (Zhang et al., 2008). A mutation in the nucleoporin-107 gene causes XX gonadal dysgenesis (Weinberg-Shukron et al., 2015). Two different homozygous nonsense gene variants of Nup188, p.Tyr96* and p.Gln113*, have been identified, and a correlation between gene variants and a severe phenotype of a new developmental syndrome having poor prognosis (due to insufficiency of nucleoporin 188 homolog proteins) is observed (Sandestig et al., 2019). ELYS is a nucleoporin required for postmitotic NPC assembly. Mutation of the zebrafish nucleoporin elys, a nucleoporin required for postmitotic NPC assembly, was reported to sensitize tissue progenitors to replication stress (Davuluri et al., 2008). We could continue listing more mutations in NPC proteins in addition to those mentioned here that might regulate the NPC channel transport mechanisms and might cause NPC channelopathies. Table 8.9 lists nucleoporin mutations in mice (Sakuma and D'Angelo, 2017). The human version of mutations are not all investigated yet. However, these mutations, if found in human NPC, the overall NPC channel transport can be expected to be altered, causing the NPC channelopathies.

## 8.4.2 MUTATIONS REGULATE ION CHANNEL TRANSLOCATION BETWEEN NUCLEAR AND PLASMA MEMBRANES

Voltage-gated potassium channel, BK or $BK_{Ca}$ (large-conductance calcium-activated potassium channels) channel, regulates gene expression by controlling nuclear calcium signaling (Li et al.,

2014). Nuclear BK channels regulate the influx of calcium from the perinuclear space into the nucleoplasm. Paxilline inhibition of nuclear BK channels was found to produce transient increases in $Ca^{2+}$ concentration in the nucleoplasm, and this nuclear calcium signaling was reported to affect calcium-dependent gene transcription, neuronal activity, and dendritic arborization.

Hutchinson–Gilford progeria syndrome (HGPS), a fatal genetic condition, was reported to manifest as premature aging caused by genetic mutations in the NE protein lamin A. Electrophysiological studies of human dermal fibroblasts, from HGPS patients, exhibited higher plasma membrane expression and larger outward $K^+$ currents than those recorded from healthy young subjects, see Figure 8.22 (Zironi et al., 2018). This suggests that as NEs are genetically disrupted, BK channels that normally partition between NEs and plasma membranes are forced into plasma membranes.

Another recent study hints toward mutation-induced regulation of BK channel translocation (Chen et al., 2020). BK channels colocalize with chromatin and nuclear lamins in electroreceptor cells. An alternatively spliced bipartite nuclear localization sequence (NLS) in kcnma1 (at site of

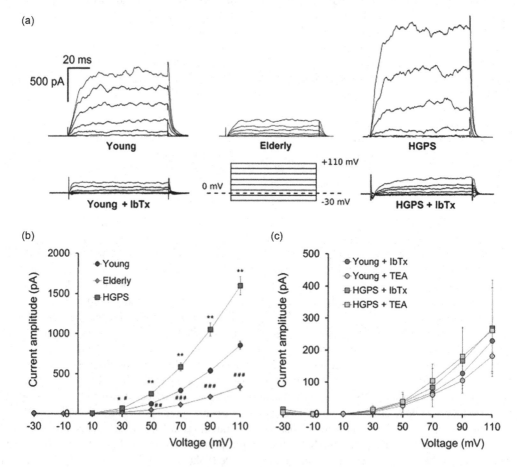

**FIGURE 8.22** Outward currents patch-clamp recorded in whole-cell configuration. (a) Representative examples of current traces recorded in hDF obtained from a young donor, an elderly, and a patient affected by HGPS. Current traces recorded after 100 nM IbTx (BK channel blocker) application and a graphical representation of the pulse protocol (holding potential at 0 mV) are also shown. (b) Average ± SEM of current–voltage relationships (I–V) recorded in hDF obtained from healthy donors (Young, n = 83; Elderly, n = 16) and patients affected by HGPS (n = 80). (c) Average ± SEM of current-voltage relationships (I–V) recorded in hDF obtained from young donors and patients affected by HGPS treated by 100 nM IbTx (n = 6) and 10 mM TEA (n = 4), a non-selective $K^+$ channels blocker Tetraethylammonium (TEA). Young vs. HGPS: *p < 0.05; **p < 0.01; Young vs. Elderly: #p < 0.05; ##p < 0.001; ###p < 0.0001. Details in Zironi et al. (2018).

mammalian STREX exon) was identified using bioinformatics sequence analysis. Skate kcnma1 wild-type cDNA transfected into HEK293 cells localized to the endoplasmic reticulum and nucleus. Mutations in the NLS (KR → AA or SVLS → AVLA) was found to independently attenuate nuclear translocation from endoplasmic reticulum. BK channel localization may thus be controlled by splicing or phosphorylation to tune electroreception and modulate gene expression (Chen et al., 2020).

### 8.4.3 NUCLEAR OUTER, INNER, AND TRANSMEMBRANE TARGETS

An array of genetic disorders and mutations encoding NE proteins have been presented earlier in this chapter (Dauer and Worman, 2009). These mutations in genes encode various proteins in NOM, NIM, and lamins of NE. Many of them are linked to ion channels of the nuclear membranes NOM and NIM. Thus, these defects in channels are branded as nuclear membrane channelopathies.

Genes with mutations, encoding nuclear lamins and associated NE proteins, are linked to multiple inherited diseases that ultimately affect tissues and organs. The diseases are referred to as laminopathies. Recent research has shown that pathogenic mutations in genes encoding NE proteins lead to defective nucleocytoplasmic connections (see Figure 8.23) that disrupt proper functioning of the linker of nucleoskeleton and cytoskeleton complex in the establishment of cell polarity. These defects, or laminopathies, are comparable to channelopathies.

The mutations in NE protein emerin has recently been investigated by many groups (e.g., Essawy et al., 2019). Emerin contributes to genome organization and cell mechanics. Through its N-terminal LAP2-emerin-MAN1 (LEM)-domain, emerin interacts with the DNA-binding protein barrier-to-autointegration (BAF). Emerin also binds to the linker of the nucleoskeleton and cytoskeleton (LINC) complex. Missense and short deletion mutations P22L, ΔK37, and T43I were recently discovered in emerin LEM-domain, in association with isolated atrial cardiac defects. The effect of mutation ΔK37 is observed to perturb emerin function within the LINC complex in response to mechanical stress.

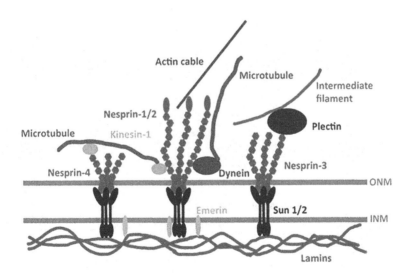

**FIGURE 8.23** Schematic diagram of linker of nucleoskeleton and cytoskeleton (LINC) complexes. Transmembrane SUN proteins of the INM/NIM interact with nesprins of the ONM/NOM within the perinuclear space. In the nucleus, SUN proteins bind to A-type lamins and interact with another NIM transmembrane protein emerin. In the cytoplasm, different nesprin proteins bind to different cytoskeletal elements. Different isoforms of nesprin-1 and nepsrin-2 can bind actin filaments directly or associate with microtubules indirectly via dynein or kinesin-1. Nesprin-3 can associate with intermediate filaments by binding to plectin. Nesprin-4 can also associate with microtubules indirectly by binding kinesin-1 (Östlund et al., 2019).

Multiple diseases are linked to mutations in the genes encoding NE Sun proteins and nesprins. Mutations in SYNE1 encoding nesprin-1 cause autosomal recessive cerebellar ataxia. These NIM and NOM protein mutations including other proteins known to form ion channels may alter membrane transport properties, and should favor channelopathies.

NOM and NIM host numerous known and unknown ion channels. Earlier in this book, we have explained crucial channel structures, functions, and their specific and nonspecific roles concerning proper functioning of nucleus inside cytoplasmic complex. Many of these nuclear membrane-based channels experience temporary or permanent alterations in their structures, and consequently in functions, mostly due to mutations occurring in certain disease conditions. These channel defects or nuclear channelopathies have recently been under scrutiny in medical research for discovering appropriate therapeutics. We might pinpoint (but not doing so due to lack of space) a lot of such channels as crucial examples that are found biophysically addressable while understanding mostly altered nuclear membrane transport properties in disease conditions.

## 8.5   CONCLUDING REMARKS

Channelopathies linked to ion channel disorders due to mutations in genes encoding the channel proteins have been found in various diseases. We have discussed several such case studies in this chapter. Cancer is perhaps most the important among these. The cancer cell hosts channels in many sections of a cell that might experience specific mutations or general alterations due to unknown reasons. These channels make good candidates for cancer channelopathies. Figure 8.24 presents a good summary for the readers.

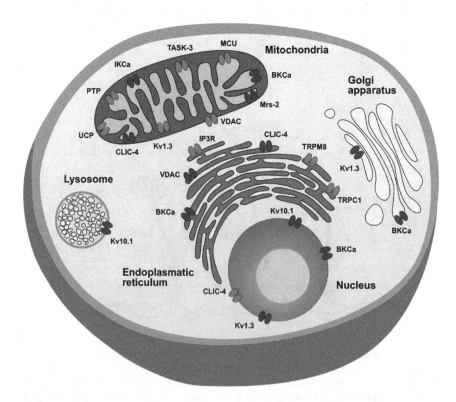

**FIGURE 8.24**   Intracellular ion channels are expressed in major subcellular compartments, for example, the nucleus, endoplasmic reticulum, lysosome, Golgi apparatus, and mitochondria. Ion channels having roles in cancer development and/or progression are stained in green (Peruzzo et al., 2016).

Once channelopathies are detected, the obvious goal is to find the right therapeutics and fight the linked diseases. Here, we wish to explain one possible research area that might address one of the biophysical aspects of understanding drug efficacy. This method may help finalize a candidate drug from a pool of chemicals designed for any target channel disorder or related mutated gene.

NIM channels may be targeted for specific purposes as their functions are individually or collectively linked to various cellular signaling and processes. Specific NIM channels require regulation (repairing or inhibition) in case those are linked to certain diseases. Agents that can successfully regulate NIM channels linked to certain disease may appear as therapeutic agents. Instead of listing or explaining these channels including drugs inhibiting them which mostly fall in physiological research, we shall try to pick a case study here to address the biophysical aspects of channel being regulated due to drugs. Potassium ion flowing potassium channels are important among those present in NIM. Here, we demonstrate how we can record the drug-induced regulation of potassium channel Kv10.1, which is a voltage-gated potassium channel found in NIM (Chen et al., 2011).

The pharmacology of the channel was characterized using astemizole, an H1 histamine receptor inhibitor, often used as a blocker of Kv10.1, because there is no report favoring the blocking of conductance other than those of the eag family including Kv10.1 at 2–5 µM concentration. In asymmetrical potassium, continuous recordings at 0 mV were made to ensure contamination by channels selective for $K^+$. The channel activity, compatible with Kv10.1, virtually disappeared due to the effects of astemizole. In cases that allowed the washout of the drug, the channel activity was recovered. Figure 8.25a depicts a similar experiment, but performed at +60 mV to increase current amplitude. In symmetrical $K^+$, the drug treatment reduced the open probability at +60 mV

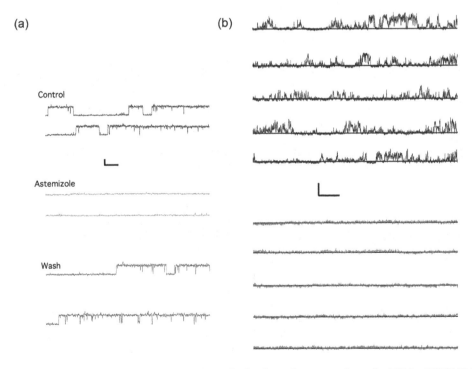

**FIGURE 8.25** Pharmacological blockade of the single channel currents from the NIM of HEK-Kv10.1. (a) Block by astemizole. Traces at the effects of +60 mV in asymmetrical potassium before (control, black color), during (astemizole, red color), and after washing (wash, blue color) 25 µM astemizole. Scale bars, 2 pA (vertical), 500 ms (horizontal). (b) Block by monoclonal antibody, mAb56. Control trace, immediately after seal formation at −80 mV in the pipette (black color), symmetrical potassium. (Red color) Recording after 20 min incubation at the same voltage. Scale bars, 0.5 pA, 500 ms. Details in ref. (Chen et al., 2011). (For color figure see eBook figure.)

membrane potential by $94\pm10\%$ (n = 5), also indicating a blockade of the channel by astemizole. As the extracellular side of the channel appeared facing the pipette, mAb56, a monoclonal antibody was used to bind to the extracellular loops of Kv10.1 and inhibit current because this antibody is commonly known to inhibit Kv10.1 channel functions (Gomez-Varela et al., 2007). The channel activity was found to decrease within 20 min of mAb56 presence, a duration found compatible with action of mAb56 on Kv10.1 whole-cell current in the plasma membrane (see Figure 8.25b). Due to the mAb56 selectivity, this result perhaps strongly supports the molecular identity of the detected NIM channel as none but Kv10.1. This channel is also proposed to have an orientation in the NIM that the intracellular C-terminus should face the nucleoplasm (Chen et al., 2011).

## REFERENCES

Ackerman, M. J., & Clapham, D. E. (1997). Ion channels-basic science and clinical disease. *The New England Journal of Medicine, 336*, 1575–1586.

Alberdi, E., Sánchez-Gómez, M. V., Cavaliere, F., Pérez-Samartín, A., Zugaza, J. L., Trullas, R., . . . Matute, C. (2010). Amyloid β oligomers induce $Ca^{2+}$ dysregulation and neuronal death through activation of ionotropic glutamate receptors. *Cell Calcium, 47*(3), 264–272. doi:10.1016/j.ceca.2009.12.010

Aon, M. A., Cortassa, S., & O'rourke, B. (2004). Percolation and criticality in a mitochondrial network. *Proceedings of the National Academy of Sciences, 101*(13), 4447–4452. doi:10.1073/pnas.0307156101

Ashrafuzzaman, M. (2018). *Nanoscale Biophysics of the Cell*. Springer International Publishing AG, Part of Springer Nature. doi:10.1007/978-3-319-77465-7

Ashrafuzzaman, M. (2020). Designing aptamers to target ceramide channels in mitochondrial membrane (Research stage)

Ashrafuzzaman, M., & Tuszynski, J. (2012a). Regulation of channel function due to coupling with a lipid bilayer. *Journal of Computational and Theoretical Nanoscience, 9*(4), 564–570. doi:10.1166/jctn.2012.2062

Ashrafuzzaman, M., & Tuszynski, J. A. (2012b). *Membrane Biophysics*. Berlin, Heidelberg: Springer-Verlag. doi:10.1007/978-3-642-16105-6

Ayyasamy, V., Owens, K. M., Desouki, M. M., Liang, P., Bakin, A., Thangaraj, K., . . . Singh, K. K. (2011). Cellular model of warburg effect identifies tumor promoting function of UCP2 in breast cancer and its suppression by genipin. *PLoS ONE, 6*(9). doi:10.1371/journal.pone.0024792

Bachmann, M, Pontarin, G., & Szabo, I. (2019). The contribution of mitochondrial ion channels to cancer development and progression. *Cellular Physiology and Biochemistry, 53*(S1), 63–78. doi:10.33594/000000198

Bacle, A., Kadri, L., Khoury, S., Ferru-Clément, R., Faivre, J., Cognard, C., . . . Ferreira, T. (2020). A comprehensive study of phospholipid fatty acid rearrangements in metabolic syndrome: Correlations with organ dysfunction. *Disease Models & Mechanisms, 13*(6), Dmm043927. doi:10.1242/dmm.043927

Baffy, G., Derdak, Z., & Robson, S. C. (2011). Mitochondrial recoupling: A novel therapeutic strategy for cancer? *British Journal of Cancer, 105*(4), 469–474. doi:10.1038/bjc.2011.245

Bichet, D. G., Tarazi, A. E., Matar, J., Lussier, Y., Arthus, M., Lonergan, M., . . . Bissonnette, P. (2012). Aquaporin-2: New mutations responsible for autosomal-recessive nephrogenic diabetes insipidus--update and epidemiology. *Clinical Kidney Journal, 5*(3), 195–202. doi:10.1093/ckj/sfs029

Bonora, M., Wieckowski, M. R., Sinclair, D. A., Kroemer, G., Pinton, P., & Galluzzi, L. (2018). Targeting mitochondria for cardiovascular disorders: Therapeutic potential and obstacles. *Nature Reviews Cardiology, 16*(1), 33–55. doi:10.1038/s41569-018-0074-0

Cavalieri, E., Bergamini, C., Mariotto, S., Leoni, S., Perbellini, L., Darra, E., . . . Lenaz, G. (2009). Involvement of mitochondrial permeability transition pore opening in α-bisabolol induced apoptosis. *FEBS Journal, 276*(15), 3990–4000. doi:10.1111/j.1742-4658.2009.07108.x

Cavalieri, E., Rigo, A., Bonifacio, M., Prati, A. D., Guardalben, E., Bergamini, C., . . . Vinante, F. (2011). Pro-apoptotic activity of α-bisabolol in preclinical models of primary human acute leukemia cells. *Journal of Translational Medicine, 9*(1), 45. doi:10.1186/1479-5876-9-45

Connors, L. H., Sam, F., Skinner, M., Salinaro, F., Sun, F., Ruberg, F. L., . . . Seldin, D. C. (2016). Heart failure resulting from age-related cardiac amyloid disease associated with wild-type transthyretin. *Circulation, 133*(3), 282–290. doi:10.1161/circulationaha.115.018852

Chen, A. L., Wu, T., Shi, L., Clusin, W. T., & Kao, P. N. (2020). Nuclear localization of calcium-activated BK channels in skate ampullary electroreceptors. doi:10.1101/2020.01.27.922161

Chen, Y., Sánchez, A., Rubio, M. E., Kohl, T., Pardo, L. A., & Stühmer, W. (2011). Functional KV10.1 channels localize to the inner nuclear membrane. *PLoS ONE, 6*(5). doi:10.1371/journal.pone.0019257

Chen, Y., & Zweier, J. L. (2014). Cardiac mitochondria and reactive oxygen species generation. *Circulation Research, 114*(3), 524–537. doi:10.1161/circresaha.114.300559

Chiara, F., & Rasola, A. (2013). GSK-3 and mitochondria in cancer cells. *Frontiers in Oncology, 3*. doi:10.3389/fonc.2013.00016

Chistiakov, D. A., Sobenin, I. A., Revin, V. V., Orekhov, A. N., & Bobryshev, Y. V. (2014). Mitochondrial aging and age-related dysfunction of mitochondria. *BioMed Research International, 2014*, 1–7. doi:10.1155/2014/238463

Cortassa, S., Aon, M. A., Winslow, R. L., & O'Rourke, B. (2004). A mitochondrial oscillator dependent on reactive oxygen species. *Biophysical Journal, 87*(3), 2060–2073. doi:10.1529/biophysj.104.041749

Dauer, W. T., & Worman, H. J. (2009). The nuclear envelope as a signaling node in development and disease. *Developmental Cell, 17*(5), 626–638. doi:10.1016/j.devcel.2009.10.016

Davuluri, G., Gong, W., Yusuff, S., Lorent, K., Muthumani, M., Dolan, A. C., & Pack, M. (2008). Mutation of the zebrafish nucleoporin elys sensitizes tissue progenitors to replication stress. *PLoS Genetics, 4*(10). doi:10.1371/journal.pgen.1000240

Deen, P., Verdijk, M., Knoers, N., Wieringa, B., Monnens, L., Os, C. V., & Oost, B. V. (1994). Requirement of human renal water channel aquaporin-2 for vasopressin-dependent concentration of urine. *Science, 264*(5155), 92–95. doi:10.1126/science.8140421

Demirbilek, H., Galcheva, S., Vuralli, D., Al-Khawaga, S., & Hussain, K. (2019). Ion transporters, channelopathies, and glucose disorders. *International Journal of Molecular Sciences, 20*(10), 2590. doi:10.3390/ijms20102590

Demuro, A., Smith, M., & Parker, I. (2011). Single-channel Ca$^{2+}$ imaging implicates Aβ1–42 amyloid pores in Alzheimer's disease pathology. *Journal of Cell Biology, 195*(3), 515–524. doi:10.1083/jcb.201104133

Doria, E., Buonocore, D., Focarelli, A., & Marzatico, F. (2012). Relationship between human aging muscle and oxidative system pathway. *Oxidative Medicine and Cellular Longevity, 2012*, 1–13. doi:10.1155/2012/830257

Duménieu, M., Oulé, M., Kreutz, M. R., & Lopez-Rojas, J. (2017). The segregated expression of voltage-gated potassium and sodium channels in neuronal membranes: Functional implications and regulatory mechanisms. *Frontiers in Cellular Neuroscience, 11*. doi:10.3389/fncel.2017.00115

Elliott, M., Ford, S., Prasad, E., Dick, L., Farmer, H., Hogg, P., & Halbert, G. (2012). Pharmaceutical development of the novel arsenical based cancer therapeutic GSAO for Phase I clinical trial. *International Journal of Pharmaceutics, 426*(1–2), 67–75. doi:10.1016/j.ijpharm.2012.01.024

Essawy, N., Samson, C., Petitalot, A., Moog, S., Bigot, A., Herrada, I., . . . Zinn-Justin, S. (2019). An emerin LEM-domain mutation impairs cell response to mechanical stress. *Cells, 8*(6), 570. doi:10.3390/cells8060570

Ferguson, M., Mockett, R., Shen, Y., Orr, W., & Sohal, R. (2005). Age-associated decline in mitochondrial respiration and electron transport in Drosophila melanogaster. *Biochemical Journal, 390*(2), 501–511. doi:10.1042/bj20042130

Fingrut, O., & Flescher, E. (2002). Plant stress hormones suppress the proliferation and induce apoptosis in human cancer cells. *Leukemia, 16*(4), 608–616. doi:10.1038/sj.leu.2402419

Flescher, E. (2005). Jasmonates—A new family of anti-cancer agents. *Anti-Cancer Drugs, 16*(9), 911–916. doi:10.1097/01.cad.0000176501.63680.80

Forrest, M. D. (2015). Why cancer cells have a more hyperpolarised mitochondrial membrane potential and emergent prospects for therapy. doi:10.1101/025197

Galluzzi, L., Morselli, E., Kepp, O., Tajeddine, N., & Kroemer, G. (2008). Targeting p53 to mitochondria for cancer therapy. *Cell Cycle, 7*(13), 1949–1955. doi:10.4161/cc.7.13.6222

Gleichmann, M., & Mattson, M. P. (2011). Neuronal calcium homeostasis and dysregulation. *Antioxidants & Redox Signaling, 14*(7), 1261–1273. doi:10.1089/ars.2010.3386

Gomez-Varela, D., Zwick-Wallasch, E., Knotgen, H., Sanchez, A., Hettmann, T., Ossipov, D., . . . Pardo, L. A. (2007). Monoclonal antibody blockade of the human Eag1 potassium channel function exerts antitumor activity. *Cancer Research, 67*(15), 7343–7349. doi:10.1158/0008-5472.can-07-0107

Grills, C., Jithesh, P. V., Blayney, J., Zhang, S., & Fennell, D. A. (2011). Gene expression meta-analysis identifies VDAC1 as a predictor of poor outcome in early stage non-small cell lung cancer. *PLoS ONE, 6*(1). doi:10.1371/journal.pone.0014635

Gulbins, E., Sassi, N., Grassmè, H., Zoratti, M., & Szabò, I. (2010). Role of Kv1.3 mitochondrial potassium channel in apoptotic signalling in lymphocytes. *Biochimica Et Biophysica Acta (BBA) - Bioenergetics, 1797*(6–7), 1251–1259. doi:10.1016/j.bbabio.2010.01.018

Guo, X. P., Zhang, X. Y., & Zhang, S. D. (1991). Clinical trial on the effects of shikonin mixture on later stage lung cancer. *Zhong Xi Yi Jie He Za Zhi, 11*, 598–599.

Han, W., Li, L., Qiu, S., Lu, Q., Pan, Q., Gu, Y., . . . Hu, X. (2007). Shikonin circumvents cancer drug resistance by induction of a necroptotic death. *Molecular Cancer Therapeutics, 6*(5), 1641–1649. doi:10.1158/1535-7163.mct-06-0511

Hanahan, D., & Weinberg, R. (2011). Hallmarks of cancer: The next generation. *Cell, 144*(5), 646–674. doi:10.1016/j.cell.2011.02.013

Hanna, M. G., Wood, N. W., & Kullmann, D. M. (1998). Ion channels and neurological disease: DNA based diagnosis is now possible, and ion channels may be important in common paroxysmal disorders. *Journal of Neurology, Neurosurgery & Psychiatry, 65*(4), 427–431. doi:10.1136/jnnp.65.4.427

Harman, D. (1982). The free-radical theory of aging. *Free Radicals in Biology, 255–275.* doi:10.1016/b978-0-12-566505-6.50015-6

Harman, D. (1992). Free radical theory of aging. *Mutation Research/DNAging, 275*(3–6), 257–266. doi:10.1016/0921-8734(92)90030-s

Harmar, A. J., Hills, R. A., Rosser, E. M., Jones, M., Buneman, O. P., Dunbar, D. R., . . . Spedding, M. (2008). IUPHAR-DB: The IUPHAR database of G protein-coupled receptors and ion channels. *Nucleic Acids Research, 37*(Suppl_1). doi:10.1093/nar/gkn728

Hashimoto, T., Harita, Y., Takizawa, K., Urae, S., Ishizuka, K., Miura, K., . . . Hattori, M. (2019). In vivo expression of NUP93 and its alteration by NUP93 mutations causing focal segmental glomerulosclerosis. *Kidney International Reports, 4*(9), 1312–1322. doi:10.1016/j.ekir.2019.05.1157

Imamura, M., Horikoshi, M., & Maeda, S. (2019). Genome-wide association study for type 2 diabetes. *Genome-Wide Association Studies, 49–86.* doi:10.1007/978-981-13-8177-5_4

Innamaa, A., Jackson, L., Asher, V., Van Shalkwyk, G., Warren, A., Hay, D., . . . Khan, R. (2013). Expression and prognostic significance of the oncogenic $K_2P$ potassium channel $KCNK_9$ (TASK-3) in ovarian carcinoma. *Anticancer Research, 33,* 1401–1408.

Jahangir, A., Ozcan, C., Holmuhamedov, E. L., & Terzic, A. (2001). Increased calcium vulnerability of senescent cardiac mitochondria: Protective role for a mitochondrial potassium channel opener. *Mechanisms of Ageing and Development, 122*(10), 1073–1086. doi:10.1016/s0047-6374(01)00242-1

Jensen, C. S., Watanabe, S., Rasmussen, H. B., Schmitt, N., Olesen, S., Frost, N. A., . . . Misonou, H. (2014). Specific sorting and post-golgi trafficking of dendritic potassium channels in living neurons. *Journal of Biological Chemistry, 289*(15), 10566–10581. doi:10.1074/jbc.m113.534495

Jones, J. L., Peana, D., Veteto, A. B., Lambert, M. D., Nourian, Z., Karasseva, N. G., . . . Domeier, T. L. (2018). TRPV4 increases cardiomyocyte calcium cycling and contractility yet contributes to damage in the aged heart following hypoosmotic stress. *Cardiovascular Research, 115*(1), 46–56. doi:10.1093/cvr/cvy156

Jones, S., Martel, C., Belzacq-Casagrande, A., Brenner, C., & Howl, J. (2008). Mitoparan and target-selective chimeric analogues: Membrane translocation and intracellular redistribution induces mitochondrial apoptosis. *Biochimica Et Biophysica Acta (BBA) - Molecular Cell Research, 1783*(5), 849–863. doi:10.1016/j.bbamcr.2008.01.009

Kamada, N., Kanaya, N., Hirata, N., Kimura, S., & Namiki, A. (2008). Cardioprotective effects of propofol in isolated ischemia-reperfused guinea pig hearts: Role of KATP channels and GSK-3β. *Canadian Journal of Anesthesia/Journal Canadien D'anesthésie, 55*(9), 595–605. doi:10.1007/bf03021433

Khan, A. U., Qu, R., Ouyang, J., & Dai, J. (2020). Role of nucleoporins and transport receptors in cell differentiation. *Frontiers in Physiology, 11.* doi:10.3389/fphys.2020.00239

Kim, J. (2014). Channelopathies. *Korean Journal of Pediatrics, 57*(1), 1. doi:10.3345/kjp.2014.57.1.1

Kim, J., Lee, S., Oh, S., Han, S., Park, H., Yoo, M., & Kang, H. (2004). Methyl jasmonate induces apoptosis through induction of Bax/Bcl-Xs and activation of caspase-3 via ROS production in A549 cells. *Oncology Reports.* doi:10.3892/or.12.6.1233

Klippel, S., Jakubikova, J., Delmore, J., Ooi, M., Mcmillin, D., Kastritis, E., . . . Mitsiades, C. S. (2012). Methyljasmonate displaysin vitroand in vivoactivity against multiple myeloma cells. *British Journal of Haematology, 159*(3), 340–351. doi:10.1111/j.1365-2141.2012.09253.x

Ko, Y. H., Verhoeven, H. A., Lee, M. J., Corbin, D. J., Vogl, T. J., & Pedersen, P. L. (2012). A translational study "case report" on the small molecule "energy blocker" 3-bromopyruvate (3BP) as a potent anticancer agent: From bench side to bedside. *Journal of Bioenergetics and Biomembranes, 44*(1), 163–170. doi:10.1007/s10863-012-9417-4

Kumar, P., Kumar, D., Jha, S. K., Jha, N. K., & Ambasta, R. K. (2016). Ion channels in neurological disorders. *Ion Channels as Therapeutic Targets, Part A Advances in Protein Chemistry and Structural Biology, 97–136.* doi:10.1016/bs.apcsb.2015.10.006

Lam, A., Karekar, P., Shah, K., Hariharan, G., Fleyshman, M., Kaur, H., . . . Rao, S. G. (2018). Drosophila voltage-gated calcium channel α1-subunits regulate cardiac function in the aging heart. *Scientific Reports, 8*(1). doi:10.1038/s41598-018-25195-0

Lan, M., Shi, Y., Han, Z., Hao, Z., Pan, Y., Liu, N., Guo, C., Hong, L., Wang, J., Qiao, T., & Fan, D. (2005). Expression of delayed rectifier potassium channels and their possible roles in proliferation of human gastric cancer cells. *Cancer Biology & Therapy, 4*(12), 1342–1347. doi:10.4161/cbt.4.12.2175

Leanza, L., Henry, B., Sassi, N., Zoratti, M., Chandy, K. G., Gulbins, E., & Szabò, I. (2012). Inhibitors of mitochondrial Kv1.3 channels induce Bax/Bak-independent death of cancer cells. *EMBO Molecular Medicine, 4*(7), 577–593. doi:10.1002/emmm.201200235

Leanza, L., Trentin, L., Becker, K. A., Frezzato, F., Zoratti, M., Semenzato, G., . . . Szabo, I. (2013). Clofazimine, Psora-4 and PAP-1, inhibitors of the potassium channel Kv1.3, as a new and selective therapeutic strategy in chronic lymphocytic leukemia. *Leukemia, 27*(8), 1782–1785. doi:10.1038/leu.2013.56

Leanza, L., Zoratti, M., Gulbins, E., & Szabo, I. (2014). Mitochondrial ion channels as oncological targets. *Oncogene, 33*(49), 5569–5581. doi:10.1038/onc.2013.578

Lemeshko, V. V. (2002). Model of the outer membrane potential generation by the inner membrane of mitochondria. *Biophysical Journal, 82*(2), 684–692. doi:10.1016/s0006-3495(02)75431-3

Lena, A., Rechichi, M., Salvetti, A., Bartoli, B., Vecchio, D., Scarcelli, V., . . . Rossi, L. (2009). Drugs targeting the mitochondrial pore act as citotoxic and cytostatic agents in temozolomide-resistant glioma cells. *Journal of Translational Medicine, 7*(1), 13. doi:10.1186/1479-5876-7-13

Li, B., Jie, W., Huang, L., Wei, P., Li, S., Luo, Z., . . . Gao, T. (2014). Nuclear BK channels regulate gene expression via the control of nuclear calcium signaling. *Nature Neuroscience, 17*(8), 1055–1063. doi:10.1038/nn.3744

Li, L., Han, W., Gu, Y., Qiu, S., Lu, Q., Jin, J., . . . Hu, X. (2007). Honokiol induces a necrotic cell death through the mitochondrial permeability transition pore. *Cancer Research, 67*(10), 4894–4903. doi:10.1158/0008-5472.can-06-3818

Lin, A., Liu, M., Ko, H., Perng, D., Lee, T., & Kou, Y. R. (2015). Lung epithelial TRPA1 transduces the extracellular ROS into transcriptional regulation of lung inflammation induced by cigarette smoke: The role of influxed $Ca^{2+}$. *Mediators of Inflammation, 2015*, 1–16. doi:10.1155/2015/148367

Liu, T., Takimoto, E., Dimaano, V. L., DeMazumder, D., Kettlewell, S., Smith, G., Sidor, A., Abraham, T. P., & O'Rourke, B. (2014). Inhibiting mitochondrial Na+/Ca 2+ exchange prevents sudden death in a guinea pig model of heart failure. *Circulation Research, 115*(1), 44–54. doi:10.1161/circresaha.115.303062

Liu, T. C.-Y., Tang, X.-M., Duan, R., Ma, L., Zhu, L., & Zhang, Q.-G. (2018). The Mitochondrial Na+/Ca2+ Exchanger is Necessary but Not Sufficient for Ca2+ Homeostasis and Viability. *Advances in Experimental Medicine and Biology*, 281–285. doi:10.1007/978-3-319-91287-5_45

Logothetis, D. E., Petrou, V. I., Adney, S. K., & Mahajan, R. (2010). Channelopathies linked to plasma membrane phosphoinositides. *Pflügers Archiv - European Journal of Physiology, 460*(2), 321–341. doi:10.1007/s00424-010-0828-y

Loonen, A. J., Knoers, N. V., Os, C. H., & Deen, P. M. (2008). Aquaporin 2 mutations in nephrogenic diabetes insipidus. *Seminars in Nephrology, 28*(3), 252–265. doi:10.1016/j.semnephrol.2008.03.006

Marchi, U. D., Sassi, N., Fioretti, B., Catacuzzeno, L., Cereghetti, G. M., Szabò, I., & Zoratti, M. (2009). Intermediate conductance $Ca^{2+}$-activated potassium channel (KCa3.1) in the inner mitochondrial membrane of human colon cancer cells. *Cell Calcium, 45*(5), 509–516. doi:10.1016/j.ceca.2009.03.014

Östlund, C., Chang, W., Gundersen, G. G., & Worman, H. J. (2019). Pathogenic mutations in genes encoding nuclear envelope proteins and defective nucleocytoplasmic connections. *Experimental Biology and Medicine, 244*(15), 1333–1344. doi:10.1177/1535370219862243

Palmieri, B., Iannitti, T., Capone, S., & Flescher, E. A. (2011). preliminary study of the local treatment of preneoplastic and malignant skin lesions using methyl jasmonate. *European Review Medical Pharmacological Science, 15*, 333–336.

Pedersen, P. L. (2012). 3-bromopyruvate (3BP) a fast acting, promising, powerful, specific, and effective "small molecule" anti-cancer agent taken from labside to bedside: Introduction to a special issue. *Journal of Bioenergetics and Biomembranes, 44*(1), 1–6. doi:10.1007/s10863-012-9425-4

Peixoto, P. M., Kinnally, K. W., & Pavlov, E. (2015). Mitochondrial channels in neurodegeneration. *The Functions, Disease-Related Dysfunctions, and Therapeutic Targeting of Neuronal Mitochondria*, 65–100. doi:10.1002/9781119017127.ch3

Pereira, G. C., Branco, A. F., Matos, J. A., Pereira, S. L., Parke, D., Perkins, E. L., . . . Oliveira, P. J. (2007). Mitochondrially targeted effects of berberine [natural yellow 18, 5,6-dihydro-9,10-dimethoxybenzo(g) -1,3-benzodioxolo(5,6-a) quinolizinium] on K1735-M2 mouse melanoma cells: Comparison with direct effects on isolated mitochondrial fractions. *Journal of Pharmacology and Experimental Therapeutics, 323*(2), 636–649. doi:10.1124/jpet.107.128017

Perier, C., & Vila, M. (2011). Mitochondrial biology and parkinson's disease. *Cold Spring Harbor Perspectives in Medicine, 2*(2). doi:10.1101/cshperspect.a009332

Perry, S. W., Norman, J. P., Barbieri, J., Brown, E. B., & Gelbard, H. A. (2011). Mitochondrial membrane potential probes and the proton gradient: a practical usage guide. *BioTechniques, 50*(2), 98–115. doi:10.2144/000113610

Peruzzo, R., Biasutto, L., Szabò, I., & Leanza, L. (2016). Impact of intracellular ion channels on cancer development and progression. *European Biophysics Journal, 45*(7), 685–707. doi:10.1007/s00249-016-1143-0

Prevarskaya, N., Skryma, R., & Shuba, Y. (2018). Ion channels in cancer: Are cancer hallmarks oncochannelopathies? *Physiological Reviews, 98*(2), 559–621. doi:10.1152/physrev.00044.2016

Quast, S., Berger, A., Buttstädt, N., Friebel, K., Schönherr, R., & Eberle, J. (2012). General sensitization of melanoma cells for TRAIL-induced apoptosis by the potassium channel inhibitor TRAM-34 depends on release of SMAC. *PLoS ONE, 7*(6). doi:10.1371/journal.pone.0039290

Ranki, H. J., Crawford, R. M., Budas, G. R., & Jovanović, A. (2002). Ageing is associated with a decrease in the number of sarcolemmal ATP-sensitive K+ channels in a gender-dependent manner. *Mechanisms of Ageing and Development, 123*(6), 695–705. doi:10.1016/s0047-6374(01)00415-8

Rao, V., Kaja, S., & Gentile, S. (2016). Ion channels in aging and aging-related diseases. *Molecular Mechanisms of the Aging Process and Rejuvenation.* doi:10.5772/63951

Rea, S. L., Ventura, N., & Johnson, T. E. (2007). Relationship between mitochondrial electron transport chain dysfunction, development, and life extension in Caenorhabditis elegans. *PLoS Biology, 5*(10). doi:10.1371/journal.pbio.0050259

Ren, Y. R., Pan, F., Parvez, S., Fleig, A., Chong, C. R., Xu, J., . . . Liu, J. O. (2008). Clofazimine inhibits human Kv1.3 potassium channel by perturbing calcium oscillation in T. lymphocytes. *PLoS ONE, 3*(12). doi:10.1371/journal.pone.0004009

Rice, R. A., Berchtold, N. C., Cotman, C. W., & Green, K. N. (2014). Age-related downregulation of the $Ca_V3.1$ T-type calcium channel as a mediator of amyloid beta production. *Neurobiology of Aging, 35*(5), 1002–1011. doi:10.1016/j.neurobiolaging.2013.10.090

Roscoe, A. K., Christensen, J. D., & Lynch, C. (2000). Isoflurane, but not halothane, induces protection of human myocardium via adenosine A1 receptors and adenosine triphosphate–sensitive potassium channels. *Anesthesiology, 92*(6), 1692–1701. doi:10.1097/00000542-200006000-00029

Rostovtseva, T. K., Gurnev, P. A., Protchenko, O., Hoogerheide, D. P., Yap, T. L., Philpott, C. C., . . . Bezrukov, S. M. (2015). α-synuclein shows high affinity interaction with voltage-dependent anion channel, suggesting mechanisms of mitochondrial regulation and toxicity in Parkinson disease. *Journal of Biological Chemistry, 290*(30), 18467–18477. doi:10.1074/jbc.m115.641746

Saeed, Y., Temple, I. P., Borbas, Z., Atkinson, A., Yanni, J., Maczewski, M., Mackiewicz, U., Aly, M., Logantha, S. J., Garratt, C. J., & Dobrzynski, H. (2018). Structural and functional remodeling of the atrioventricular node with aging in rats: The role of hyperpolarization-activated cyclic nucleotide–gated and ryanodine 2 channels. *Heart Rhythm, 15*(5), 752–760. doi:10.1016/j.hrthm.2017.12.027

Sakuma, S., & D'Angelo, M. A. (2017). The roles of the nuclear pore complex in cellular dysfunction, aging and disease. *Seminars in Cell & Developmental Biology, 68*, 72–84. doi:10.1016/j.semcdb.2017.05.006

Sandestig, A., Engström, K., Pepler, A., Danielsson, I., Odelberg-Johnsson, P., Biskup, S., . . . Stefanova, M. (2019). *NUP188* biallelic loss of function may underlie a new syndrome: Nucleoporin 188 insufficiency syndrome? *Molecular Syndromology, 10*(6), 313–319. doi:10.1159/000504818

Santulli, G., & Marks, A. (2015). Essential roles of intracellular calcium release channels in muscle, brain, metabolism, and aging. *Current Molecular Pharmacology, 8*(2), 206–222. doi:10.217 4/1874467208666150507105105

Scaduto, R. C., & Grotyohann, L. W. (1999). Measurement of mitochondrial membrane potential using fluorescent rhodamine derivatives. *Biophysical Journal, 76*(1), 469–477. doi:10.1016/s0006-3495(99)77214-0

Schilling, T., & Eder, C. (2014). Microglial K+ channel expression in young adult and aged mice. *Glia, 63*(4), 664–672. doi:10.1002/glia.22776

Shoshan, M. C. (2012). 3-bromopyruvate: Targets and outcomes. *Journal of Bioenergetics and Biomembranes, 44*(1), 7–15. doi:10.1007/s10863-012-9419-2

Shoshan-Barmatz, V., & Mizrachi, D. (2012). VDAC1: From structure to cancer therapy. *Frontiers in Oncology, 2.* doi:10.3389/fonc.2012.00164

Simamura, E., Shimada, H., Hatta, T., & Hirai, K. (2008). Mitochondrial voltage-dependent anion channels (VDACs) as novel pharmacological targets for anti-cancer agents. *Journal of Bioenergetics and Biomembranes, 40*(3), 213–217. doi:10.1007/s10863-008-9158-6

Simms, B., & Zamponi, G. (2014). Neuronal voltage-gated calcium channels: Structure, function, and dysfunction. *Neuron, 82*(1), 24–45. doi:10.1016/j.neuron.2014.03.016

Spillane, J., Kullmann, D. M., & Hanna, M. G. (2015). Genetic neurological channelopathies: Molecular genetics and clinical phenotypes. *Journal of Neurology, Neurosurgery & Psychiatry.* doi:10.1136/jnnp–2015-311233

Srivastava, S., Cohen, J. S., Vernon, H., Barañano, K., Mcclellan, R., Jamal, L., . . . Fatemi, A. (2014). Clinical whole exome sequencing in child neurology practice. *Annals of Neurology, 76*(4), 473–483. doi:10.1002/ana.24251

Strickland, M., Yacoubi-Loueslati, B., Bouhaouala-Zahar, B., Pender, S. L., & Larbi, A. (2019). Relationships between ion channels, mitochondrial functions and inflammation in human aging. *Frontiers in Physiology, 10*. doi:10.3389/fphys.2019.00158

Szabó, I., & Zoratti, M. (1991). The giant channel of the inner mitochondrial membrane is inhibited by cyclosporin A. *Journal of Biological Chemistry, 266*, 3376–3379.

Thiffault, I., Speca, D. J., Austin, D. C., Cobb, M. M., Eum, K. S., Safina, N. P., . . . Sack, J. T. (2015). A novel epileptic encephalopathy mutation in KCNB1 disrupts Kv2.1 ion selectivity, expression, and localization. *Journal of General Physiology, 146*(5), 399–410. doi:10.1085/jgp.201511444

Torkamani, A., Bersell, K., Jorge, B. S., Bjork, R. L., Friedman, J. R., Bloss, C. S., . . . Kearney, J. A. (2014). De novoKCNB1mutations in epileptic encephalopathy. *Annals of Neurology, 76*(4), 529–540. doi:10.1002/ana.24263

Truong, A. H., Murugesan, S., Youssef, K. D., & Makino, A. (2016). Mitochondrial ion channels in metabolic disease. In P. I. Levitan & M. D. P. A. Dopico (Eds.), *Vascular Ion Channels in Physiology and Disease*. Cham: Springer International Publishing, pp. 397–419

Venetucci, L., Denegri, M., Napolitano, C., & Priori, S. G. (2012). Inherited calcium channelopathies in the pathophysiology of arrhythmias. *Nature Reviews Cardiology, 9*(10), 561–575. doi:10.1038/nrcardio.2012.93

Wang, Y., Zeng, W., Xiao, X., Huang, Y., Song, X., Yu, Z., . . . Xu, T. (2013). Intracellular ASIC1a regulates mitochondrial permeability transition-dependent neuronal death. *Cell Death & Differentiation, 20*(10), 1359–1369. doi:10.1038/cdd.2013.90

Weinberg-Shukron, A., Renbaum, P., Kalifa, R., Zeligson, S., Ben-Neriah, Z., Dreifuss, A., . . . Zangen, D. (2015). A mutation in the nucleoporin-107 gene causes XX gonadal dysgenesis. *Journal of Clinical Investigation, 125*(11), 4295–4304. doi:10.1172/jci83553

Williams, G. S., Boyman, L., Chikando, A. C., Khairallah, R. J., & Lederer, W. J. (2013). Mitochondrial calcium uptake. *Proceedings of the National Academy of Sciences, 110*(26), 10479–10486. doi:10.1073/pnas.1300410110

Wolf, A., Agnihotri, S., Micallef, J., Mukherjee, J., Sabha, N., Cairns, R., . . . Guha, A. (2011). Hexokinase 2 is a key mediator of aerobic glycolysis and promotes tumor growth in human glioblastoma multiforme. *The Journal of Experimental Medicine, 208*(2), 313–326. doi:10.1084/jem.20101470

Worman, H. J., & Bonne, G. (2007). "Laminopathies": A wide spectrum of human diseases. *Experimental Cell Research, 313*(10), 2121–2133. doi:10.1016/j.yexcr.2007.03.028

Yagoda, N., Rechenberg, M. V., Zaganjor, E., Bauer, A. J., Yang, W. S., Fridman, D. J., . . . Stockwell, B. R. (2007). RAS–RAF–MEK-dependent oxidative cell death involving voltage-dependent anion channels. *Nature, 447*(7146), 865–869. doi:10.1038/nature05859

Zaydman, M. A., Silva, J. R., & Cui, J. (2012). Ion channel associated diseases: Overview of molecular mechanisms. *Chemical Reviews, 112*(12), 6319–6333. doi:10.1021/cr300360k

Zhang, X., Chen, S., Yoo, S., Chakrabarti, S., Zhang, T., Ke, T., . . . Wang, Q. K. (2008). Mutation in nuclear pore component NUP155 leads to atrial fibrillation and early sudden cardiac death. *Cell, 135*(6), 1017–1027. doi:10.1016/j.cell.2008.10.022

Zironi, I., Gavoçi, E., Lattanzi, G., Virelli, A., Amorini, F., Remondini, D., & Castellani, G. (2018). BK channel overexpression on plasma membrane of fibroblasts from Hutchinson-Gilford progeria syndrome. *Aging, 10*(11), 3148–3160. doi:10.18632/aging.101621

Zorova, L. D., Popkov, V. A., Plotnikov, E. Y., Silachev, D. N., Pevzner, I. B., Jankauskas, S. S., Babenko, V. A., Zorov, S. D., Balakireva, A. V., Juhaszova, M., Sollott, S. J., & Zorov, D. B. (2018). Mitochondrial membrane potential. *Analytical Biochemistry, 552*, 50–59. doi:10.1016/j.ab.2017.07.009

# 9 Bioinformatics of Ion channels
## *Artificial Intelligence, Machine Learning, and Deep Learning*

Bioinformatics methods are capable of modeling known biological structures and predicting unknown ones. Additionally, versatile bioinformatics techniques can store the information processed in various biological and biophysical studies in the created databank, and call and utilize the information from the databank in pinpointing crucial molecular processes of an individual system or collective systems. Thus, these techniques help establish scientific links between various mechanisms and processes and produce concluding evidence that is otherwise unattainable using conventional theoretical and experimental techniques. Although just 2% of experimentally identified structures are transmembrane proteins, genomic studies suggest that these proteins make up to 30% of all coded proteins. Bioinformatics methods enable modeling these unknown protein structures, functions, transmembrane location, and ligand binding. Current in silico modeling tools use various computational methods, which are capable of providing results that may mimic nearly biologically relevant functionality. General understanding of genetics, the gene-based mutations, and emergence of disease, as well as information on evolution that concern ion channel structures and functions including both normal and abnormal biological systems status quo may be addressed using bioinformatics techniques. Bioinformaticians and biophysicists with necessary expertise and interest in computer science techniques have recently started covering various ion channel aspects using artificial intelligence (AI), machine learning (ML), and deep learning (DL) algorithms and models. With the help of a few example studies, we shall provide an introduction to these novel research trends.

## 9.1 BIOINFORMATICS PREDICTIONS OF ION CHANNEL STRUCTURES AND FUNCTIONS

X-ray crystallography and nuclear magnetic resonance (NMR) data on transmembrane proteins are generally used to predict the optimal protein structures. These techniques require the use of extremely expensive necessary ingredients and a tuned laboratory setup. Bioinformatics modeling utilizing appropriate techniques that may promote in silico mechanics and energetics of the protein structure considering the underlying mechanisms is often popularly considered in the biophysical studies of proteins. Membrane proteins are generally studied specifically to address their ion channel-forming potency. Bioinformatics techniques play crucial roles when important molecular actions are to be inspected to explain experimental facts obtained in in vitro studies, such as their imaging in the interface of hydrophobic/hydrophilic regions and electrophysiology record of currents across membranes hosting the proteins. Molecular dynamics (MD) simulations are important computational techniques to detect energetics underlying biomolecular interactions. We have been successful in biophysical addressing, using MD simulations, of the channel energetics involving channel subunits and membrane lipids for small channels, such as gramicidin A, alamethicin, and chemotherapy drug channels in model membrane systems (Ashrafuzzaman and Tuszynski, 2012a; 2012b; Ashrafuzzaman et al., 2012; 2014; 2020a; 2020b). In these publications, we could establish a single fundamental fact that the channel stability inside the membrane is due to nothing but molecular mechanisms depending on charge-based screened Coulomb interaction energetics among functional charge groups in the ion channel complex involving channel subunit peptides or drugs and membrane lipids. Our computational in silico assays (numerical computations and MD simulations) simply supported the experimental findings in the distance and time-dependent channel subunit–lipid interaction energetics theoretically. We calculated the binding energies and evaluated the binding energetics in the channel complex, and thus, determine the statistical mechanical nature in the channel stability in a biological thermodynamic environment. The readers are invited to read directly from these articles to gain further insights.

In addition to various computational assays addressing the general structure and function of channel proteins, bioinformatics templates that draw information from various databanks on the channel protein structures, genomics of the proteins, and mutations in genes of the ion channel proteins provide crucial information about channel functions in both healthy cells and mutated (disease) conditions.

The aspects addressing the ion channel protein genetics and mutations are presented later in this chapter using a few example case studies. Here, we wish to address the general aspects of ion channel structures and functions using bioinformatics techniques including various computational assays and in silico modeling. e.g. among large number of sources see refs. (Kurczynska et al., 2016, Kurczynska and Kotulska, 2018). Table 9.1 presents a set of ion channels that are addressed using various in silico computational techniques (Maffeo et al., 2012).

A two-decade old review analyzed, in combination with MD simulations and associated calculations and modeling, and provided approaches to understanding structure/function relationships in human ion channels (Capener, 2002). Here, the modeling techniques were analyzed for two classes of potassium channels: voltage-gated (Kv) and inward rectifier (Kir) channels. It was clarified how the transmembrane pore region could be modeled based on NMR structures of the pore-lining M2 helix.

What matters most in understanding the ion channel function are based on mostly two aspects: (i) ion channel pore region geometry, and (ii) energetics that control the pore opening/closing phenomena. Direct and indirect experimental techniques can address them phenomenologically, but underlying mechanisms largely rely on modeling of the channel using bioinformatics techniques (Maffeo et al., 2012).

The protein family involved in constructing any specific type of channel is highly variable and closely related to other channels as well, thus making it very difficult to identify new types of channel sequences. Taking potassium channel as an example case, Heil and colleagues (2006) introduced an interesting bioinformatics method, the so-called "Property Signature Method" (PSM), to address the issue of identification of the channel sequences. PSM is based directly on the physicochemical properties of amino acids rather than on the amino acids themselves. A signature for the pore region, including the selectivity filter, has been created, which represents the most common physicochemical properties of known potassium channels. This string enables the genome-wide screening for sequences with similar features despite a very low degree of amino acid similarity within a protein family.

While developing PSM, the dataset used comprises 461 potassium channel α-subunits representing different families, see Figure 9.1 (Heil et al., 2006). For cross-validation, only potassium channel sequences with a pairwise sequence similarity of <80% were used (187 sequences). The set contains additional 957 non-α-subunits, thus providing false positives. These sequences include closely related ion channels, proteins binding to potassium channels, and other randomly chosen proteins. The latter were included to ensure that the signature discriminates between potassium channels and other proteins, and not only between potassium channels and related sequences. All the sequences used here were extracted from Swiss-Prot (Bairoch and Apweiler, 2000).

A potassium channel pore region profile was created using the dataset. This profile is not used to describe the conserved amino acid positions in this region, but it describes all variations found in the different potassium channel families.

This profile is then translated into a descriptor which describes the different properties of the sequence region. The amino acids at each profile position are analyzed, and the properties whose

**TABLE 9.1**

**Ion Channel Modeling and Simulation Studies**

| | Implicit | | All-atom MD | Hybrid | CG | Others (QM) |
|---|---|---|---|---|---|---|
| System | Continuum | Solvent MD | | | | |
| Gramicidins | 8–15 | 16, 17 | 18–51 | 52, 53 | | 54 |
| Outer membrane porins | 55–61 | 55, 62–64 | 55, 65–87 | 55 | | |
| α-hemolysin | 88–93 | 88, 90, 94, 95 | 90, 93, 96–99 | 90, 93, 100–102 | | |
| K+ channels | 103–109 | 110–117 | 29, 88, 111–113, 116–197, 425, 595, 596, 602–604 | 107, 198–202 | 203–206 | 207–211 |
| nAChR | | | 212–220 | | | |
| MscL/MscS | 221–228 | 225, 229 | 230–248, 248–257 | 225, 258 | 222, 258 | |
| Anion channels (VDAC,ClC) | 259 | | 260–264 | 265–268 | | |
| Aquaporins | | | 269–274 | | | |
| NH$_{+4}$ transporters | | | 275–278 | | | |
| Other channel | 279–310 | 311–318 | 299, 312, 319–348 | 302, 330, 337, 349–356 | 357, 358 | |
| Synthetic nanopores | 359–370 | 371–374 | 375–391 | 350, 382, 383, 392 | 393 | 394 |

*Source:* Reprinted (adapted) with permission from Maffeo et al. (2012). Copyright (2012) American Chemical Society.
The references quoted in the table are readily found as referenced in article (Maffeo et al., 2012). Here the general area of ion channels are organized according to the system type and computational models employed.

| | | |
|---|---|---|
| Potassium channels | Voltage-gated | 208 (80) |
| | Inward-rectifier | 87 (29) |
| | Double-pore (2+2) | 59 (30) |
| | Double-pore (6+2) | 1 (1) |
| | Calcium-dependent (SK/IK) | 16 (6) |
| | Calcium-dependent (BK) | 39 (5) |
| | Kcsa + MthK | 2 (2) |
| | Kch | 14 (11) |
| | Hyperpolarization-activated | 32 (20) |
| | unclassified | 3 (3) |
| | Σ | 461 (187) |
| other | Potassium-channel associated | 188 |
| | Calcium channel | 169 |
| | other Channels | 9 |
| | unspecified | 591 |
| | Σ | 957 |

**FIGURE 9.1** Channel families – composition of the dataset. All sequences were extracted from Swiss-Prot (Bairoch and Apweiler, 2000). The potassium channels represent the different families and topologies of known potassium channels. For the cross-validation, nonpotassium channels were used as false positives, and all sequences with >80% sequence similarity were removed from the potassium channels (number of remaining channels in brackets). The (2 + 2) double-pore channels consist of two α-subunits with four transmembrane-domains each, the α-subunits of (6 + 2) double-pore channels possess eight transmembrane domains. The three unclassified potassium channels cannot be unambiguously classified. The unspecified proteins were randomly chosen from Swiss-Prot.

Bit string representation of amino acid properties (groups: side chain type [aliphatic, aromatic, sulfur, imino, amide, hydroxyl, acidic, basic], tertiary structure preference [exposed, buried, intermediate], functional properties [acidic, basic, hydrophobic, polar], secondary structure preference [α-helix, β-strand, coil], size [tiny, small, medium, large, very large]):

| | Amino acid | bit string |
|---|---|---|
| A | Alanine | 10000000001001010010000 |
| C | Cysteine | 00100000001000101001000 |
| D | Aspartate | 00000101001000000100100 |
| E | Glutamate | 00000101001000010000010 |
| F | Phenylalanine | 01000000010001001000001 |
| G | Glycine | 10000000001000100110000 |
| H | Histidine | 00000011000001010000010 |
| I | Isoleucine | 10000000010001001000100 |
| K | Lysine | 00000011000100010000010 |
| L | Leucine | 10000000010001010000100 |
| M | Methionine | 00100000010001010000010 |
| N | Asparagine | 00001000100000100100100 |
| P | Proline | 00010000010001001001000 |
| Q | Glutamine | 00001000100001100000010 |
| R | Arginine | 00000001100010010000001 |
| S | Serine | 00000100001000100100100 |
| T | Threonine | 00000100001000101001000 |
| V | Valine | 10000000010001001001000 |
| W | Tryptophan | 01000000001001001000001 |
| Y | Tyrosine | 01000000001000101000001 |

**FIGURE 9.2** Amino acid properties. Bit string representation of the amino acids composed of 23 properties. The relative frequency of occurrence in such a state was converted into binary values by majority vote. With respect to size, the amino acids were categorized according to their molecular weight: tiny when ≤71 Da, small when ≤103 Da, medium when ≤115 Da, large when ≤137 Da, and very large when >137 Da.

absence or presence are conserved are used to describe this position. Hits are ranked according to the number of properties found in both the property descriptor and the target sequence. The screening algorithm was implemented in programming language C++.

The PSM uses an amino acid representation via a binary signature derived from various physicochemical properties. A total of 23 properties were used and combined into five groups: side chain type, functional properties, secondary and tertiary structure preference, and size, see Figure 9.2 (Heil et al., 2006). Each amino acid is represented by a binary string created in which a bit is set to 1 if the corresponding property applies to the amino acid. Five bits are set, one for each property group. The remaining bits are set to 0. This property encryption results in 20 unique bit strings, one for each amino acid, used in the algorithm. The method is divided into two steps: (i) a profile of the aligned pore domains is created which includes all amino acids present in at least 3% of the 461 potassium channels; (ii) this profile is translated into a string representing the physicochemical properties of the sequences.

Large-scale analysis of potassium channel sequences confirms the requirement of identifying the potassium channel α-subunit proteins (Harte and Ouzounis, 2002). As the potassium channel family is highly diverse and closely related to other ion channels using amino acids to classify potassium channels in PSM has been found to be imprecise. PSM is found to be superior over the Markov models and BLASTp (Altschul et al., 1997; Harte and Ouzounis, 2002; Moulton et al., 2003). In addition, they use the potassium channel motifs from the PRINTS Database (Attwood et al., 1997). These approaches use multiple methods to overcome a single method's limitations of recognizing sequences of a certain subset of the potassium channel family. These issues are indeed resolved in PSM that can detect properties which are representative for all subsets of the potassium channel family. Moreover, PSM analyzes the physicochemical properties of amino acids and enables a more sensitive extraction of information coded in the amino acid sequences. For details, readers may consult the original article (Heil et al., 2006).

*Saccharomyces cerevisiae* genome was screened using PSM (Heil et al., 2006). Both of the only two hits found were the pore domains of the two-pore potassium channel TOK1, the only known *S. cerevisiae* potassium channel. Despite the close relationship and high homology of the two potassium transporters, TRK1 and TRK2, to potassium-selective pore domains of TOK1, these two have been classified as nonpotassium channels.

Heil et al. also performed another test with *Caenorhabditis elegans* having complete genome sequence (Hodgkin et al., 1998). Its genome is well understood regarding potassium channel sequences; about 40 two-pore-domain potassium channels are annotated. All were recovered using PSM, and additionally, one new potential pore domain was identified.

Regarding the potassium channel signature, a summary of the conserved properties at 60% and 80% conservation threshold is presented in Figure 9.3. Despite the high divergency in the sequence set, 63 properties are conserved at the 60% significance level and 19 properties are conserved at the 80% significance level. Not shown are the unusual properties coded in the signature – about 350 properties at 60% significance level and 330 properties at 80% significance level. These properties contribute significantly to the specificity of the method.

PSM is superior to conventional methods for the search for sequences with a very low conservation level. PSM has a main advantage which is that the signature describes, for each amino acid position, the frequently selected properties and uncommon properties in the potassium channel α-subunits. Using position-bound properties in the signature has another advantage, that is, the results interpretation appears quite simple. Next to the number of missing and unusual properties, the method returns (for each sequence) a vector that displays which sequence positions contain the missing and the untypical residues, respectively, thus facilitating fast analysis of the sequence.

## 9.2 ION CHANNEL GENOMES TRACK EARLY ANIMAL EVOLUTION

A comparative genomics study provided a new window into the past that may be applied for understanding the early evolution of animal nervous systems (Liebeskind et al., 2015). An important controversy is whether nervous systems evolved just once or independently in different animal lineages. Liebeskind and colleagues explored the history of the ion channel gene families that are most central to nervous system function. They tracked when these gene families expanded in animal evolution and found that they radiated on several occasions and, in some cases, underwent periods of contraction. The multiple origins of these gene families may signify large-scale convergent evolution of nervous system complexity.

| signature position | # | conserved properties | |
|---|---|---|---|
| | | 60% | 80% |
| [DGNRST]$_1$ | 6 | polar, loop | - |
| [FILVWY]$_2$ | 6 | internal, hydrophobic, β-strand | hydrophobic, β-strand |
| [AFIL STVW]$_3$ | 9 | hydrophobic | - |
| [ADEGHIST]$_4$ | 8 | - | - |
| [ACGS]$_5$ | 4 | no tertiary., polar | no tertiary. |
| [FILMVY]$_6$ | 6 | internal, hydrophobic, β-strand | internal, hydrophobic |
| [FLWY]$_7$ | 4 | aromatic, hydrophobic, β-strand, very large | - |
| [FKLWY]$_8$ | 5 | aromatic, hydrophobic, β-strand, very large | - |
| [ACGILSTV]$_9$ | 8 | aliphatic, no tertiary. | - |
| [FILMSTV]$_{10}$ | 7 | internal, hydrophobic | - |
| [EISTV]$_{11}$ | 5 | β-strand, small | - |
| [HSTV]$_{12}$ | 4 | polar, small | - |
| [EFILMQV]$_{13}$ | 7 | internal, hydrophobic | - |
| [ALSTV]$_{14}$ | 5 | aliphatic, no tertiary., hydrophobic, small | - |
| [CST]$_{15}$ | 3 | hydroxyl, no tertiary., polar, β-strand, small | no tertiary., polar, small |
| [ILTV]$_{16}$ | 4 | aliphatic, internal, hydrophobic, β-strand | - |
| [G]$_{17}$ | 1 | aliphatic, no tertiary., polar, loop, very small | aliphatic, no tertiary., polar, loop, very small |
| [FLY]$_{18}$ | 3 | aromatic, internal, hydrophobic, β-strand, very large | - |
| [G]$_{19}$ | 1 | aliphatic, no tertiary., polar, loop, very small | aliphatic, no tertiary., polar, loop, very small |
| [DFNRSY]$_{20}$ | 6 | - | - |
| [IKLMQRVY]$_{21}$ | 8 | α-helical | - |
| [ACHRSTVY]$_{22}$ | 8 | no tertiary., polar | - |
| [AI V]$_{23}$ | 4 | aliphatic, hydrophobic | hydrophobic |
| [EGHIKLNQSTVY]$_{24}$ | 12 | - | - |
| [DEGNQST]$_{25}$ | 7 | polar | - |
| Σ properties | | 63 | 19 |

**FIGURE 9.3** Conservation of properties at 60% and 80% significance level. A scheme of the secondary structure is drawn left of the signature positions. Despite the low amino acid conservation, there are properties which are conserved in 80% of the sequences. As expected from the analysis of potassium channel pores (Doyle et al., 1998), hydrophobic residues dominate the pore, with a few polar residues to decrease the energetic barrier for the charged K$^+$ ions (Heil et al., 2006).

Ancestral gene content reconstruction was used to track the timing of gene family expansions for the major families of ion channel proteins that are known to drive nervous system function. Animals with nervous systems were found to have broadly similar complements of ion channel types, but these complements might evolve independently. Ion channel gene family evolution was found to experience large loss events, two of which were immediately followed by rounds of duplication. Ctenophores, cnidarians, and bilaterians have been found to undergo independent bouts of gene expansion in channel families involved in synaptic transmission and action potential shaping, suggesting a genomic signature of expanding nervous system complexity. Ancestral nodes, wherefrom nervous systems are hypothesized to have originated, was found not to experience large expansions, suggesting that the possible origin of nerves did not experience an immediate burst of complexity, instead, the nervous system evolution complexity experienced a slow fuse in stem animals, followed by independent gene gains and losses.

A custom bioinformatics pipeline (Liebeskind et al., 2015) was used to collect and annotate predicted proteins from 16 ion channel families, see Table 9.2 for 41 broadly sampled opisthokonts (the group that includes animals, fungi, and related protists) and an apusozoan outgroup. The ion channel families play diverse roles in nervous systems – some families (such as the voltage-gated families) are almost solely associated with nervous system function in animals, whereas others (such as P2X receptors) play more diverse roles, with only some isoforms being expressed in nervous systems. This dataset was then used to infer ancestral genome content and the timing of gene duplications using EvolMap (Sakarya et al., 2008).

These gene families were found to be ancient (Kai et al., 2011; Moroz et al., 2014), with all but two [Cys-loop receptor (LIC) and acid sensing channel (ASC)] being found in the most recent common ancestor (MRCA) of the taxa examined here (Liebeskind et al., 2015). Only the ASC family was found to be metazoan-specific. All the families were pulled together and net gains and percent losses on the species tree were plotted, see Figure 9.4 (Liebeskind et al., 2015). The animal lineage has been dominated by gains and the fungal lineage by losses. Two major loss events, however,

## TABLE 9.2
## Ion Channel Families (Liebeskind et al., 2015)

| Abbreviation | Full Names | Function |
|---|---|---|
| Ano | Anoctamin, $Ca^{2+}$ activated $Cl^-$ | Smooth muscle, excitability |
| ASC | Epithelial (ENaC), acid sensing channel (ASIC) | Osmoregulation, synaptic transmission |
| CNG/HCN | Cyc. nucleotide gated | Sensory transduction, heart |
| $Ca_v$ | Voltage-gated $Ca^+$ channel | AP, muscle contraction, secretion |
| ClC | Voltage-gated $Cl^-$ channel | Muscle membrane potential, kidney |
| GIC | Glutamate receptor (iGluR) | Synaptic transmission |
| LIC | Ligand-gated, Cys-loop receptor | Synaptic transmission |
| $K_v$ | Voltage-gated $K^+$ channel | AP, membrane potential regulation |
| $Na_v$ | Voltage-gated $Na^+$ channel | AP propagation |
| Leak | Sodium leak (NALCN), yeast calcium channel (Cch1) | Regulation of excitability |
| P2X | Purinurgic receptor | Vascular tone, swelling |
| PCC | Polycystine, Mucolipin | Sensory transduction, kidney |
| RyR | Ryanodine receptor, $IP_3$ receptor | Intracellular, muscle contraction |
| Slo | Voltage and ligand-gated $K^+$ | AP, resting potential |
| TPC | Two-pore channel | Intracellular, NAADP signaling |
| TRP | Transient receptor potential | Sensory transduction |

The channels play various roles. Some are almost exclusively associated with nervous system function, whereas others have additional roles outside the nervous system.

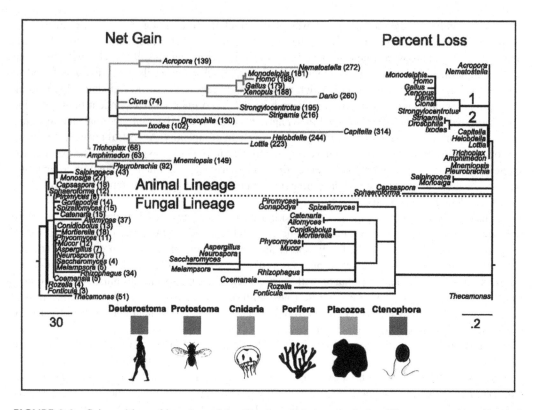

**FIGURE 9.4** Gain and loss of ion-channel families in opisthokont evolution. The two trees have identical topologies. The branch lengths of the tree on the left are the net gain (gains minus losses), and the branch lengths of the tree on the right represent percent loss (losses minus gains as a percentage of parent copy number). Total numbers of ion channels in each taxon are shown on the left-hand tree. Two branches in animals that had large loss events are labeled: the common ancestors of deuterostomes and ecdysozoans.

occurred, just before major gene family expansions (exception), in the common ancestors of deuterostomes and ecdysozoans.

In the phylogenetic pattern of gain and loss for each of the 16 ion channel families (Table 9.2), large expansions of the LIC, glutamate-gated channel (GIC), and voltage-gated potassium channel (Kv) families at several places on the tree were found (Liebeskind et al., 2015). These gene-family expansions occurred independently in the branches leading to the MRCAs of bilaterians, vertebrates, and cnidarians. The vertebrate gene family expansions occurred after the loss event in the branch leading to the MRCA of deuterostomes. The loss events involved reductions in several families, with the largest families, such as LIC, having the largest losses. The branch that gave rise to the MRCA of ctenophores underwent an expansion resembling the expansions in bilaterians and cnidarians, but the LIC family was lost in ctenophores. No expansions were reported in the branches leading to the MRCA of cnidarians plus bilaterians or to the MRCA of animals – two places where nervous systems have been hypothesized to have evolved (Moroz et al., 2014).

Ecdysozoans and lophotrocozoans were found to have large expansions of LIC, GIC, and Kv channels, as well as large expansions of the ASC family, see Figure 9.5a (Liebeskind et al., 2015). These expansions occurred mostly in the terminal lineages leading to each species, see Figure 9.5. Figure 9.5a shows ion channel family counts from representative species from each major lineage. All taxa with nervous systems, with the notable exception of the tunicate *Ciona*, were enriched for similar gene families. The two taxa without nervous systems, *Trichoplax* and *Amphimedon*, had smaller ion channel complements. The MRCAs of chordates, cnidarians plus bilaterians, and

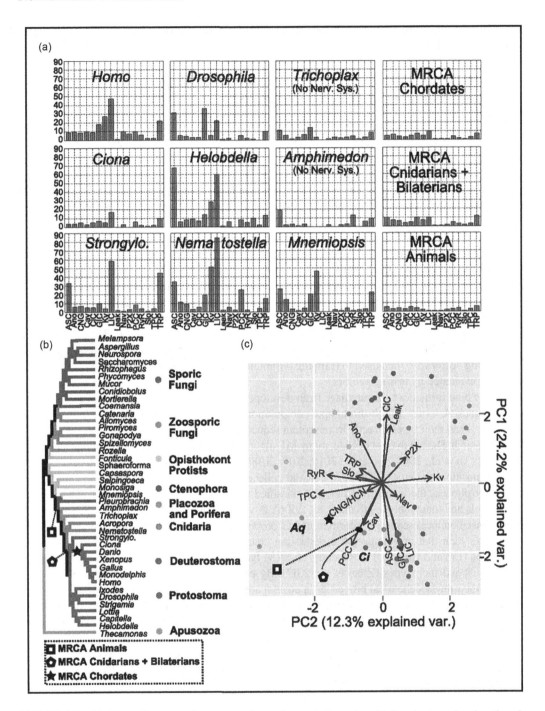

**FIGURE 9.5** (a) Channel counts of extant species and ancestral species. (b) Species tree showing the relationships of extant taxa and the location of key ancestral nodes. (c) PCA of normalized ion channel gene contents for all tips and three ancestral nodes. Proximity in the space of the two PCs indicates similar gene contents. Loadings of the ion channel families are shown as vectors in the two axes. The size and direction of the loading vector indicates its correlation with the two components. Loading arrows point to regions where that gene family is in high relative abundance. Labeled species are *Amphimedon* (Aq) and *Ciona* (Ci).

animals each had ion channel complements that resembled extant animals without nervous systems more than animals with nervous systems.

## 9.3  BIOINFORMATICS PREDICTIONS OF ION CHANNEL GENES AND CHANNEL CLASSIFICATION

Ion channels are directly or indirectly associated with various types of cellular disorders leading to specific diseases. Therefore, ion channels are therapeutic and diagnostic targets of many drugs. About 700 drugs are known so far to act upon ion channels (Han et al., 2019). Knowledge of ion channel genes and their mutations is key to understanding diseases and drug discovery. Bioinformatics techniques may be helpful in understanding the roles of ion channels in diseases through analysis of genetics-based classifications (Han et al., 2019), as well as genetic mutations (Klassen et al., 2011; Xu et al., 2019) of ion channels.

**Artificial Intelligence Techniques Help Predicting Ion Channel Genes**. ML, a subset of AI, was used recently to extract the feature vectors of various ion channels (Han et al., 2019). The SVMProt and the k-skip-n-gram methods were used, which helped obtain 188- and 400-dimensional features, respectively. SVMProt is a web-based support vector machine (SVM) software developed for the functional classification of a protein from its primary sequence (Cai, 2003). The structural class of a protein is inconsiderably correlated with the constituent amino acid composition, and the SVM was found to be a powerful computational tool for predicting the structural classes of proteins (Cai et al., 2002). In the k-skip-n-gram method, each protein sequence is transferred into a vector. The training vectors are then used to train the parameters of random forest. The performance of the method is evaluated by testing vectors.

Various bioinformatics software have been developed to predict the identification of ion channels in membranes. A series of high-throughput computational tools are now available which help predict ion channels and their types directly from protein sequences, helping in ion channel-targeted drug discovery research. During the last decade, many ML algorithm-based computational methods have been developed (Yu et al., 2014; Zou et al., 2016; 2017; 2018a; 2018b; Qu et al., 2019), which may be used in drug repositioning (Yu et al., 2016). Saha and colleagues used amino acid composition and dipeptide composition as the feature vectors, and classified them using an SVM to predict voltage-gated ion channels and their subtypes (Saha et al., 2006). A voltage-gated potassium channel identification method based on local sequence information was proposed (Liu et al., 2006; 2014), which was found to be better than that of voltage-gated potassium channel identification based on global sequence information (Lin and Ding, 2011). Recently, an SVM-based model was constructed to quickly predict ion channels and their types (Zhao et al., 2017). By considering the residue sequence information including their physicochemical properties, a novel feature-extracted method which combined dipeptide composition with the physicochemical correlation between two residues was employed. A model based on SVM to search for predicted ion channels and their subfamilies using the sequence similarity search feature of the basic local alignment search tool was recently developed (Gao et al., 2016).

In a recent article, a briefing in the application of ML methods in ion channel has been made (Lin and Chen, 2015). This review focused on the development of prediction methods for ion channels in terms of the following issues:

   i. datasets of ion channel proteins,
  ii. ML methods to predict ion channels,
 iii. feature selection techniques to obtain optimal features for ion channel predictions, and
  iv. prospect of bioinformatics methods prediction of ion channels using appropriate and available tools.

Han and colleagues (2019) used SVM and random forest classifiers to identify ion channels and further classify them. The maximum-relevance-maximum-distance (MRMD) method was introduced for feature selection to improve the prediction accuracy. Three steps were followed to predict and classify ion channels. First, a protein sequence was detected to determine if it belonged to an ion channel. If the test results appeared positive, the protein sequence was classified as either a voltage-gated ion channel or a ligand-gated ion channel. Finally, if the protein sequence was found to belong to a voltage-gated ion channel, the sequence were classified as belonging to potassium (K+), sodium (Na+), calcium (Ca$^{2+}$), or anion voltage-gated ion channel class.

The flowchart shows the stepwise adopted basic processes that Han and colleagues considered for gene detection and channel classification, see Figure 9.6 (Han et al., 2019). We avoid explaining how they introduce the dataset, feature extraction method, dimension reduction method, and classifier used in this study, but readers may find them in the original article.

The data used for the prediction model were collected from Lin and Ding (2011). The ion channel sequences were collected from the Universal Protein Resource (UniProt) and ligand-gated ion channel databases (Marco et al., 2006). A total of 148 voltage-gated ion channels (81 potassium channels, 29 calcium channels, 12 sodium channels, 26 anion channels) and 150 ligand-gated ion channels were finally extracted. In total, 300 protein sequences were randomly selected from the UniProt as nonion channels, with the consistency of these nonion channel sequences at <40%. Two feature extraction ML methods, the SVMProt 188-D feature extraction method (which is based on protein composition and physicochemical properties), and the k-skip-n-gram 400-D feature extraction

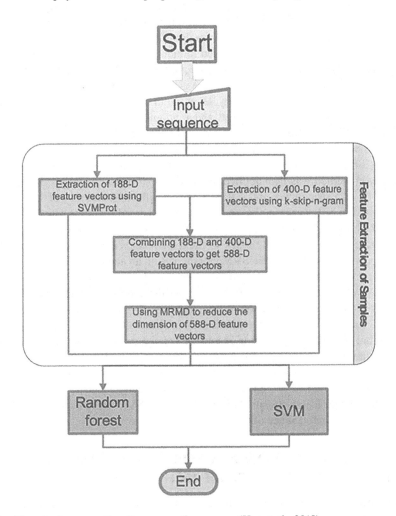

**FIGURE 9.6**   Flowchart representing the proposed processes (Han et al., 2019).

**TABLE 9.3**

**Prediction Results of Ion Channels and Nonion Channels**

| Method | Ion Channel (%) | Nonion Channel (%) | OA (%) |
|---|---|---|---|
| Random forest (188D) | 90.3 | 77.2 | 83.7793 |
| SVM (188D) | 87.0 | 78.5 | 82.7759 |
| Random forest (400D) | 87.7 | 77.5 | 82.6087 |
| SVM (400D) | 86.6 | 83.7 | 85.1171 |
| Random forest (588D) | 77.5 | 190 | 83.7793 |
| SVM (588D) | 83.2 | 80 | 81.6054 |
| Random forest (587D) | 77.2 | 89.7 | 83.4448 |
| SVM (587D) | 77.2 | 83.3 | 80.2676 |

method were used. These two feature representation methods were combined to form a new feature vector containing more than one feature. SVM and random forest classifiers were used to classify the new feature vector set. The dimensionality reduction method based on MRMD (http://lab. malab.cn/soft/MRMD/index_en.html) was then employed to reduce the dimensionality of the generated feature vectors (Xu et al., 2016). MRMD selects the feature with the highest correlation and least redundancy by calculating the maximum relevance and maximum distance. Here, a random forest classifier was used to build a model. As random forest is a classifier that uses multiple trees to train and predict samples, it is popularly used in bioinformatics research where applicable (Pan et al., 2018). It is a well-performing tool using especially the random forest algorithm (Buntine and Niblett, 1992) in many practical fields, such as the classification and regression of gene sequences, action recognition, face recognition, anomaly detection in data mining, and metric learning.

The predictive effects of the SVM-based and random forest-based methods on both ion and nonion channels in different dimensions were compared in this study (Table 9.3) (Han et al., 2019). The 10-fold cross-validation results of the 188-dimensional features, 400-dimensional features, and mixed features (188-dimensional features combined with 400-dimensional features) are listed in the table. The MRMD method was then applied to reduce the dimensions of the 588-dimensional features to obtain 587-dimensional features. The average classification accuracy of the 587-dimensional features is reported to be lower than that of the 400-dimensional features. The SVM classifier was reported to be the best method for classifying the 400-dimensional features, with an average overall accuracy (OA) rate of 85.1%. Overall, 86.6% of the ion channels and 83.7% of the nonion channels could be appropriately identified by the SVM classifier, with an accuracy of 85.1%. The feature vectors of the 188- and 400- dimensional features yielded good prediction results.

The accuracy was evaluated on the 188-dimensional features, 400-dimensional features, and mixed features (188-dimensional features combined with 400-dimensional features), and the 88-dimensional features obtained after the dimensional reduction using the MRMD method for discriminating between the classification results of voltage-gated and ligand-gated ion channels. The results are summarized in Table 9.4 for these two classes and in Table 9.5 for ion specificity in voltage-gated ion channels (Han et al., 2019). Overall, 93.9% of the voltage-gated ion channels and 86.0% of the ligand-gated ion channels could be correctly identified using the random forest method. The random forest classifier is better than the SVM classifier in some cases and can improve the prediction performance of the model.

## 9.4  DETECTION OF ION CHANNEL GENETIC MUTATIONS USING ARTIFICIAL INTELLIGENCE TECHNIQUES

Mutations in genes are generally known to be responsible for diseases. Genetic mutations involving ion channel subunits or proteins are also often responsible for various diseases. AI techniques may

**TABLE 9.4**
**Compare the Results between Voltage-Gated and Ligand-Gated Ion Channels**

| Method | Voltage-Gated Ion Channels (%) | Ligand-Gated Ion Channels (%) | OA (%) |
|---|---|---|---|
| Random forest (188D) | 93.9 | 86.0 | 89.9329 |
| SVM (188D)) | 91.9 | 86.7 | 89.2617 |
| Random forest (400D) | 88.5 | 82.7 | 85.5705 |
| SVM (400D) | 82.4 | 83.3 | 82.8859 |
| Random forest (588D) | 89.2 | 86.0 | 87.5839 |
| SVM (588D)) | 91.9 | 86.7 | 89.2617 |
| Random forest (188D) | 92.6 | 86.7 | 89.5973 |
| SVM (188D) | 91.9 | 86.7 | 89.2617 |

**TABLE 9.5**
**Prediction Results for Four Types of Voltage-Gated Ion Channels**

| Method | K (%) | Ca (%) | Na (%) | Anion (%) | OA (%) | AA (%) |
|---|---|---|---|---|---|---|
| Random forest (188D) | 97.5 | 37.9 | 50 | 46.2 | 72.973 | 57.9 |
| SVM (188D) | 96.3 | 48.3 | 58.3 | 69.2 | 79.0541 | 68.0 |
| Random forest (400D) | 97.5 | 6.9 | 50 | 23.1 | 62.8378 | 44.4 |
| SVM (400D) | 85.2 | 62.1 | 50 | 73.1 | 75.6757 | 67.6 |
| Random forest (588D) | 97.5 | 34.5 | 50 | 57.7 | 74.3243 | 59.9 |
| SVM (588D) | 96.3 | 48.3 | 53.3 | 69.2 | 79.0541 | 60.2 |
| Random forest (424D) | 98.8 | 34.5 | 58.3 | 46.2 | 73.6486 | 59.5 |
| SVM (424D) | 96.3 | 48.3 | 58.3 | 69.2 | 79.0541 | 68.0 |

be applied to establish such evidence in bioinformatics explorations. We use a few case studies to address this phenomenon for certain diseases.

### 9.4.1 Ion Channel Genetic Variants in Epilepsy

Ion channel mutations are important causes of rare Mendelian disorders affecting the heart, brain, and other tissues. Mendelian mutations are linked with single-channel defects causing familial episodic and degenerative excitability disorders in the cardiovascular (Demolombe et al., 2005), nervous (Catterall et al., 2008), neuroendocrine (Hiriart and Aguilar-Bryan, 2008), and immune surveillance systems (Cahalan and Chandy, 2009).

Klassen and colleagues performed parallel exome sequencing of 237 channel genes in a human sample and compared variant profiles of unaffected individuals to those with the most common neuronal excitability disorder, sporadic idiopathic epilepsy (Klassen et al., 2011). Rare missense variation in known Mendelian disease genes is prevalent in both groups at similar complexity. This study reveals that even deleterious ion channel mutations confer uncertain risk to an individual depending on the other variants with which they are combined.

Comparisons were made on exomic single nucleotide polymorphism (SNP) profiles, including the type, relative burden, and pattern of variants within a large ion channel candidate gene set between healthy (unaffected) individuals and those with severe neurological excitability disease to evaluate personal genetic liability. Table 9.6 summarizes SNPs (Klassen et al., 2011).

This study found SNPs in both groups for every targeted gene; of the validated SNPs, 1,355 were unique to either population, and the majority (1,740) were shared. The data have expanded the

**TABLE 9.6**

**SNPs in 237 Ion Channel Genes in Those with Idiopathic Generalized Epilepsy and Neurologically Normal Individuals**

| Type/Location of SNP | Number of Validated SNPs[a] | Percent of Validated Dataset (%) | Number of Novel SNPs Discovered | Number of Validated SNPs per Megabases Sequenced[f] | | |
|---|---|---|---|---|---|---|
| | | | | Cases Only (n=152) | Controls Only (n=139) | SNPs in Both Cases and Controls (n=291) |
| Promoter[b] | 80 | 2.6 | 18 | 0.4 | 0.1 | 0.4 |
| 5′ UTR | 79 | 2.6 | 7 | 0.2 | 0.1 | 0.5 |
| 3′ UTR | 461 | 14.9 | 62 | 1.4 | 0.6 | 3.0 |
| Synonymous (sSNP) | 936 | 30.2 | 351 | 5.1 | 2.2 | 4.2 |
| Nonsynonymous (nsSNP) | 668 | 21.6 | 415 | 4.9 | 2.2 | 1.9 |
| Nonsense/Stop codon | 9 | <1 | 9 | 0.1 | 0.03 | 0 |
| Splice site SNP[c] | 12 | <1 | 9 | 0.1 | 0.03 | 0.02 |
| Splice region SNP[d] | 90 | 2.9 | 13 | 0.3 | 01 | 0.6 |
| Intron SNP | 737 | 23.8 | 101 | 2.3 | 1.0 | 4.7 |
| Undefined[e] | 23 | <1 | 4 | 0.1 | 0 | 0.2 |
| **Total** | **3095** | **100.0** | **989** | **14.6** | **6.3** | **15.6** |

[a] Validated SNPs are SNPs that were confirmed through a combination of (1) visual validation; (2) previous discovery (dbSNP ID), (3) detected on a custom MIP chip, (4) Biotage and/or 454 sequencing.

[b] SNPs in promoter regions are reported as such if the promoter for the gene is known and defined in our gene models.

[c] Splice site (+2 to −2 bp from defined exon boundary at/near splice junction).

[d] Splice region (−2 to −15 bp from defined exon boundary, located in the intron, from splice junction).

[e] Undefined SNPs are SNP detected in regions that are not defined in our gene models.

[f] The number of individual SNPs per megabase sequenced does not reflect the SNP frequency.

known channel SNP lists in dbSNP. This addition also confirms the existence of rare allelic variation across a broad spectrum of ion channel genes. The rich variation is found to agree with that emerging from whole genome sequencing of individuals (Durbin et al., 2010), and from >2,100 cases that are screened for variants in a subset of clinically important cardiac channel genes (Kapplinger et al., 2009). An individual's channotype is apparently unique. In this cohort, it is found that no individuals were free of SNPs, and no two channotypes among 291 individuals were identical, see Figure 9.7 (Klassen et al., 2011). Across both groups, this study found an overlapping variety of SNP types, including sSNPs, nsSNPs, and SNPs in promoter, coding, UTR, and intronic regions. Nonsense SNPs were observed in both of these populations.

A total of 300 missense channel variants in 139 unaffected individuals were found. Of these, 23 were in human epilepsy (hEP) genes signaling that the allelic penetrance in channelopathy is underappreciated, see Figure 9.8 (Klassen et al., 2011). The R393H nsSNP in the ion-selective pore of the SCN1A gene is believed to cause severe myoclonic epilepsy of infancy (Claes et al., 2001). It was also detected once, and only in the control population. The in vitro functional studies of this mutation, however, failed to produce measurable sodium current (Ohmori et al., 2006), indicating that deleterious alterations in protein structure in a known hEP gene are insufficient to predict the risk of epilepsy. The finding of missense mutations in known hEP genes in the control cohort supports a biophysical model that other subunits, as yet unrelated to epilepsy, may constitute genetic excitability modifiers, thus providing direct evidence validating the multigenic basis for complex inheritance of channelopathy phenotypes.

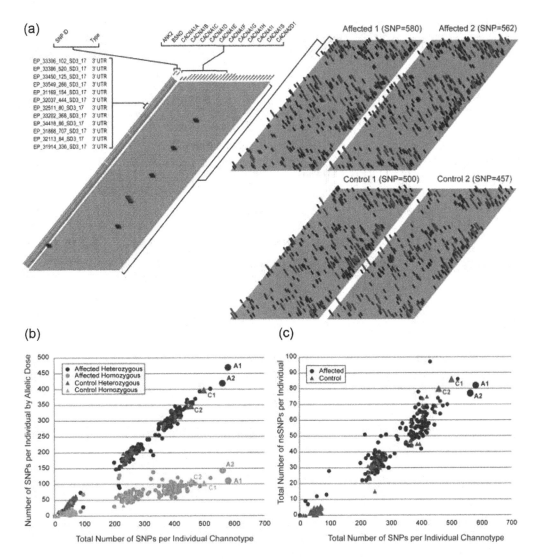

**FIGURE 9.7** Multiple variants render each individual channotype unique. (a) Low-resolution (gray background) 3D representations illustrating the most extreme channotypes present in the study cohort (two cases and two controls, each with >450 SNPs). The columns list the channel subunit genes in alphabetical order (ANK – SCN) and the rows list validated individual SNP identifiers organized alphabetically by type (3'UTR – promoter). An enlargement at left in teal is presented for clarity and scale. The gene dosage of the minor (variant) allele for a SNP is denoted by a bar (tall red = homozygous minor allele; short blue = heterozygous minor allele). Sparsely populated regions present in all four channotypes reflect low frequency novel SNPs. (b) Histogram of all individuals by cohort with the total number of SNPs in the individual plotted against the total number of SNPs in the channotype that are heterozygous or homozygous. The affected and control cohorts show similar allelic dosages with increasing SNP count. (c) Histogram of all individuals within each cohort showing the total number of SNPs per individual plotted against number of nsSNPs contained in the channotype. The number of nsSNPs in a channotype increases with increasing total SNP count in both populations. The individual channotypes profiled in (a). (A1, A2, C1, C2) are indicated in the histograms. (For color figure see eBook figure.)

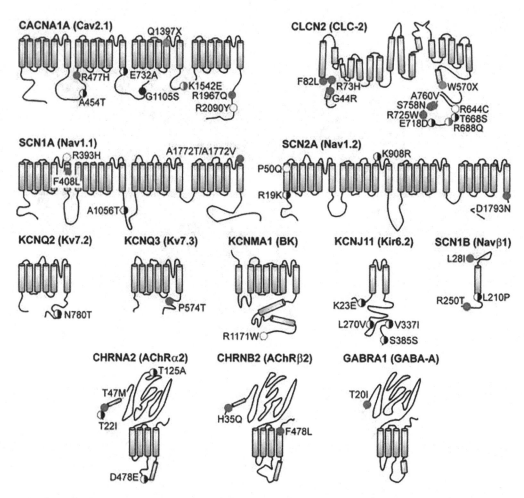

**FIGURE 9.8** Known monogenic human epilepsy (hEP) genes populated with missense and nonsense variants found in the cohort. The protein products of 12 ion channel genes known to cause monogenic epilepsy are shown schematically. Validated missense and nonsense SNPs discovered by profiling are represented by circles marking the nearest amino acid location as determined by comparative multiple alignment. Presence of a SNP denoted by the fill pattern (filled circle = in affected only; open circle = in controls only; half filled circle = SNP is present in both groups). The nsSNPs in dbSNP are colored black, novel nsSNPs from our study are red, and the nonsense SNPs are colored blue. (For color figure see eBook figure.)

The study also explored the value of computational models to assist in personal risk prediction (Klassen et al., 2011). Epilepsy with no known cause (idiopathic epilepsy, IE) was found to be an ideal condition to study the impact of sporadic genetic channel variation on the cortical function, as seizure disorders are found to affect 1%–2% of the population, and analyses of the rare Mendelian forms reveal that ion channels are in fact major determinants of the phenotype, as 17/20 confirmed monogenic syndromes arise in individuals heterozygous for an SNP in a channel subunit gene (Reid et al., 2009). Thus, the study observed considerable genetic complexities and overlapping patterns of both rare and common variants in known excitability disease genes across both populations.

Understanding of the genetic mutations in ion channels using bioinformatics techniques are expected to help largely in drug discovery. In epilepsy, almost one-third of the patients are refractory to current antiepileptic drug treatments, which, with few exceptions, target ion channels. The detected sequence variants that alter access to drug-binding sites are obvious candidate mechanisms for pharmacoresistance, and variant profiles may even personalize the treatment by identifying ineffective drugs in epilepsy and other excitability disorders concerning channel modulation is clinically useful.

The variant discovery via large-scale sequencing efforts is only the first step in illuminating the complex allelic architecture underlying personal disease risk. In silico modeling of channel variations in cell and network models will be crucial to future strategies that might be able to assess mutation profile pathogenicity and drug response in individuals having a broad spectrum of excitability disorders related to diseases.

### 9.4.2  Ion Channel Genetic Variants in Alzheimer's Disease

Alzheimer's disease (AD) is a heterogeneous genetic disorder characterized by the early hippocampal atrophy and cerebral Aβ peptide deposition. Using Tissue Info to screen for genes that are found to preferentially express in the hippocampus and located in AD-linkage regions, a novel gene on 10q24.33 was identified, CALHM1 (Dreses-Werringloer et al., 2008). CALHM1 was shown to encode a multipass transmembrane glycoprotein that controls cytosolic $Ca^{2+}$ concentrations and Aβ levels. CALHM1 homomultimerizes, shares considerable sequence similarities with the NMDA receptor's selectivity filter, and generates a considerable $Ca^{2+}$ conductance across the plasma membrane. It was determined that the CALHM1 P86L polymorphism (rs2986017) is significantly associated with AD in independent case-control studies of 3,404 participants (allele-specific OR = 1.44, P = $2 \times 10^{-10}$). The P86L polymorphism was found to increase Aβ levels by interfering with CALHM1-mediated $Ca^{2+}$ permeability. Thus, it was concluded that CALHM1 encodes an essential component of a novel cerebral $Ca^{2+}$ channel that controls Aβ levels and susceptibility to AD.

Dreses-Werringloer and colleagues showed that a region of CALHM1 structure shares sequence similarities with the selectivity filter of NMDAR and that the N72 residue is a key determinant in the control of cytosolic $Ca^{2+}$ levels by CALHM1. Electrophysiological analyses in CALHM1-expressing Xenopus oocytes and CHO cells were found to demonstrate that CALHM1 induced a novel plasma membrane $Ca^{2+}$-selective cation current, suggesting that CALHM1 might be a novel pore-forming ion channel (Dreses-Werringloer et al., 2008). In a subsequent study, however, rare genetic variants in CALHM1 were reported to lead to $Ca^{2+}$ dysregulation and predicted to contribute to the risk of EOAD through a mechanism independent from the classical Aβ cascade (Rubio-Moscardo et al., 2013). Here, the role of CALHM1 variants in early-onset AD (EOAD) was investigated. They sequenced all CALHM1 coding regions in three independent series comprising 284 EOAD patients and 326 controls. Two missense mutations in patients (p.G330D and p.R154H) and one (p.A213T) in a control individual were identified. Calcium imaging analyses revealed that while the mutation found in a control (p.A213T) behaved as wild-type CALHM1 (CALHM1-WT), a complete abolishment of the $Ca^{2+}$ influx was associated with the mutations found in EOAD patients (p.G330D and p.R154H). The CALHM1 P86L polymorphism was found in another study associated with elevated cerebrospinal fluid (CSF) Aβ in normal individuals at risk for AD (Koppel et al., 2011), which support that CALHM1 controls Aβ metabolism in vitro in cell lines (Dreses-Werringloer et al., 2008) and in vivo in human CSF (Kauwe et al., 2010). Despite having crucial molecular-level understanding in mentioned findings, we wish to elaborate on understanding the genetic mutations in ion channels concerning AD utilizing bioinformatics techniques (Dreses-Werringloer et al., 2008).

In the study (Dreses-Werringloer et al., 2008), the human genome with Tissue Info was studied to annotate human transcripts with tissue expression levels derived from the expressed sequence tag database (dbEST) (Campagne and Skrabanek, 2006). Out of 33,249 human transcripts, the Tissue Info screen was found to identify 30 transcripts, corresponding to 12 genes, with expression restricted to the hippocampus, see Table 9.7 (Dreses-Werringloer et al., 2008). These transcripts were found to match either one or two ESTs sequenced from the hippocampus. Among these genes, one of unknown function, previously annotated as FAM26C, matched two hippocampal ESTs and mapped to the AD locus on 10q24.33. This gene, referred to as CALHM1 (calcium homeostasis

**TABLE 9.7**

**Tissue Info Expression Screen[a] (Dreses-Werringloer et al., 2008)**

| Chromosome | Band | Ensembl Transcript ID | Hit(s) | Hit(s) in Hippocampus[b] | Tissue Summary | Gene Name/ Other ID |
|---|---|---|---|---|---|---|
| 1 | p34.3 | ENST00000319637 | 2 | 2 | Hippocampus | EPHA10 |
| 2 | p21 | ENST00000306078 | 2 | 1 | Hippocampus | KCNG3 |
| 2 | q37.1 | ENST00000313064 | 2 | 1 | Hippocampus | C2orf52 |
| 6 | q15 | ENST00000303726 | 3 | 1 | Hippocampus | CNR1 |
| 6 | q25.3 | ENST00000308254 | 1 | 1 | Hippocampus | Retired in Ensembl 46 |
| 6 | q27 | ENST00000322583 | 1 | 1 | Hippocampus | NP_787118.2 (Link) |
| 9 | q21.33 | ENST00000298743 | 3 | 1 | Hippocampus | GAS1 |
| 10 | q24.33 | ENST00000329905 | 3 | 2 | Hippocampus | CALHM1 (FAM26C) |
| 11 | q24.1 | ENST00000354597 | 3 | 1 | Hippocampus | OR8B3 |
| 17 | q25.3 | ENST00000326931 | 2 | 1 | Hippocampus | Q8N8L1_HUMAN |
| 19 | p12 | ENST00000360885 | 1 | 1 | Hippocampus | Retired in Ensembl 46 |
| X | q27.2 | ENST00000298296 | 1 | 1 | Hippocampus | MAGEC3 |

[a]  One transcript is shown for each gene identified in the screen. Genomic location and number of hit(s) in dbEST are reported for each transcript.

[b]  Hit(s) in hippocampus indicates how many ESTs matching the transcript were sequenced from a cDNA library made from the hippocampus.

*Source:*  https://www.ncbi.nlm.nih.gov/protein/NP_787118.2.

modulator 1), encodes an open reading frame (ORF) of 346 amino acids and is predicted to contain four hydrophobic domains (HDs; TMHMM prediction) and two N-glycosylation motifs (NetNGlyc 1.0 prediction) (Figure 9.9a). No significant amino acid sequence homology to other functionally characterized proteins was found. Sequence database searches identified five human homologs of CALHM1 (collectively identified as the FAM26 gene family). Two homologs of human CALHM1 with broader tissue expression profiles were clustered next to CALHM1 in 10q24.33 and designated CALHM2 (26% protein sequence identity, previously annotated as FAM26B) and CALHM3 (39% identity, FAM26A) (Figure 9.9a). CALHM1 is conserved across at least 20 species, including mouse and *C. elegans* (Figure 9.9a and b).

As CALHM1 maps to a chromosomal region associated with susceptibility for LOAD, we tested whether CALHM1 SNPs could be associated with the risk of developing the disease.

Two nonsynonymous SNPs have been found in databases, rs2986017 (+394 C/T; P86L) and rs17853566 (+927 C/A; H264N). Dreses-Werringloer and colleagues sequenced the entire CALHM1 ORF using genomic DNA from 69 individuals, including 46 autopsy-confirmed AD cases and 23 age-matched normal controls. The rs17853566 SNP was not observed in this group, and the presence of the rs2986017 SNP was confirmed, with a potential over-representation of the T allele in AD subjects (AD=36%, controls=22%), see Table 9.8 (Dreses-Werringloer et al., 2008). They next assessed the impact of rs2986017 on the risk of developing AD in four other independent case-control populations (2,043 AD cases and 1,361 controls combined, see Table 9.8). The T allele distribution was increased in AD cases compared to controls in all the studies, with odds ratios (ORs) ranging from 1.29 to 1.99 (OR=1.44, $P=2 \times 10^{-10}$ in the combined population). This association was highly homogeneous among the different case-control studies (test for heterogeneity, $P=0.59$, $I2=0\%$). We also observed that the T allele frequency in autopsy-confirmed AD cases was similar to that observed in probable AD case populations (see Table 9.8). In the combined population, the CT or TT genotypes were both associated with an increased risk of developing AD (ORCT vs. CC ranging from 1.18 to 1.64, OR=1.37, $P=3 \times 10^{-5}$ in the combined population and ORTT vs.

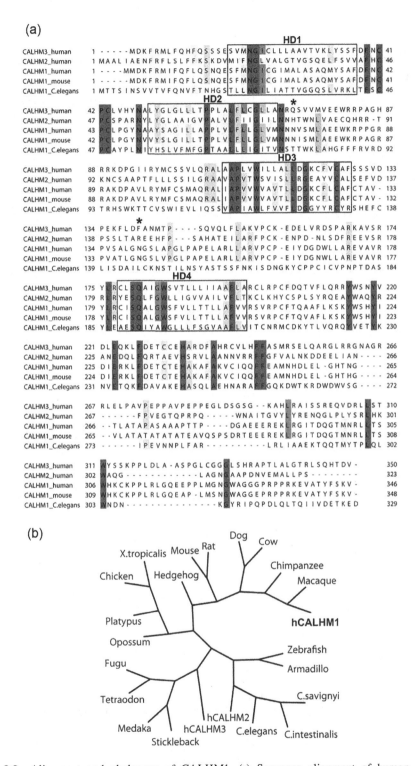

**FIGURE 9.9** Alignment and phylogeny of CALHM1. (a) Sequence alignment of human CALHM3, CALHM2, and CALHM1, and of murine and *C. elegans* CALHM1. Conserved sequences are highlighted in blue and sequence conservation is mapped in a color gradient, with the darkest color representing sequences with absolute identity and lighter colors representing sequences with weaker conservation. Boxes denote hydrophobic domains 1–4 (HD1–4). Stars, predicted N-glycosylation sites on human CALHM1. (b) Phylogenetic tree including human CALHM1 (hCALHM1). (For color figure see eBook figure.)

**TABLE 9.8**

**Allele and Genotype Distributions of the CALHM1 P86L Polymorphism (rs2986017) in AD Case and Control Populations**

| | n | Allele Distribution (%) | | Genotype Distribution (%) | | |
|---|---|---|---|---|---|---|
| | | C | T | CC | CT | TT |
| **USA Screening Sample[a,b]** | | | | | | |
| Controls | 23 | 36 (0.78) | 10 (0.22) | 14 (0.61) | 8 (0.35) | 1 (0.04) |
| Autopsied AD cases | 46 | 59 (0.64) | 33 (0.36) | 20 (0.44) | 19 (0.40) | 7 (0.16) |
| **France I[c,d]** | | | | | | |
| Controls | 565 | 907 (0.80) | 223 (0.20) | 370 (0.65) | 167 (0.30) | 28 (0.05) |
| AD cases | 710 | 1,051 (0.74) | 369 (0.26) | 410 (0.58) | 231 (0.32) | 69 (0.10) |
| **France II[e,f]** | | | | | | |
| Controls | 483 | 716 (0.74) | 250 (0.26) | 271 (0.56) | 174 (0.36) | 38 (0.08) |
| AD cases | 645 | 888 (0.69) | 402 (0.31) | 303 (0.47) | 282 (0.44) | 60 (0.09) |
| **UK[g,h]** | | | | | | |
| Controls | 205 | 320 (0.78) | 90 (0.22) | 127 (0.62) | 66 (0.32) | 12 (0.06) |
| AD cases | 365 | 504 (0.69) | 226 (0.31) | 193 (0.53) | 118 (0.32) | 54 (0.15) |
| Autopsied AD cases | 127 | 169 (0.66) | 85 (0.34) | 57 (0.45) | 55 (0.43) | 15 (0.12) |
| **Italy[i,j]** | | | | | | |
| Controls | 85 | 131 (0.77) | 39 (0.23) | 52 (0.61) | 27 (0.32) | 6 (0.07) |
| AD cases | 150 | 210 (0.70) | 90 (0.30) | 74 (0.49) | 62 (0.41) | 14 (0.09) |
| **Combined Studies[k,l]** | | | | | | |
| Controls | 1,361 | 2,110 (0.77) | 612 (0.23) | 834 (0.61) | 442 (0.32) | 85 (0.06) |
| AD cases | 2,043 | 2,881 (0.71) | 1,205 (0.29) | 1,057 (0.52) | 767 (0.37) | 219 (0.11) |

[a]  $P=0.10$;

[b]  $P=$ns;

[c]  $P=0.0002$;

[d]  $P=0.001$;

[e]  $P=0.006$;

[f]  $P=0.01$;

[g]  $P=0.0002$;

[h]  $P=0.00002$;

[i]  $P=0.10$;

[j]  $P=$ns;

[k]  $P=2\times10^{-10}$;

[l]  $P=7\times10^{-9}$

OR (CT vs. CC)=1.37, 95% CI [1.18–1.59], $P=3\times10^{-5}$.

OR (CT vs. CC)=1.27, 95% CI [1.08–1.50], $P=0.004$ adjusted for age, gender, APOE status, and center.

OR (TT vs. CC)=2.03, 95% CI [1.56–2.65], $P=2\times10^{-7}$.

OR (TT vs. CC)=1.77, 95% CI [1.33–2.36], $P=9\times10^{-5}$ adjusted for age, gender, APOE status, and center.

ns, nonsignificant.

CC ranging from 1.44 to 4.02, OR = 2.03, P = $2 \times 10^{-7}$ in the combined population; Table 2). All these observations were independent of the APOE status (see Table 9.8 and P for interaction = 0.26).

In the report, compelling evidence is provided that the rs2986017 SNP in CALHM1, which results in the P86L substitution, is associated with an increased risk for LOAD and a significant dysregulation of $Ca^{2+}$ homeostasis and APP metabolism. The P86L polymorphism was found to impair the plasma membrane $Ca^{2+}$ permeability, reduce cytosolic $Ca^{2+}$ levels, affect sAPPα production, and concomitantly derepress the effect of CALHM1 on Aβ accumulation. Indeed, these results help in understanding AD involving ion channel malfunctions due to specific genetic mutations owing partially to bioinformatics techniques, establishment of various databases, and development of advanced algorithms.

## 9.5 DEEP LEARNING MODELS EXPLAIN ION CHANNEL FEATURES

Earlier we have addressed how ML can help understand crucial ion channel aspects. Here we wish to discuss the role of another popular technique DL in understanding ion channels. Application of ML algorithms (e.g., in ion channel understanding) almost always require structural (e.g., ion channel protein) data, while DL networks rely on layers of artificial neural networks. Both ML and DL are actually forms of AI, although DL is considered a specific type of ML. Both these AI techniques start with training and test data and a model and go through an optimization process to ultimately find the weights that make the model best fit the data. In this section, we wish to see how DL may assist us in understanding ion channels. We must keep in mind that understanding ion channels using this new AI technique is just beginning. Hence, readers may not get any fully conclusive scenario related to crucial ion channel structural and functional aspects.

### 9.5.1 DEEP LEARNING MODEL IDEALIZES SINGLE MOLECULAR ACTIVITY OF ION CHANNELS

A DL model based on convolutional neural networks and long short-term memory architecture has just been reported which can automatically idealize complex single molecule activity more accurately and faster than traditional methods (Celik et al., 2020). The critical first step in understanding electrophysiology technique recorded ion channel current traces is event detection, so called "idealization," where noisy raw data are turned into discrete records of protein movement (Neher and Sakmann, 1976; Hamill et al., 1981). However, till today there have been practical limitations in patch-clamp data idealization; high-quality idealization is typically laborious and becomes infeasible and subjective with complex biological data containing many distinct native single-ion channel proteins gating simultaneously. In the DL model of Celik and colleagues, there are no parameters to set, for example, baseline, channel amplitude, or numbers of channels. This DL model may be useful in unsupervised automatic detection of single-molecule transition events.

Both fluorescence resonance energy transfer (FRET) and patch-clamp electrophysiology on single-molecule research provide high-resolution data on the molecular state of proteins in real time. However, the analyses of the produced data are usually time consuming and laborious requiring expert supervision. Celik and colleagues have demonstrated that the deep neural network, Deep-Channel, combining recurrent and convolutional layers can detect events in single-channel patch-clamp data automatically. Deep-Channel is completely unsupervised and adds objectivity to single-channel data analyses. With complex data, Deep-Channel also outperforms traditional manual threshold crossing both in terms of speed and accuracy. This method displays very high accuracy across various input datasets.

A hybrid recurrent convolutional neural network (RCNN) (Wardah et al., 2019; Celik et al., 2020) model is introduced to idealize ion channel records, with up to five ion channel events occurring simultaneously. To train and validate models, an analogue synthetic ion channel record generator system was developed and it has been found that the our Deep-Channel model, involving long short-term

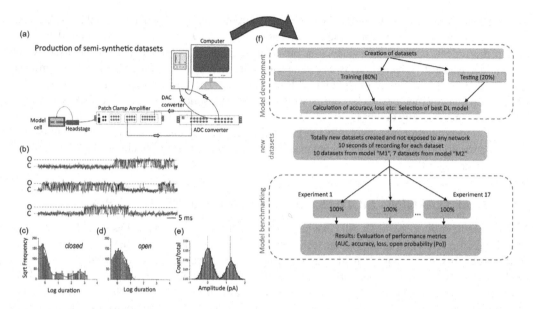

**FIGURE 9.10** Workflow diagram: generation of artificial analog datasets. (a) For training, validation and benchmarking, data were generated first as fiducial records with authentic kinetic models in MATLAB (Figure 9.11); these data were then played out through a CED digital-to-analog converter to a patch-clamp amplifier that sent this signal to a model cell and recorded the signal back (simultaneously) to a hard disk with CED Signal software via a CED analog to digital converter. The degree of noise could be altered simply by moving the patch-clamp headstage closer to or further from the PC. In some cases, drift was added as an additional challenge via a separate Matlab script. Raw single-channel patch-clamp data produced by these methods are visually indistinguishable from genuine patch-clamp data. To illustrate this point, a standard analysis work-up for one such experiment is shown with (b) raw data, followed by analyses with QuB: kinetic analyses of (c) channel open and (d) closed dwell times. Finally, we show (e) all points amplitude histogram. The difference between this and standard ion channel data is that here we have a perfect fiducial record with each experimental dataset, which is impossible to acquire without simulation. (f) Illustrates our overall model design and testing workflow. The Supplementary Information in the study by Celik et al. (2020) includes training metrics from the initial validation and the main text here shows performance metrics acquired from 17 experiments with entirely new datasets. The training datasets typically contained millions of sample points, and the 17 benchmarking experiments were sequences of 100,000 samples each.

memory (LSTM) and convolutional neural network (CNN) layers, rapidly and accurately idealizes or detects experimentally observed single molecule events without the need for human supervision. To our knowledge, this work is the first DL model designed for the idealization of patch-clamp single molecule events.

Figure 9.10 illustrates the data generation workflow and Figure 9.11 illustrates the Deep-Channel architecture (Celik et al., 2020). In training and model development, it was found that while LSTM models display good performance, the combination with a time-distributed CNN improved performance, a so-called RCNN which was called Deep-Channel by Celik and colleagues. After training and model development (Celik et al., 2020), 17 newly generated datasets were used, previously unseen by Deep-Channel, and thus uninvolved with the training process. Authentic ion channel data (Figure 9.10b) were generated as described in the methods from two kinetic schemes, the first (M1) with low channel open probability, and the second (M2) with a high open channel probability, and thus an average of approximately three channels open at a time (Figure 9.12b). Across the datasets they included data from both noisy, difficult-to-analyze signals and low noise (high signal to ratio samples) as would be the case in any patch-clamp project. Examples of these data, together with ground truth and Deep-Channel idealization, are shown in Figure 9.13 (Celik et al., 2020). All of the

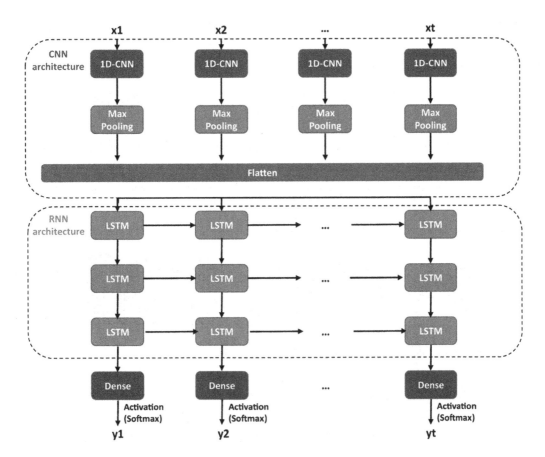

**FIGURE 9.11** The input time series data were fed to the 1D Convolution layer (1D-CNN) which includes both 1D convolution layers and max pooling layers. After this, data were flattened to the shape of the next network layer, which is an LSTM. Three LSTM layers were stacked and each contains 256 LSTM units. Dropout layers were also appended to all LSTM layers with the value of 0.2 to reduce overfitting. This returned features from the stacked LSTM layers. The updated features were then forwarded to a regular dense layer with a SoftMax activation function giving an output representing the probability of each class (e.g., the probability × channels being open at each time step). In postnetwork processing, the most likely number of channels open at a given time was calculated simply as the class with the highest probability at a given instant (Argmax).

Deep-Channel results described here have been achieved with a single DL model script "capacity to detect a maximum of five channels," requiring no human intervention beyond giving the script the correct filename or path. Hence, there was no need to set baseline, channel amplitude, or channel number.

For channels having a low opening probability (see stochastic gating model M1, Figure 9.12a), the data idealization process becomes close to a binary detection problem (see Figure 9.13a), with ion channel events type closed or open, labels "0" and "1," respectively. In this classification, the receiver operating characteristic (ROC) curve for ion channel event classification for both open and closed event detection exceeds 96%. In low open probability experiments, Deep-Channel returned a macro-F1 of 97.1 % ± 0.02%, whereas the segmented-$k$ means (SKM) method in a traditional software package QuB resulted a macro-F1 of 95.5% ± 0.025%, and 50% threshold method in QuB gave a macro-F1 of 84.7% ± 0.05%, n = 10.

For datasets including highly active channels (from model M2), it becomes a multiclass comparison problem, hence Deep-Channel outperformed both 50% threshold-crossing and SKM methods in QuB considerably. The Deep-Channel macro-F1 for such events was 0.87 ± 0.07, however SKM

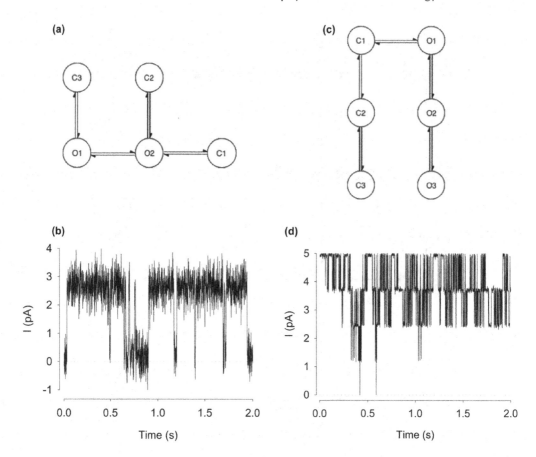

**FIGURE 9.12** "Patch-clamp" data were produced from two different stochastic models. (a, c) The Markovian models used for simulation of ion channel data. Ion channels typically move between Markovian states that are either closed (zero conductance) or open (unitary conductance, g). The current passing when the channel is closed is zero (aside from recording offsets and artefacts), whereas when open the current (i) passing is given by i = g × V, where V is the driving potential (equilibrium potential for the conducting ion minus the membrane potential). In most cases, there are several open and closed states ("O1"–"O3", or "C1"–"C3", respectively). The central dogma of ion channel research is that the g will be the same for O1, O2, or O3. Although substates have been identified in some situations, these are beyond the scope of our current work. a Model M1; the stochastic model from Davies et al. (2010) and its output. (b) This model has a low open probability, and so the data is mostly a representation of zero or one channel open. (c) Model M2; the stochastic model from ref. O'Brien and Barrett-Jolley (2018) and its output data. (d) Since open probability is high, the signal is largely composed of three or more channels simultaneously open.

macro-F1 in QuB, without manual baseline correction, dropped sharply to 0.57 ± 0.15, and 50% threshold-crossing macro-F1 fell to 0.47 ± 0.37 (Student's paired *t*-test between methods, p = 0.0052).

### 9.5.2 Deep Learning to Classify Ion Transporters and Channels from Membrane Proteins

Recently, an article was published proposing a DL method for automatically classifying ion transporters/pumps and ion channels from membrane proteins (Taju and Ou, 2019). This technique is proposed by training deep neural networks and using the position-specific scoring matrix profile as an input.

From structural and behavioral perspectives, ion channels differ significantly from ion transporters, see Figure 9.14 (Taju and Ou, 2019). The DL method of Taju and Ou is dedicated to classifying

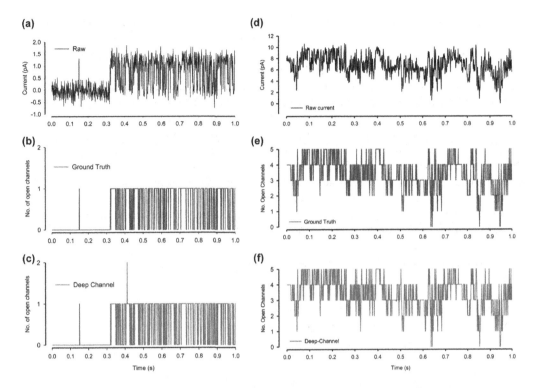

**FIGURE 9.13** Qualitative performance of Deep-Channel with previously unseen data. (a–c) Representative example of Deep-Channel classification performance with low activity ion channels (data from model M1, Figure 9.12 a, b): (a) The raw semi-simulated ion channel event data (black). (b) The ground truth idealization/ annotation labels (blue) from the raw data above in (a). (c) The Deep-Channel predictions (red) for the raw data above (a). (d–f) Representative example of Deep-Channel classification performance with five channels opening simultaneously (data from model M2, Figure 9.12c, d). (d) The semi-simulated raw ion channel event data (black). (e) The ground truth idealization/annotation labels (blue) from the raw data above in (d). (f) The Deep-Channel label predictions (red) for the raw data above (d).

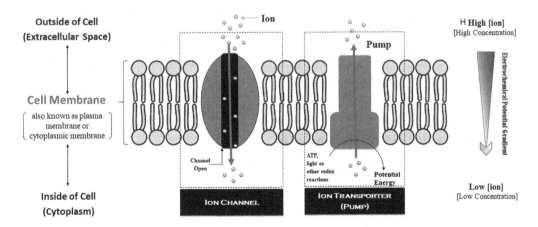

**FIGURE 9.14** Schematic representation (and mutual comparison) of an ion channel and ion transporter.

these two structural events. Three-stage approaches have been adopted, in which five techniques for data normalization are used; next three imbalanced data techniques are applied to the minority classes, and then, six classifiers are compared with the proposed method (Taju and Ou, 2019). We shall present here a brief of the results and interpretations.

The goal is to propose a method for automatically classifying ion transporters and ion channels from membrane proteins via training deep neural networks (DNNs) using a CNN as a selected algorithm which can capture hidden pattern information inside the dataset. The hidden feature extraction from the position-specific scoring matrix (PSSM) of the proteins dataset is, therefore, the best feature to produce relevant evolutionary information of protein sequences. More importantly, this feature can be applied to various problems in bioinformatics and ML with promising results when compared to other feature extraction methods. First, the representation of all protein data in FASTA (stands for fast-all) format is changed into PSSM profiles format. Second, the DL method is demonstrated by using this representation to be able to accurately classify some of the proteins that are separated from the training data. Finally, to validate the approach, fivefold cross validation is used to test the model of the proposed method.

The guidelines of the five-step rule (Chou, 2011) are followed to make the following five steps very clear:

- how to construct or select a valid benchmark dataset to train and test the predictor;
- how to formulate the biological sequence samples with an effective mathematical expression that can truly reflect their intrinsic correlation with the target to be predicted;
- how to introduce or develop a powerful algorithm (or engine) to operate the prediction;
- how to properly perform cross-validation tests to objectively evaluate the anticipated accuracy of the predictor; and
- how to establish a user-friendly web server for the predictor that is accessible to the public.

For details on all these five steps, readers may refer to Taju and Ou (2019). Figure 9.15 presents a schematic representation of the membrane protein classification prediction steps. The dataset used were collected from the UniProtKB database (UniProt, 2016) (see Table 9.9).

We avoid elaborating on the detailed techniques here. To represent the input data, PSSM-based feature extractor was applied here, and a $20 \times 20$ matrix was produced. Initially, the Position-Specific Iterative Basic Local Alignment Search Tool (PSI-BLAST) (Altschul et al., 1997) against the National Center for Biotechnology Information (NCBI)'s nonredundant (nr) protein database (ftp://ftp.ncbi.nih.gov/blast/db/FASTA/) was utilized to generate the PSSM profiles. PSI-BLAST is a protein sequence profile search method, and a PSSM is a matrix generated using a protein query that performs PSI-BLAST search to find its similarity from the standard biological databases and creates a position-specific matrix. For each protein query, the PSSM produces $N \times 20$ matrix with a profile component where N represents the length of protein sequence and the columns shows the protein's amino acid substitution scores. The PSSM features got generated by summing the same amino acid rows in the PSSM profile and then divided by the length of the sequence and finally scaled by some feature normalization techniques (Taju and Ou, 2019).

**TABLE 9.9**
**The Datasets Used in This Experiment**

| Datasets | Original Data | Similarity <20% | Testing Data | Training Data |
|---|---|---|---|---|
| Ion channels | 845 | 301 | 60 | 241 |
| Ion transporters | 1,051 | 351 | 70 | 281 |
| Membrane proteins | 8,295 | 4,263 | 850 | 3,413 |
| **Total proteins** | **10,191** | **4,915** | **980** | **3,935** |

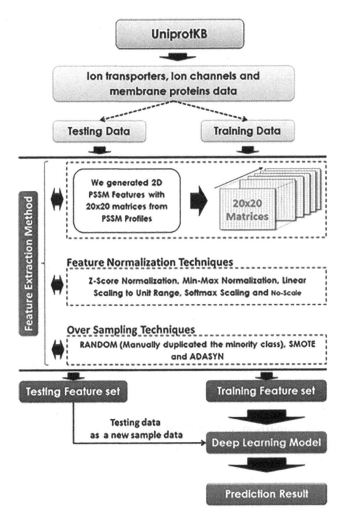

**FIGURE 9.15** Schematic representation of the steps for ion transporters and ion channels predictions from membrane proteins.

The composition of 20 amino acids analysis, n-gram analysis, and the visualization of important sequence motif using word cloud technique were shown. The variance of 20 amino acid residues was computed among the three classes of protein datasets. The experiment compared the performance of the proposed DL model against five different data normalization techniques and three oversampling techniques. The k-fold cross-validation as a model validation technique was applied to evaluate the model. Then, the best model performance was compared with some classifiers such as perceptron Gaussian naïve Bayes, random forest, nearest neighbors, SVM, and nearest centroid classifiers using independent test data to examine the effect of different algorithms. The sequence analyses were performed on training data to find some information about amino acids and the based pair of residue patterns at important motifs in the datasets. Figure 9.16a shows that the amino acids with letter Ala (A), Gly (G), Leu (L), Ser (S), and Val (V) are dominant and important, as shown in the figure of amino acid composition or the occurrence frequency of amino acids for all proteins. The variance of 20 amino acid residues across the ion channels, ion transporters, and membrane proteins were then computed, Figure 9.16b. Variance analysis was used to measure how far the data are spread out from the average value. Amino acids Leu (L), Ser (S), Ala (A), Val (V), Gly (G), Glu (E), Ile (I), Arg (R), and Thr (T) with high frequencies in analysis top frequent motifs had a variance

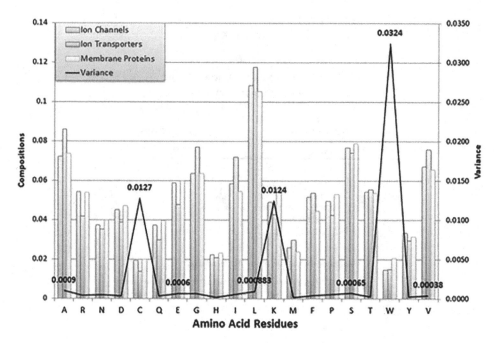

**FIGURE 9.16** (a) Amino acid composition of ion channels (blue) and ion transporters (red) compared to membrane proteins (green) and (b) variance of 20 amino acid residues. (For color figure see eBook figure.)

---

**TABLE 9.10**

**Performance Comparison on Classifying Ion Transporters and Ion Channels from Membrane Proteins Using Fivefold Cross Validation Technique**

| Datasets | Sen | Spec | Acc | MCC |
|---|---|---|---|---|
| Ion channels (class A) | 89.20 | 84.89 | 87.05 | 0.75 |
| Ion transporters (class B) | 86.76 | 88.23 | 87.49 | 0.75 |
| Other proteins (class C) | 92.50 | 96.19 | 94.35 | 0.89 |

---

lower than 0.005, and the amino acids Cys (C), Lys (K), and Trp (W) had a variance higher than 0.005 of 0.013, 0.012, and 0.032, respectively.

The datasets were then classified to distinguish three classes, namely, ion channels (class A), ion transporters (class B), and other proteins (class C) using the following techniques (Taju and Ou, 2019):

- Comparative results on different feature normalizations techniques.
- Comparative results on different imbalanced dataset techniques.

To evaluate the performance of the predictor model, fivefold cross-validations were applied in the training datasets. Table 9.10 reports the results of the fivefold cross-validation technique that was applied in the training data, which is a challenging step to find the best model prediction of independent test sets. The performance reached the highest sensitivity (89.20%), specificity (84.89%), overall accuracy (87.05%), and MCC (0.75) for class A. Class B achieves sensitivity (86.76%), specificity (88.23%), overall accuracy (87.49%), and MCC (0.75), and performance of sensitivity (92.50%), specificity (96.19%), overall accuracy (94.35%), and MCC (0.89) were seen for class C.

The application of all these Deep-Channel algorithms and models has been found possible, though with limitations, for the case of biological data on ion channels. We have presented here two example studies where DL algorithms have been utilized to demonstrate various ion channel features. The effectiveness of Deep-Channel, an artificial deep neural network to detect events in single molecule datasets has been demonstrated. The method is exclusively applicable not only for patch-clamp data but has the potential for deep learning convolution/LSTM networks to tackle other related biological data analysis problems. The ion channel, ion pump, and other membrane protein classification using DL algorithms and modeling has been found quite impressive and time and resource-saving initiative. Over the next decade, we may see an exponential increase in use of AI, ML, and DL in understanding natural status and mutated conditions of ion channels of biological cells.

# REFERENCES

Altschul, S., Madden, T. L., Schäffer, A. A., Zhang, J., Zhang, Z., Miller, W., & Lipman, D. J. (1997). Gapped BLAST and PSI-BLAST: A new generation of protein database search programs. *Nucleic Acids Research, 25*(17), 3389–3402. doi:10.1093/nar/25.17.3389

Ashrafuzzaman, M., Tseng, C., Duszyk, M., & Tuszynski, J. A. (2012). Chemotherapy drugs form ion pores in membranes due to physical interactions with lipids. *Chemical Biology & Drug Design, 80*(6), 992–1002. doi:10.1111/cbdd.12060

Ashrafuzzaman, M., Tseng, C., & Tuszynski, J. (2014). Regulation of channel function due to physical energetic coupling with a lipid bilayer. *Biochemical and Biophysical Research Communications, 445*(2), 463–468. doi:10.1016/j.bbrc.2014.02.012

Ashrafuzzaman, M., Tseng, C., & Tuszynski, J. (2020a). Charge-based interactions of antimicrobial peptides and general drugs with lipid bilayers. *Journal of Molecular Graphics and Modelling, 95*, 107502. doi:10.1016/j.jmgm.2019.107502

Ashrafuzzaman, M., Tseng, C., & Tuszynski, J. (2020b). Dataset on interactions of membrane active agents with lipid bilayers. *Data in Brief, 29*, 105138. doi:10.1016/j.dib.2020.105138

Ashrafuzzaman, M., & Tuszynski, J. (2012a). Regulation of channel function due to coupling with a lipid bilayer. *Journal of Computational and Theoretical Nanoscience, 9*(4), 564–570. doi:10.1166/jctn.2012.2062

Ashrafuzzaman, M., & Tuszynski, J. A. (2012b). *Membrane Biophysics*. Berlin, Heidelberg: Springer-Verlag. doi:10.1007/978-3-642-16105-6

Attwood, T. K., Beck, M. E., Bleasby, A. J., Degtyarenko, K., Michie, A. D., & Parry-Smith, D. J. (1997). Novel developments with the PRINTS protein fingerprint database. *Nucleic Acids Research, 25*(1), 212–216. doi:10.1093/nar/25.1.212

Bairoch, A., & Apweiler, R. (2000). The SWISS-PROT protein sequence database and its supplement TrEMBL in 2000. *Nucleic Acids Research, 28*(1), 45–48. doi:10.1093/nar/28.1.45

Buntine, W., & Niblett, T. (1992). A further comparison of splitting rules for decision-tree induction. *Machine Learning, 8*(1), 75–85. doi:10.1007/bf00994006

Cahalan, M. D., & Chandy, K. G. (2009). The functional network of ion channels in T. lymphocytes. *Immunological Reviews, 231*(1), 59–87. doi:10.1111/j.1600-065x.2009.00816.x

Cai, C. (2003). SVM-Prot: Web-based support vector machine software for functional classification of a protein from its primary sequence. *Nucleic Acids Research, 31*(13), 3692–3697. doi:10.1093/nar/gkg600

Cai, Y., Liu, X., Xu, X., & Chou, K. (2002). Prediction of protein structural classes by support vector machines. *Computers & Chemistry, 26*(3), 293–296. doi:10.1016/s0097-8485(01)00113-9

Campagne, F., & Skrabanek, L. (2006). Mining expressed sequence tags identifies cancer markers of clinical interest. *BMC Bioinformatics, 7*(1). doi:10.1186/1471-2105-7-481

Capener, C. E. (2002). Ion channels: Structural bioinformatics and modelling. *Human Molecular Genetics, 11*(20), 2425–2433. doi:10.1093/hmg/11.20.2425

Catterall, W. A., Dib-Hajj, S., Meisler, M. H., & Pietrobon, D. (2008). Inherited neuronal ion channelopathies: New windows on complex neurological diseases. *Journal of Neuroscience, 28*(46), 11768–11777. doi:10.1523/jneurosci.3901-08.2008

Celik, N., O'Brien, F., Brennan, S., Rainbow, R. D., Dart, C., Zheng, Y., . . . Barrett-Jolley, R. (2020). Deep-channel uses deep neural networks to detect single-molecule events from patch-clamp data. *Communications Biology, 3*(1). doi:10.1038/s42003-019-0729-3

Chou, K. (2011). Some remarks on protein attribute prediction and pseudo amino acid composition. *Journal of Theoretical Biology, 273*(1), 236–247. doi:10.1016/j.jtbi.2010.12.024

Claes, L., Del-Favero, J., Ceulemans, B., Lagae, L., Broeckhoven, C. V., & Jonghe, P. D. (2001). De Novo mutations in the sodium-channel gene SCN1A cause severe myoclonic epilepsy of infancy. *The American Journal of Human Genetics, 68*(6), 1327–1332. doi:10.1086/320609

Davies, L. M., Purves, G. I., Barrett-Jolley, R., & Dart, C. (2010). Interaction with caveolin-1 modulates vascular ATP-sensitive potassium (KATP) channel activity. *The Journal of Physiology, 588*(17), 3255–3266. doi:10.1113/jphysiol.2010.194779

Demolombe, S., Marionneau, C., Lebouter, S., Charpentier, F., & Escande, D. (2005). Functional genomics of cardiac ion channel genes. *Cardiovascular Research, 67*(3), 438–447. doi:10.1016/j.cardiores.2005.04.021

Doyle, D. A., Cabral, J. M., Pfuetzner, R. A., Kuo, A., Gulbis, J. M., Cohen, S. L., . . . MacKinnon, R. (1998). The structure of the potassium channel: Molecular basis of $K^+$ conduction and selectivity. *Science, 280* (5360), 69–77. doi:10.1126/science.280.5360.69

Dreses-Werringloer, U., Lambert, J., Vingtdeux, V., Zhao, H., Vais, H., Siebert, A., . . . Marambaud, P. (2008). A polymorphism in CALHM1 influences $Ca^{2+}$ homeostasis, Aβ levels, and Alzheimer's disease risk. *Cell, 133*(7), 1149–1161. doi:10.1016/j.cell.2008.05.048

Durbin, R. M., Abecasis, G. R., Altshuler, D. L., Auton, A., Brooks, L. D., Gibbs, R. A., . . . Consortium, G. P. (2010). A map of human genome variation from population-scale sequencing. *Nature, 467*, 1061–1073. doi:10.1038/nature09534

Gao, J., Cui, W., Sheng, Y., Ruan, J., & Kurgan, L. (2016). PSIONplus: Accurate sequence-based predictor of ion channels and their types. *PLoS ONE 11*(4), e0152964. doi:10.1371/journal.pone.0152964

Hamill, O. P., Marty, A., Neher, E., Sakmann, B., & Sigworth, F. J. (1981). Improved patch-clamp techniques for high-resolution current recording from cells and cell-free membrane patches. *Pflügers Archiv - European Journal of Physiology, 391*(2), 85–100. doi:10.1007/bf00656997

Han, K., Wang, M., Zhang, L., Wang, Y., Guo, M., Zhao, M., . . . Wang, C. (2019). Predicting ion channels genes and their types with machine learning techniques. *Frontiers in Genetics, 10*. doi:10.3389/fgene.2019.00399

Harte, R., & Ouzounis, C. A. (2002). Genome-wide detection and family clustering of ion channels. *FEBS Letters, 514*(2–3), 129–134. doi:10.1016/s0014-5793(01)03254-9

Heil, B., Ludwig, J., Lichtenberg-Frate, H., & Lengauer, T. (2006). Computational recognition of potassium channel sequences. *Bioinformatics, 22*(13), 1562–1568. doi:10.1093/bioinformatics/btl132

Hiriart, M., & Aguilar-Bryan, L. (2008). Channel regulation of glucose sensing in the pancreatic β-cell. *American Journal of Physiology-Endocrinology and Metabolism, 295*(6). doi:10.1152/ajpendo.90493.2008

Hodgkin, J., Robert Horvitz, H., Jasny, B. R., & Kimble, J. (1998). C. Elegans: Sequence to biology. *Science, 282*(5396), 2011–2011. doi:10.1126/science.282.5396.2011

Kai, W., Kikuchi, K., Tohari, S., Chew, A. K., Tay, A., Fujiwara, A., Hosoya, S., Suetake, H., Naruse, K., Brenner, S., Suzuki, Y., & Venkatesh, B. (2011). Integration of the genetic map and genome assembly of fugu facilitates insights into distinct features of genome evolution in teleosts and mammals. *Genome Biology and Evolution, 3*, 424–442. doi:10.1093/gbe/evr041

Kapplinger, J. D., Tester, D. J., Salisbury, B. A., Carr, J. L., Harris-Kerr, C., Pollevick, G. D., . . . Ackerman, M. J. (2009). Spectrum and prevalence of mutations from the first 2,500 consecutive unrelated patients referred for the FAMILION® long QT syndrome genetic test. *Heart Rhythm, 6*(9), 1297–1303. doi:10.1016/j.hrthm.2009.05.021

Kauwe, J. S., Cruchaga, C., Bertelsen, S., Mayo, K., Latu, W., Nowotny, P., . . . Goate, A. M. (2010). $O_2$-07-03: Validating predicted biological effects of Alzheimer's disease associated SNPs using cerebrospinal fluid biomarker levels. *Alzheimer's & Dementia, 6*. doi:10.1016/j.jalz.2010.05.348

Klassen, T., Davis, C., Goldman, A., Burgess, D., Chen, T., Wheeler, D., . . . Noebels, J. (2011). Exome sequencing of ion channel genes reveals complex profiles confounding personal risk assessment in epilepsy. *Cell, 145*(7), 1036–1048. doi:10.1016/j.cell.2011.05.025

Koppel, J., Campagne, F., Vingtdeux, V., Dreses-Werringloer, U., Ewers, M., Rujescu, D., . . . Marambaud, P. (2011). CALHM1 P86L polymorphism modulates CSF Aβ levels in cognitively healthy individuals at risk for Alzheimer's disease. *Molecular Medicine, 17*(9–10), 974–979. doi:10.2119/molmed.2011.00154

Kurczynska, M., & Kotulska, M. (2018). Automated method to differentiate between native and mirror protein models obtained from contact maps. *PLOS ONE, 13*(5). doi:10.1371/journal.pone.0196993

Kurczynska, M., Kania, E., Konopka, B. M., & Kotulska, M. (2016). Applying PyRosetta molecular energies to separate properly oriented protein models from mirror models, obtained from contact maps. *Journal of Molecular Modeling, 22*(5). doi:10.1007/s00894-016-2975-3

Liebeskind, B. J., Hillis, D. M., & Zakon, H. H. (2015). Convergence of ion channel genome content in early animal evolution. *Proceedings of the National Academy of Sciences, 112*(8). doi:10.1073/pnas.1501195112

Lin, H., & Chen, W. (2015). Briefing in application of machine learning methods in ion channel prediction. *The Scientific World Journal, 2015*, 1–7. doi:10.1155/2015/945927

Lin, H., & Ding, H. (2011). Predicting ion channels and their types by the dipeptide mode of pseudo amino acid composition. *Journal of Theoretical Biology, 269*(1), 64–69. doi:10.1016/j.jtbi.2010.10.019

Liu, L.-X., Li, M.-L., Tan, F.-Y., Lu, M.-C., Wang, K.-L., Guo, Y.-Z., Wen, Z.-N., & Jiang, L. (2006). Local sequence information-based support vector machine to classify voltage-gated potassium channels. *Acta Biochimica Et Biophysica Sinica, 38*(6), 363–371. doi:10.1111/j.1745-7270.2006.00177.x

Liu, W.-X., Deng, E.-Z., Chen, W., & Lin, H. (2014). Identifying the subfamilies of voltage-gated potassium channels using feature selection technique. *International Journal of Molecular Sciences, 15*(7), 12940–12951. doi:10.3390/ijms150712940

Maffeo, C., Bhattacharya, S., Yoo, J., Wells, D., & Aksimentiev, A. (2012). Modeling and simulation of ion channels. *Chemical Reviews, 112*(12), 6250–6284. doi:10.1021/cr3002609

Marco, D., Marie-Ange, D., & Nicolas, L. (2006). LGICdb: A manually curated sequence database after the genomes. *Nucleic Acids Research, 34*(90001). doi:10.1093/nar/gkj104

Moroz, L. L., Kocot, K. M., Citarella, M. R., Dosung, S., Norekian, T. P., Povolotskaya, I. S., . . . Kohn, A. B. (2014). The ctenophore genome and the evolutionary origins of neural systems. *Nature, 510*(7503), 109–114. doi:10.1038/nature13400

Moulton, G., Attwood, T. K., Parry-Smith, D. J., & Packer, J. C. (2003). Phylogenomic analysis and evolution of the potassium channel gene family. *Receptors and Channels, 9*(6), 363–377. doi:10.3109/714041017

Neher, E., & Sakmann, B. (1976). Single-channel currents recorded from membrane of denervated frog muscle fibres. *Nature, 260*(5554), 799–802. doi:10.1038/260799a0

O'Brien, F., & Barrett-Jolley, R. (2018). CVS role of TRPV: from single channels to HRV assessment with Artificial Intelligence. *FASEB Journal, 32*, 732.6.

Ohmori, I., Kahlig, K. M., Rhodes, T. H., Wang, D. W., & George, A. L. (2006). Nonfunctional SCN1A is common in severe myoclonic epilepsy of infancy. *Epilepsia, 47*(10), 1636–1642. doi:10.1111/j.1528-1167.2006.00643.x

Pan, G., Jiang, L., Tang, J., & Guo, F. (2018). A novel computational method for detecting DNA methylation sites with DNA sequence information and physicochemical properties. *International Journal of Molecular Sciences, 19*(2), 511. doi:10.3390/ijms19020511

Qu, K., Guo, F., Liu, X., Lin, Y., & Zou, Q. (2019). Application of machine learning in microbiology. *Frontiers in Microbiology, 10*. doi:10.3389/fmicb.2019.00827

Reid, C. A., Berkovic, S. F., & Petrou, S. (2009). Mechanisms of human inherited epilepsies. *Progress in Neurobiology, 87*(1), 41–57. doi:10.1016/j.pneurobio.2008.09.016

Rubio-Moscardo, F., Setó-Salvia, N., Pera, M., Bosch-Morató, M., Plata, C., Belbin, O., . . . Clarimón, J. (2013). Rare variants in Calcium Homeostasis Modulator 1 (CALHM1) found in early onset Alzheimer's disease patients alter calcium homeostasis. *PLoS ONE, 8*(9). doi:10.1371/journal.pone.0074203

Sakarya, O., Kosik, K. S., & Oakley, T. H. (2008). Reconstructing ancestral genome content based on symmetrical best alignments and Dollo parsimony. *Bioinformatics, 24*(5), 606–612. doi:10.1093/bioinformatics/btn005

Saha, S., Zack, J., Singh, B., & Raghava, G. P. S. (2006). VGIchan: Prediction and classification of voltage-gated ion channels. *Genomics, Proteomics & Bioinformatics, 4*(4), 253–258. doi:10.1016/s1672-0229(07)60006-0

Taju, S. W., & Ou, Y. (2019). DeepIon: Deep learning approach for classifying ion transporters and ion channels from membrane proteins. *Journal of Computational Chemistry, 40*(15), 1521–1529. doi:10.1002/jcc.25805

UniProt: The universal protein knowledgebase. (2016). *Nucleic Acids Research, 45*(D1). doi:10.1093/nar/gkw1099

Wardah, W., Khan, M., Sharma, A., & Rashid, M. A. (2019). Protein secondary structure prediction using neural networks and deep learning: A review. *Computational Biology and Chemistry, 81*, 1–8. doi:10.1016/j.compbiolchem.2019.107093

Xu, L., Liang, G., Liao, C., Chen, G., & Chang, C. (2019). K-Skip-n-Gram-RF: A random forest based method for Alzheimer's disease protein identification. *Frontiers in Genetics, 10*. doi:10.3389/fgene.2019.00033

Xu, Y., Guo, M., Liu, X., Wang, C., Liu, Y., & Liu, G. (2016). Identify bilayer modules via pseudo-3D clustering: Applications to miRNA-gene bilayer networks. *Nucleic Acids Research*. doi:10.1093/nar/gkw679

Yu, G., Smith, D. K., Zhu, H., Guan, Y., & Lam, T. T. Y. (2016). ggtree: An r package for visualization and annotation of phylogenetic trees with their covariates and other associated data. *Methods in Ecology and Evolution, 8*(1), 28–36. doi:10.1111/2041-210x.12628

Yu, G., Wang, L.-G., Yan, G.-R., & He, Q.-Y. (2014). DOSE: an R/Bioconductor package for disease ontology semantic and enrichment analysis. *Bioinformatics, 31*(4), 608–609. doi:10.1093/bioinformatics/btu684

Zhao, Y.-W., Su, Z.-D., Yang, W., Lin, H., Chen, W., & Tang, H. (2017). IonchanPred 2.0: A tool to predict ion channels and their types. *International Journal of Molecular Sciences, 18*(9), 1838. doi:10.3390/ijms18091838

Zou, Q., Chen, L., Huang, T., Zhang, Z.G., & Xu, Y.G. (2017). Machine learning and graph analytics in computational biomedicine. *Artificial Intelligence in Medicine* 83, 1. doi: 10.1016/j.artmed.2017.09.003

Zou, Q., Li, J. J., Song, L., Zeng, X. X., & Wang, G. H. (2016). Similarity computation strategies in the microRNA-disease network: A survey. *Briefings in Functional Genomics 15*, 55–64. doi: 10.1093/bfgp/elv024

Zou, Q., Lin, G., Jiang, X., Liu, X., & Zeng, X. (2018a). Sequence clustering in bioinformatics: An empirical study. *Briefings in Bioinformatics*. bby090. doi: 10.1093/bib/bby090

Zou, Q., Qu, K. Y., Luo, Y. M., Yin, D. H., Ju, Y., & Tang, H. (2018b). Predicting diabetes mellitus with machine learning techniques. *Front. Genet.* 9, 515. doi: 10.3389/fgene.2018.00515

# 10 Quantum Mechanics of Ion Channels

Ion channels are stable over a wide range of time scales in the order of millisecond (ms) to microsecond (µs) or even smaller time range. This time scale is based on measurements using available experimental techniques. The same time scale is indeed followed by channel-carried charges while getting transported across the host membrane. This concept of gross time-dependent motion is valid when we consider ion channels to be classical mechanical systems following related formalism regarding their construction, transport, and other localized static and dynamic phenomena. However, ion channel gating-charge movement for activating a channel is predicted to occur in a few brief packets of charges wherein each packet carries a small proportion of electron charges. The movement of such a small-scale charge in a small channel length involves local physiological environmental fluctuations, hinting toward the presence of some kind of quantum mechanical states. In quantum mechanics, when a wave function explains the state of a charge or particle, two aspects are considered, namely, the probability amplitude and the square of its modulus providing the probability density of finding the electron in a certain position in space. Ion channels are predicted to host processes involving ion movement to follow important quantum mechanical principles.

We predict ion channels to host quantum mechanical effects in regard to the charge movement inside and across the channels. However, the general understanding of channel structure and function has so far been made using classical mechanical treatments. Going beyond common concepts, as we inspect some of the scales involved in ion channel structures, functions, and the underlying energetics, we may find clues about the parameters that need to be targeted to explain some features of ion channels using quantum mechanical wave function. Ion channels transport charges across low dimension cross-sections and lengths on an ultrafast time scale. The highest length of channel is equivalent to the membrane thickness of ~3–7 nanometer (nm) or even longer in case the channel constructing proteins extend beyond the membrane thickness. However, the lowest cut-off value for a channel length is unclear. Is it close to zero, close to the order of the length of a couple of hydrogen bonds, or is it any theoretically limited (due to relative hydrophobic/hydrophilic constrained) specific value? Modeling lipid-lined, toroidal-type ion pore suggests that the membrane thickness vanishes at the pore opening (Ashrafuzzaman et al., 2012). Hence, the channel length is specific to the channel type. The types of channels range from general class, such as toroidal and barrel-stave, to specific class, such as sodium and potassium. Specifically, as an example, the passage of sodium ions occurs through the channel protein exhibiting the shape of a truncated cone, 14 nm in height. One end has a diameter of 12 nm and is asymmetric, while the other is more symmetric and has a diameter of 7–10 nm (Sato et al., 1998). The estimated hydrodynamic radius of potassium ($K^+$) is ~4 Å (Díaz-Franulic et al., 2015; Moldenhauer et al., 2016). This group estimated a diameter of 10 Å for the internal entrance of the shaker channel, while 10–12 Å for Kv1.2 was reported by Long et al. (2005). This narrow pore is expected to contribute higher electrical resistance in small conductance channels, making the electric field significantly drop away from the selectivity filter. The estimated external access resistance is measured to be ~1.9 GΩ (Díaz-Franulic et al., 2015).

All the scales we have mentioned here on which ion channels work are quite extreme. This encourages us to look for quantum mechanical effects as we know that quantum mechanics is the branch of physics relating to the very small world where objects may exist in a haze of probability and lose certainty. This loss of certainty in the coordinates and amount of free charges in ion channels, especially at the channel's preopen states, may draw real attention to the treatment of the problem using quantum mechanics.

DOI: 10.1201/9781003010654-10

Figure 10.1 models different states that may be considered generally available in ion channels. Here, we have proposed a model representation on channel energetics, energetics of transitions between various stable or unstable states that altogether may explain the channel phenomenology. We may hypothesize that the channel exists in a combination of classical and quantum energy states (explained in Chapters 1 and 6). Occasionally, channels, while active, may concomitantly hold both classical and quantum phenomena. However, using conventional experimental techniques, such as electrophysiology record of channel currents, we may only end up detecting the classical mechanical states (CMSs) and the transitions thereof, see Figure 10.1 (AOB section: A ↔ O, O ↔ B, A ↔ B). In addition to preopen state (A) and postopen state (B), we may also consider two more independent energy states, quantum mechanical state (QMS) (A′) and stochastic state (SS) (B′). QMS has its

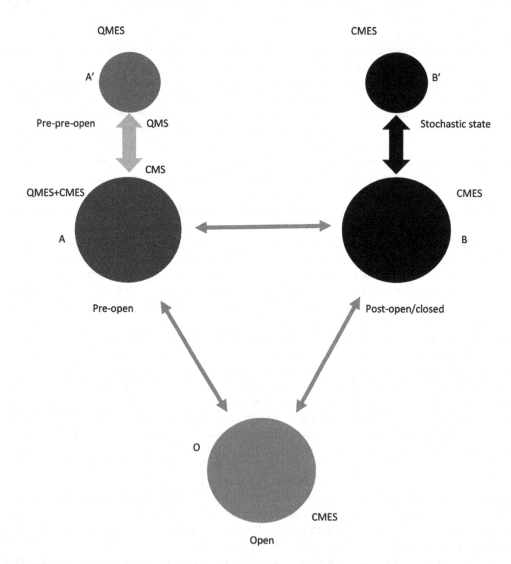

**FIGURE 10.1**   Model demonstration of ion channel transitions between various energy states. Preopen state is considered to contain both classical and quantum mechanical states. Conceptually, as the preopen states initiate to translate toward open states, it starts losing QMESs and gaining the control of CMESs. During this process, the quantal charges start to lose the quantum mechanical wave functions and initiate the process of sinking in the classical charge bath. At the opening of the channel, the quantum mechanical states are vanished while classical mechanical states appear active.

quantum mechanical energy state (QMES) condition, while the SS holds classical mechanical energy state (CMES) condition. Transitions A $\leftrightarrow$ A′ and B $\leftrightarrow$ B′ may occur due to modest energetic fluctuations that often go unnoticed in the classical treatment of ion channels using conventional techniques which deal mainly with the transitions A $\leftrightarrow$ O, O $\leftrightarrow$ B requiring heavy amount of free energetics.

We may need to work more in the inactivation (let's call it preopen or postopen) state of the channel's (gate) structure to pinpoint the quantum effects. Because when the channel enters into open/active state, a bulk amouth of charges concomitantly enters into the state, destroying the quantum energy state to ensure the treatment to be largely (or perhaps even only) classical. From this hypothesis, only at preopen or certain postopen states, we may observe the quantal charge effects close to the ground state. At these extreme conditions, the fluctuations in the coordinates and values of (quantal) charges may occur due to modest change in the ground state energy (is the lowest energy state in quantum mechanical system) due to known or unknown physical reasons. One reason may be transient polarization occurring due to the fluctuations in free charge densities in the biomolecules involved in constructing ion pores (explained in earlier chapters in detail). As the channel enters into the open state, due to the presence of super high charge density, the ground state is destroyed and anything we measure corresponds to a series of excited states, comparable to the energy states of classical systems. However, at low ion density, especially in ion channel selectivity filter (SF), the location of ions may not be always absolutely deterministic at one exact position at a given time. This uncertainly needs to be dealt using quantum mechanical principles. Therefore, although the ion conduction across ion pores is a classical phenomenon, there are special cases where the classical mechanical approaches may fail and quantum mechanics may need to emerge to fully understand the channel conduction mechanisms.

We wish to address some aspects related to the ion channel structures, functions, and energetics where uncertainly is evident using quantum mechanical treatments (see also Chapter 6). In conventional electrophysiology record, we can never measure the conduction or movement of quantal charges at extremely low scales. Determination of a quantum mechanical wave function may help address the coordinates and probabilities of quantal charges associated with certain energy states of ion channels.

## 10.1   ION CHANNEL GATING CURRENTS FROM QUANTAL CHARGE MOVEMENTS

Three decades ago, Conti and Stühmer recorded the asymmetric displacement currents ($I_g$) associated with the gating of nerve sodium channels in cell-attached macropatches of *Xenopus laevis* oocytes injected with exogenous mRNA coding for rat-brain-II sodium channels (Conti and Stühmer, 1989). This type of measurements on gating currents are perhaps not so surprising today, but at that time, it was a great achievement. However, we point at another aspect, apparently a crucial one. The investigators went on measuring the $I_g$ fluctuations to ascertain the discreteness of the conformational changes which precede the ion channel opening. The variance of the fluctuations was reported to indicate that most of the gating-charge movement that accompanies the activation of a single sodium channel occurs in 2–3 brief packets, each carrying an equivalent of about 2.3 e⁻, e⁻ is electron charges.

According to Conti and Stühmer (1989), the quantal nature of the gating current may be revealed by the analysis of macroscopic fluctuations. A general theory has been proposed. This study predicted generalization to transient phenomena in gating mechanism involving a finite total charge transfer. It also predicted gating currents of low pico ampere (pA) order resulting from quantal charge movements having current fluctuations of the order of femto ampere (fA) when recorded with Bessel filter band width B = 10 kHz. These fluctuations are comparable to the background noise of a good patch-clamp recording system. The experimental results are shown in Figures 10.2 and 10.3.

Almost immediately after the publication of Conti and Stühmer's study on the detection of the quantal nature of the sodium channel gating current (Conti and Stühmer, 1989), another important

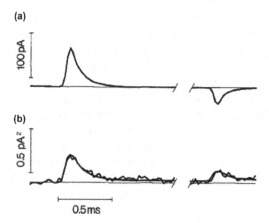

**FIGURE 10.2**    The relationship between mean gating currents and their mean squared fluctuation in repeated measurements. (a) Average of 1.730 $I_g$ responses to the same test depolarization at +20 mV from a holding potential of −100 in V. (b) Noisy trace: mean square fluctuation of I recordings, offset by the mean baseline variance (0.42 pA²); smooth trace: expected variance of the shot-noise from gating-charge quanta with an estimated valence $z_a = 2.2$ (Eq. (4); see also Figure 3 in Conti and Stühmer (1989). Notice that the same $z_a$, but different Qs (13.3 and 4.3 fC, respectively), were used to draw the smooth line for the pulse and the tail data. In A and B the break in the time axis represents an interval of about 2 ms. Recording bandwidth: B = 8 kHz. Temperature: T = 15°C. Reproduced from Conti and Stühmer (1989).

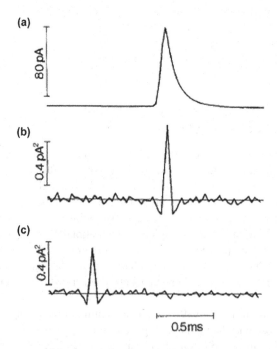

**FIGURE 10.3**    (a) Time course of the average gating current, Ig, from the same experiment illustrated in Figure 2 in Conti and Stühmer (1989), shown on an expanded time scale. (b) Autocovariance of the fluctuations during the development of the Ig response. (c) Autocovariance of the baseline fluctuations. Each fluctuation autocovariance function was measured as the average product of the fluctuation at the respective reference time and the fluctuations at any other time of the stimulation cycle. The decay phase of Ig is well fitted by a single exponential with a time constant of 100 ps, whereas both autocovariances in B and C drop to negligible values within one sampling interval (25 μs). Reproduced from Conti and Stühmer (1989).

paper was published by Greeff and Forster who measured the quantal gating charge of sodium channel inactivation (Greeff and Forster, 1991). Sodium channel open↔inactivation transition was found at voltages above −10 up to +40 mV and the quantal gating charge 1.21 e⁻ was obtained associated to the transition at temperatures 5°C and 15°C. The quantal nature in both channel gating and open ↔ inactivation transition currents raises an open question whether the underlying gating and transition mechanisms may be theoretically addressable using quantum mechanical calculations. To gain an in-depth knowledge about the biochemical, molecular biological, physiological, and structural biological approaches that elucidated the structure and function of sodium channels at the atomic level, the readers may consider reading a review by Catterall (2014). Sodium channels gating pore, inactivation, and drug binding regions are geometrically modeled in low nanometer (nm)-scale resolution. The quantal charge involvement in these gating and inactivation phases requires a better understanding using both classical and quantum mechanical approaches.

The quantum mechanical properties of a particle is described by a wave function. As we find a quantal charge being involved in channel state determination and interstate transitions, we may think of some solutions addressed by quantum treatment of the problem. The solution to the Schrödinger equation, the wave function, may be determined for the quantal charge particle and may describe the quantum mechanical properties of the quantal particle on microscopic scales. The quantal charged particle position, momentum, and energy may be derived from the wave function.

## 10.2 QUANTUM TUNNELING OF ION CHANNELS

Theoretically, tunneling is possible whenever quantum states are separated by a potential energy barrier, where the potential energy is a function of some state variable(s) of the system. The probability of tunneling transitions, relative to transitions caused by thermal fluctuations, depends on the height and width of the barrier and the total energy of the system. High and narrow barriers favor tunneling; low and wide barriers favor classical transitions dependent on thermal fluctuations. Figure 10.4 models the classical movement and quantum tunneling of particles across potential barriers.

Quantum tunneling between closed and open states of sodium channels (see Figure 10.5 for the channel's primary structure) was addressed by Chancey et al. (1992) almost three decades ago. They measured the sodium channel activation time of tunneling. They calculated the Coulomb interactions between the S4 α-helix and negative residues on the nearest-neighbor helices and had included longer range interactions in terms of an effective background interaction. The complete theoretical calculations are presented by Chancey et al. (1992).

Periodic pairing of charges between the S4 and adjacent helices in the model (Catterall, 2014) causes the resting and depolarized states of the channel to correspond to local minima in the S4 potential energy curve. Harmonic potentials closely fit the energy curves around each of the two minima and the energy barrier between them is closely modeled by a parabola (see Figure 10.6). These approximations allow a semi-classical calculation of the S4 helix's tunneling rate. At 37°C, for an inter-helix axial spacing of 1 nm, tunneling times in the range of 1 μs to a few ms were computed for a single S4 segment, depending on the equilibrium temperature of the cell membrane (see Figure 10.7).

Tunneling is an important process in chemical reaction dynamics (Klinman, 1989), including electron transport along molecular bonds (Conti and Stühmer, 1989), transitions in biological systems involving electrons (Jortner, 1976) and protons (Guy, 1990; Stühmer et al., 1989), and transitions between conformational states of biopolymers (Bell, 1980). The quantum tunneling in sodium channel, as presented by Chancey et al. (1992), is an important addition to the understanding of quantum mechanical effects in biological systems.

## 10.3 QUANTUM COHERENCE AND DECOHERENCE OF ION CHANNELS

A system is considered coherent if there exists a definite phase relation between different states. Coherence generally describes properties that correlate between physical quantities of a single wave

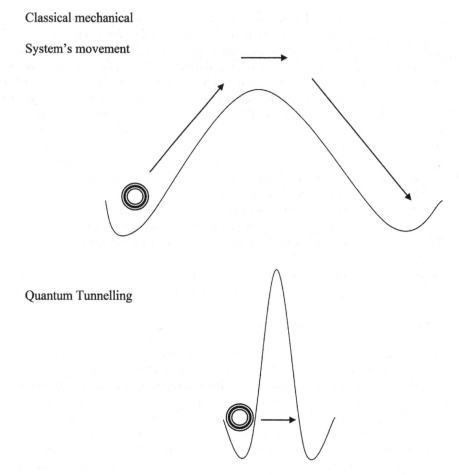

**FIGURE 10.4** Comparison of the movement of particles (e.g., electrons or quanta of charges) while crossing an energy barrier landscape (arrow indicates the direction of movement). **Top panel**, classical mechanical systems cross over the energy landscape. In this case, the barrier width is considerably wide. The particle's location is deterministic in this classical mechanical treatment, and the probability of the particle to be at its location (where ever that may be) is always 1, leaving no possibility for the particle to be elsewhere at the same time. **Bottom panel**, in quantum mechanics, particles are allowed to tunnel through the energy barrier landscape. In this case, the landscape width must be very narrow. Due to quantum tunneling, although the particle is hypothetically seen (in this model diagram) on the left side of the landscape, there is always a finite probability for the particle to be found on the right side of the landscape. Therefore, the probability of finding the particle on either side of the landscape is always less than 1, specifically between 0 and 1.

or between several waves or wave packets. In quantum mechanics, coherence is an important phenomenon. Quantum coherence relies on the concept that all quantum objects have wave-like properties and that any quantum state simultaneously consists of multiple states. Quantum decoherence acts opposite to the coherence. It is simply the loss of quantum coherence.

Quantum coherence in biological processes is already predicted, or evident. Just an example, electronic coherence between excitons as well as between exciton and charge-transfer states has been revealed in photosynthesis (Romero et al., 2014). The conversion of solar to chemical energy in photosynthesis occurs in which the absorbed excitation energy is converted into a stable charge-separated state by ultrafast electron transfer events. Here, electron coherence is predicted to be maintained by vibrational modes. Earlier, direct visualization of coherent nuclear motion in the excited state of the primary electron donor in the bacterial RC was made (Vos et al., 1993).

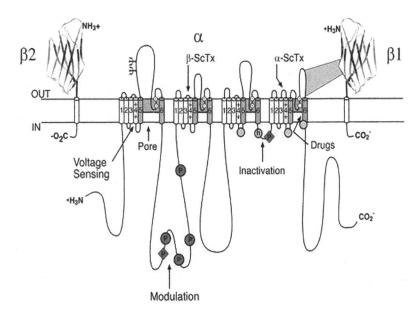

**FIGURE 10.5** The primary structures of the subunits of the voltage-gated sodium channels. Cylinders represent alpha helical segments. Bold lines represent the polypeptide chains of each subunit with length approximately proportional to the number of amino acid residues in the brain sodium channel subtypes. The extracellular domains of the $\beta 1$ and $\beta 2$ subunits are shown as immunoglobulin-like folds. □, sites of probable N-linked glycosylation; P in red circles, sites of demonstrated protein phosphorylation by PKA (circles) and PKC (diamonds); green, pore-lining segments; white circles, the outer (EEEE) and inner (DEKA) rings of amino residues that form the ion selectivity filter and the tetrodotoxin binding site; yellow, S4 voltage sensors; h in blue circle, inactivation particle in the inactivation gate loop; blue circles, sites implicated in forming the inactivation gate receptor. Sites of binding of $\alpha$- and $\beta$-scorpion toxins and a site of interaction between $\alpha$ and $\beta 1$ subunits are also shown. Reproduced from Catterall (2014). (For color figure see eBook figure.)

Here, a simpler version of the photosystem II reaction center (PSII RC) suggested a functional role for these motions in the primary electron transfer reaction, a view that is theoretically modeled by Parson and Warshel (2004). A long-lived coherence in photosynthetic complexes has been observed in light-harvesting (Savikhin et al., 1997; Brixner et al., 2005; Zigmantas et al., 2006; Engel et al., 2007; Calhoun et al., 2009; Collini et al., 2010; Schlau-Cohen et al., 2012; Hildner et al., 2013) and in oxidized bacterial RCs (Lee et al., 2007; Westenhoff et al., 2012).

Quantum coherence has been revealed in ion channels. A comparison of a classical and a quantum mechanical calculation of the motion of $K^+$ ions in the highly conserved KcsA SF motive of voltage-gated ion channels has recently been made by Summhammer and co-investigators (Summhammer et al., 2018). SF is a protein structural motif forming a tunnel inside the ion channel. The filter selects a specific type of ion and transfers it to ensure selectivity in ion transfer through a channel type. Figure 10.8 details the $K^+$ channel SF (Doyle et al., 1998).

Summhammer et al. (2018) showed that the de Broglie wavelength of thermal ions is not much smaller than the periodic structure of Coulomb potentials in the nanopore model of the SF. This implies that the location of an ion may no longer be totally deterministic at one exact position at a given time. This uncertainty can be described by a quantum mechanical wave function. They demonstrate solutions of a nonlinear Schrödinger model providing insight into the role of short-lived (~1 ps) coherent ion transition states and attribute an important role to subsequent decoherence and the associated quantum to classical transition for permeating ions. Short coherences are not just beneficial but also necessary to explain the fast-directed permeation of ions through the potential barriers of the filter.

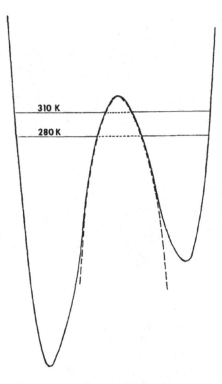

**FIGURE 10.6**    The potential energy barrier separating the unshifted and shifted positions of the α-helix. The barrier region is closely modeled by the parabola (- - -). The highest point of the barrier is taken as the vertex of the parabola. The two horizontal levels show the mean oscillation energy, E, of the α-helix's center of mass when the equilibrium temperature of the cell membrane is 280 K (lower) and 310 K (upper). E is measured from the left well's potential minimum. Reproduced from Chancey et al. (1992).

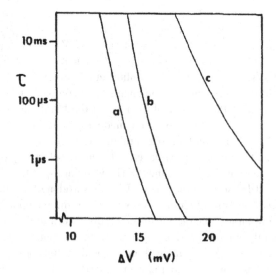

**FIGURE 10.7**    The characteristic tunneling time τ as a function of membrane depolarization for three equilibrium temperatures of the membrane: (a) 310 K, (b) 305 K, and (c) 300 K. τ is the time interval between membrane depolarization and the α-helical shift (gate activation). Reproduced from Chancey et al. (1992).

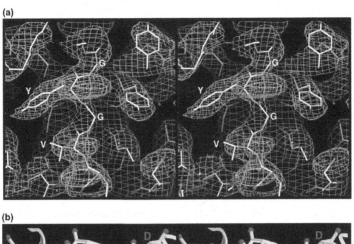

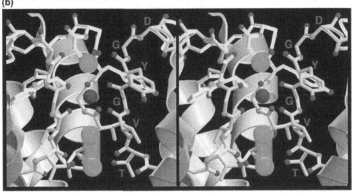

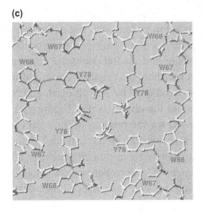

**FIGURE 10.8** Detailed views of the K⁺ channel selectivity filter. (a) Stereo view of the experimental electron density (green) in the selectivity filter. The map was calculated with native-sharpened amplitudes and multiple isomorphous replacement (MIR)-solvent-flattened-averaged phases. The selectivity filter of three subunits is shown as a stick representation with several signature sequence residues labeled. The Rb⁺ difference map (yellow) is also shown. (b) Stereo view of the selectivity filter in a similar orientation to (a) with the chain closest to the viewer removed. The three chains represented comprise the signature sequence amino acids Thr, Val, Gly, Tyr, Gly running from bottom to top, as labeled in single-letter code. The Val and Tyr side chains are directed away from the ion conduction pathway, which is lined by the main chain carbonyl oxygen atoms. Two K⁺ ions (green) are located at the opposite ends of the selectivity filter, roughly 7.5 Å apart, with a single water molecule (red) in between. The inner ion is depicted as in rapid equilibrium between adjacent coordination sites. The filter is surrounded by inner and pore helices (white). Although not shown, the model accounts for hydrogen bonding of all amide nitrogen atoms in the selectivity filter except for that of Gly77. (c) A section of the model perpendicular to the pore at the level of the selectivity filter and viewed from the cytoplasm. The view highlights the network of aromatic amino acids surrounding the selectivity filter. Tyrosine-78 from the selectivity filter (Y78) interacts through hydrogen bonding and van der Waals contacts with two Trp residues (W67, W68) from the pore helix. Reproduced from Doyle et al. (1998). (For color figure see eBook figure.)

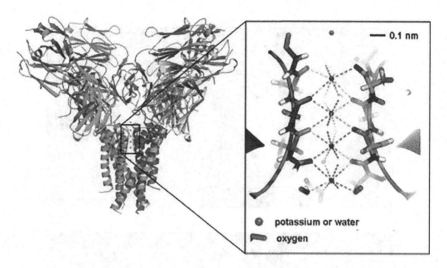

**FIGURE 10.9** Schematic illustration of the KcsA potassium channel after PDB 1K4C. KcsA protein complex with four transmembrane subunits (left) and selectivity with four axial trapping sites formed by the carbonyl oxygen atoms in which a potassium ion or a water molecule can be trapped. Reproduced from Vaziri and Plenio (2010).

The KcsA SF consists of four ion binding sites (Doyle et al., 1998), where each site is physically represented by a minimum potential energy. The axial separation between these potential minima is about 0.24 nm. The height of the potential barrier varies between ~1.7 and 8.0 $k_B$T depending on both protein thermal vibrations and ion position in the channel, confirmed by molecular dynamics (MD) simulation (Gwan and Baumgaertner, 2007). $k_B$ is the Boltzmann's constant, and T is the absolute temperature. It is shown that ion translocation is based on the collective hopping of ions and water molecules which is mediated by the flexibly charged carbonyl groups lining the backbone of the pore. Pairwise translocations are evident, where one ion and one water molecule form a bound state. The water molecules are predicted to act as rectifiers during the hopping of ion–water pairs.

A coherent process behind ion transmission is predicted by March et al. (2018). Earlier Vaziri and Plenio (2010) suggested that the SF of potassium ion channels (see Figures 10.8 and 10.9) may exhibit quantum coherence, which might be relevant for the process of ion selectivity and conduction. We avoid presenting the bulk amount of theories written in the article. Technology wise, they showed that quantum resonances could provide an alternative approach to ultrafast two-dimensional (2D) spectroscopy to probe these quantum coherences. The emergence of resonances in the conduction phase of ion channels that are modulated periodically by time-dependent external electric fields can serve as signatures of quantum coherence in the system.

The chains of ions and water molecules in the SF, each coordinated by eight oxygen atoms of carbonyl groups, were treated as one-dimensional coupled quantum harmonic oscillators connected to a source and a sink and subject to dephasing-type noise and thermal excitations (see Figure 10.10). It is demonstrated that such a system, if driven by an external varying potential, exhibits quantum resonances that are absent in classical rate equation-type models (see Figure 10.11). Such resonances can be used to quantify the amount of quantum coherence in a system. If such coherences could indeed be detected using valid experimental techniques, their possible biological relevance would be highly important. As they would be short-lived, it might lead to a new understanding of biomolecular functions relying on the interplay between quantum coherence and environmentally induced decoherence.

The conductivity of an externally driven channel exhibits resonances where the conductivity is strongly suppressed. This resonance is known to decrease in the presence of dephasing noise, and

**FIGURE 10.10** A source drives excitations into the first site of a chain where the nearest neighbors are coupled by a hopping interaction. The excitations leave the chain via the last site (Vaziri and Plenio, 2010).

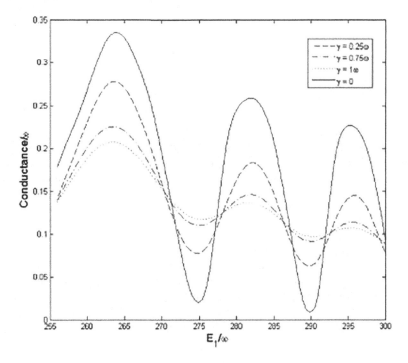

**FIGURE 10.11** Induced quantum resonances in the selectivity filter of the potassium channel in the presence of an external driving field with (dashed lines) and without (solid line) decoherence (Vaziri and Plenio, 2010).

it can be related to the amount of quantum coherence in the channel. Thus, the depth of resonances may be taken as a measure of the presence of quantum coherence. Besides, incoherence is also found to be proportional to conductivity (Vaziri and Plenio, 2010).

The flux rate of ions through ion pore at a given concentration is fairly constant. The width of the expected quantum resonances in Figure 10.11 is not very sharp. These are the reasons we may expect to find the resonances experimentally. Single-channel biophysics studies with techniques being slightly modified can help in experimental investigations of the quantum resonances in the SF of potassium channels (Vaziri and Plenio, 2010).

Recently, regarding the superposition states of potassium ion in SF of KcsA channel, Salari et al. (2015) has calculated decoherence time using MD simulations based on models explained in Figure 10.12 in reference to earlier studies (Doyle et al., 1998). Superposition states of potassium ions in this study indicate that decoherence times are in the order of picoseconds (Figure 10.13). These times are 10–100 million times bigger than the order calculated by Tegmark two decades ago (Tegmark, 2000). This mentioned difference in decoherence time perhaps lies in using the different models by these two groups, as explained in Figure 10.12 (Salari et al., 2015). The ionic superposition becomes decohered as a consequence of environmental scattering. This decoherence time can be considered adequate for quantum states of cooled ions in the filter to leave their quantum traces on the filter and action potentials.

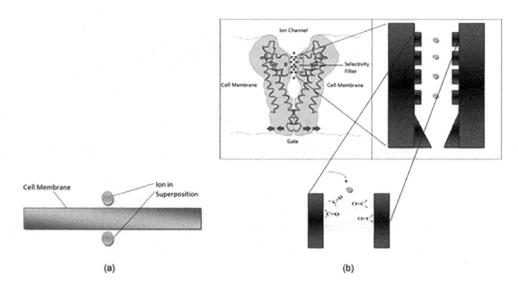

(a)                                                                (b)

**FIGURE 10.12** Schematic difference between Tegmark's model (2000) and the real model, in which Tegmark has not considered the ion channel structure. (a) Tegmark's model (2000), in which two ions are in a superposition state of "inside" and "outside" of the membrane. The superposition distance is the thickness of the membrane. Tegmark has assumed 10 nm for the thickness and therefore the superposition distance. (b) The real model, in which quantum superposition states occur in the selectivity filter of the ion channel. The superposition distance here is 0.3 nm. The 3.4-nm-long KcsA channel comprises a 1.2-nm-long selectivity filter that is composed of four P-loop monomers. Each P-loop is composed of five amino acids linked by peptide units (H–N–C=O), where N–C=O is an amide group and C=O is a carbonyl group. If a positively charged ion, such as sodium or potassium, enters the selectivity filter, the ion can be transiently trapped by Coulomb interactions, with the negative charges provided by the oxygen ions.

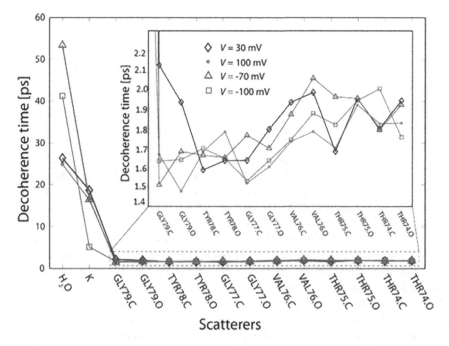

**FIGURE 10.13** Decoherence time of ionic superposition states versus scatterers in the selectivity filter of an ion channel. The results indicate that the main scatterers are carbonyls of amino acids yielding decoherence times of the order of picoseconds for superposition states (Salari et al., 2015).

Recently, Salari et al. (2017) simulated two neighboring ion channels on a cell membrane with the double-slit experiment in physics and investigated to inspect whether there is any possibility of matter-wave interference of ions via movement through ion channels. The model is presented in Figure 10.14: two neighboring ion channels on a neural cell membrane with a double-slit experiment have been simulated.

The possibility of matter-wave interference through the slits has been investigated. An estimation has then been made on the plausible distances between the ion channels to produce ionic interference. The effect of environmental decoherence on quantum states of ions inside and outside the slits has been addressed, and accordingly the coherence lengths of ions for making interference was obtained.

Simulation of Salari et al. (2017) is based on Gaussian wave packet as a simulation for potassium ions which move through two neighboring ion channels in two dimensions. The use of Gaussian wave packet is sufficiently general because it includes the limit case of plane waves. Here, the macroscopicity method is used for obtaining interference patterns using a dimensionless form of the Schrödinger equation, in which a new dimensionless parameter $\hbar$(proportional to the de Broglie wavelength, normalized by a characteristic length) appears showing quantitatively the quantum behavior of the system (Figure 10.15). We avoid presenting the details, but the detailed theoretical analysis can be found in Salari et al. (2017).

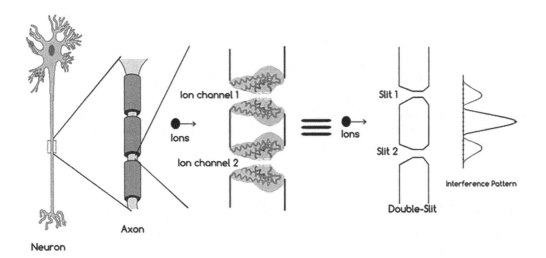

**FIGURE 10.14**   (Left) Ions passing through two neighboring ion channels on a cell membrane in a neuron, (Right) Simulation of two neighboring ion channels as a double-slit interferometer.

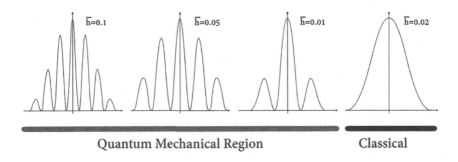

**FIGURE 10.15**   The value of $\hbar$ determines the quantum mechanical or classical behavior. In the classical regime, the interference pattern cannot be formed, while in quantum regime, there are possibility of obtaining the interference pattern (Salari et al., 2017).

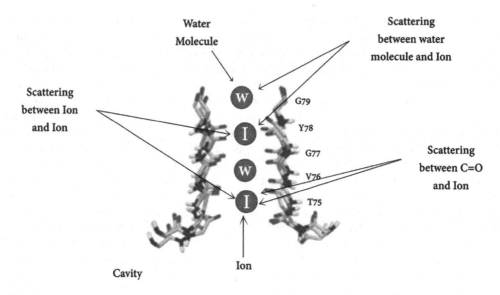

**FIGURE 10.16** The quantum states of ions in the selectivity filter may be decohered due to scattering with environmental particles and biomolecules in biological temperature (Salari et al., 2017).

The obtained decoherence timescales indicate that the quantum states of ions can only survive for short times, i.e., ≈100 picoseconds in each channel and ≈17–53 ps outside the channels. We may now assume that the SF is a cavity with volume V containing N particles (see Figure 10.16). Here the ion is the system that can be scattered by the particles in the cavity. The main scattering can happen between the ion and the particles in the cavity such as C=O bonds, water molecules, and other ions (see Figure 10.16). Thus, the results hint that the quantum interference of ions seems unlikely due to environmental decoherence.

## 10.4 QUANTUM ENTANGLEMENT OF ION CHANNELS

In ion channels, the gross conduction state refers to the flow of large number of charged particles across the channel. This phenomenon of the movement of large amount of particles is classical in nature, which needs to be explained using classical physics principles. However, in the initiation phase of the ion conduction or in any closed localized environment where ion channel gating mechanism primarily occurs, or when the gating mechanism gets broken, it may be possible to treat individually charged particles independently or pairwise or in a small localized complex of particles in highly specialized physical condition. The quantum state of the ions in the SF (Figure 10.16) is such physical process where even a very special phenomenon "quantum entanglement" may occur.

When particles as pairs or in groups are generated, interact, or share spatial proximity within a localized region in ways such that the quantum state of individual particle cannot be described independently disregarding the state of the others, irrespective even of their mutual separation, quantum entanglement may occur. The state of particles in channel (Figure 10.16) may be explained better using the movement of their ($K^+$ and water molecules) coordination, as presented in Figure 10.17 (Morais-Cabral et al., 2001).

All possible states of the SF have been considered in which there can exist zero, one, or two ions, and in which ion pairs are separated by at least a single water molecule (see Figure 10.18). Between states B and C having different electron density profiles, correspondingly between configurations 1,3 and 2,4 in Figure 10.17, $K^+$ energy states are equal (Morais-Cabral et al., 2001). Thus,

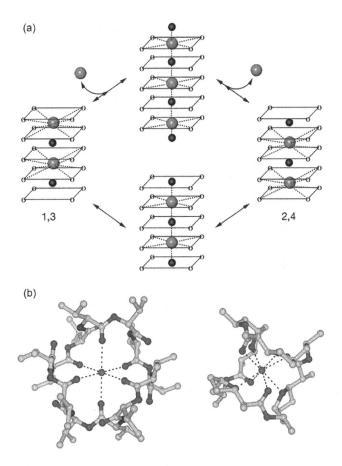

**FIGURE 10.17** (a), Detailed description of the cycle. The selectivity filter is depicted as five sets of four in-plane oxygen atoms (the top is outside the cell), with $K^+$ ions and water molecules shown as green and red spheres, respectively. $K^+$ ions undergo coordination by eight oxygen atoms when in the 1,3 and 2,4 configurations. Movement along either the concentration-independent path (bottom) or the concentration-dependent path (top) would involve octahedral coordination by six oxygen atoms, two provided by the intervening water molecules. (b) The atomic structures of $K^+$-selective antibiotics nonactin (right) and valinomycin (left) exhibit coordination by eight and six oxygen atoms, respectively. (For color figure see eBook figure.)

the difference in energies of these two conformations 1,3 and 2,4 for $K^+$ (Figure 10.17) is 0, so the transition between these two conformations require no free energy. This raises important questions like are these back-and-forth transitions due to events such as quantum tunneling or are there some kind of quantum entanglement occurring?

Electron density profiles (I4 crystal form, 2.4 Å Bragg spacings) for various $K^+$ concentrations, 3–200 mM $K^+$, are shown in Figure 10.19, that exhibit important features. The 3 mM $K^+$ profile shows a predominant peak on each end of the SF and a small peak near position 3. Therefore, $Rb^+$ and $K^+$ seem to be similar when only a single ion is present in the filter. As the $K^+$ concentration is raised, two peaks are seen to appear at positions 2 and 3, yielding a profile with four peaks having roughly equal heights and widths. Adjacent peaks are separated by about 3.2 Å. As a potassium ion has a diameter of 2.7 Å, they could, in principle, fit in the filter side by side. However, due to possible electrostatic repulsion, this would seem to be an unstable binding configuration. Two $K^+$ ions only very rarely occur with a separation distance of less than 3.5 Å. Therefore, two subsequent $K^+$ ions are expected to be separated by an intervening water molecule (Figures 10.17a and 10.18).

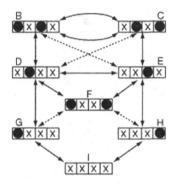

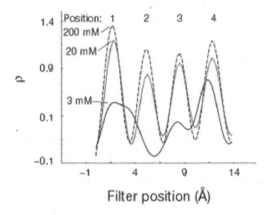

**FIGURE 10.18** Analysis of a potassium channel conduction-state diagram constructed on the basis of the electron density profiles (Morais-Cabral et al., 2001). Each set of four boxes (positions 1–4 from left to right) represents a state of the selectivity filter (left side outside the cell) with a specified configuration of ions (black circles) and water (crosses). The eight states, B–I, are the configurations of 0, 1, or 2 ions, assuming that ions are separated by at least a single water molecule. Arrows show the transition paths that connect the states through concerted movements of the ion–water queue (solid lines) or by ion–water exchange at the ends of the filter (dashed lines). Two paths connecting states B and C represent a concentration-independent movement of the ions from state B (1,3 configuration) to state C (2,4 configuration) and a concentration-dependent transition due to entry of a third ion on one side, causing exit of an ion on the opposite side.

**FIGURE 10.19** Electron density ($\rho$) profile of potassium channel selectivity filter (Morais-Cabral et al., 2001).

At high concentration, the electron density for $K^+$ seems to be more or less evenly distributed over the filter and that the distribution of $K^+$ in the filter occurs following certain hydration principle leaving the possibility of distribution of $K^+$ over a spectrum of hydration. There are two hypothesis we wish to consider:

i. As a potassium ion has a diameter of 2.7 Å, they could, in principle, fit in the filter side by side (defying the possible electrostatic repulsion), in which case the successive $K^+$ ions exist at dehydrated environment, accounting for a physiological or biological environment having a very low dielectric permittivity equivalent to that of hydrophobic environment. If we hypothetically consider the inter-potassium interactions to extend across the SF length (~1.2 nm) which is closed from the surrounding environment, then there would be 2–4 potassium ions accommodable inside the filter.

ii. As the energy states of 1,3 and 2,4 configurations (Figure 10.17) are exactly the same (Morais-Cabral et al., 2001), we may consider the potassium ions to exist in either of these configurations with equal probability. That's more like tunneling between two energy states without losing energy, or two states are entangled.

Thus, the problem can be treated quantum mechanically as we treat the interactions among ions in almost a closed vacuum, where all interaction energy states have same values. Bernroider and Roy (2005) provided a general quantum mechanical calculation on this issue. They proposed a quantum information scheme that builds on the interference properties of entangled ion states transiently confined by local potentials within the permeation path of voltage-gated ion channels. It is understood that the submolecular organization of parts of the ion channel constructing protein carries a logical coding potency that goes beyond the pure catalytic function of the channel, subserving the transmembrane crossing of an electrodiffusive barrier. The KcsA channel protein can transiently stabilize three $K^+$ energy states, two within the permeation pathway and one within the "water cavity" located toward the intracellular side of the permeation path (Doyle et al., 1998; Guidoni and Carloni, 2002). The overall state structure are summarized a bit differently using a chemical scheme, as shown in Figure 10.20 (Bernroider and Roy, 2005).

In Figure 10.20, the channel is shown to consist of possible three ion modes, schematically aligned along one (z) axis, the longitudinal section of the protein core, with the cells interior cytoplasmic site to the left. The gate states appear with two possibilities, either closed (top panel) or open (bottom panel). Ions are coordinated by their electrostatic Coulomb interactions with either carbonyl-derived oxygens in the permeation path ($O_i - K^+$) or water (W)–ion interaction ($W_i - K^+$). The subscript i denotes the "in-plane" number, along the perpendicular x, y plane of molecules (water) or atoms (oxygens) coordinating the ion. The interior of the permeation pore is delineated and the Coulomb domains giving rise to eight-fold coordination states under the closed gate conditions (a, b) are shaded. Note: with double ion occupancy of the permeation path, there are two "confined" ion configurations in the closed gate state. Using the originally proposed z-coordinate number (numbers 0–4), these are the configurations 2,4 (a) and 1,3 (b). Only short-range electrostatic interaction terms are given, leaving out (i) water-damped dipole interactions from surrounding short pore helices in the water cavity, and (ii) side chain interaction terms from the ionizable residues near the SF.

The maximum conduction of $K^+$ requires the energy difference between the 1,3 and 2,4 configurations to approach to zero, dedicated to the protein function optimization (Morais-Cabral et al., 2001). A very important message is found here. The highest conductance condition requires two energy states to exist in exactly identical energy landscapes. Here, we may draw the clue of a hidden mechanisms, which is the transition probability between two energy landscapes is maximum, or the probability of ions to be in either state is equal, or even the transition time between two energy landscapes is instantaneous. Isn't the concept close to the quantum mechanical treatment of problems?

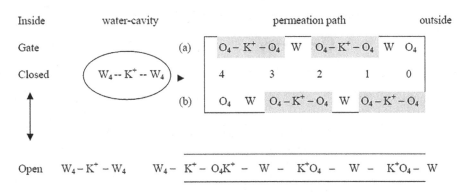

**FIGURE 10.20** Potassium channel state configurations (presented differently) that mimic Figures 10.17 and 10.18.

The so-called "energetic optimization hypothesis of ion channels" suggests that "conspecific" ions in the channel are energetically balanced, whereas "heterospecific ions" such as rubidium (Rb$^+$) ions in a K$^+$ channel, are energetically imbalanced. Anomalous distribution of Rb$^+$ was reported compared with K$^+$ in the SF (Morais-Cabral et al., 2001). The states of the joint two K$^+$ ion configurations 1,3 and 2,4 may, therefore, be assumed as degenerate, that is, their energy difference $\Delta E = 0$. Degeneracy in quantum mechanical system is quite common. Two or more distinguishable states of any quantum mechanical system are considered as degenerate if the states appear with same energy value. As we detect degeneracy among two conformational states 1,3 and 2,4 for K$^+$ ions in a KcsA channel, we may predict some sort of quantum mechanical principles to be active in these ion conductance states. Of course, the prediction needs to be at least theoretically analyzed, if not possible to easily validate experimentally. An energetics of the conformatiuonal changes of the system, considering the classical distribution, has been graphically represented in Figure 10.21 (Bernroider and Roy, 2005).

Bernroider and Roy (2005) proposed a theoretical analysis to explain the ions' transitions between conformational states. If we wish to find the ions in one of these two energetically indistinguished states 1,3 and 2,4 as a function of time, we may apply the time-dependent perturbation concepts, in particular Rabis formula. This allows us to estimate the probability ($P_2$) of occupying, for example, state 2,4, if a constant perturbation is switched on at $t = 0$ and the system sets out from state 1,3. Bernroider and Roy (2005) calculated the value of $P_2$.

$$P_2 = \left[ \left( 4V^2 \right) \middle/ \left( \omega_0^2 + 4V^2 \right) \right] \cdot \sin^2 \tfrac{1}{2} \left( \omega_0^2 + 4V^2 \right)^{1/2} t$$

For degeneracy, $\Delta E = 0$, we also have $\omega_0 = 0$. So $P_2 = \sin^2 Vt$. This predicts a complete state transition for degenerate systems, with $P \to 1$ after a time $t = \pi/2V$. t is estimated in MD simulations for KcsA channel transitions to scale along $6.10^{-11}$s (Guidoni and Carloni, 2002), we can estimate the perturbation strength $V \approx \pi/2.\ 10^{11}$s$^{-1}$. The "trapping frequencies" above can be recovered from the

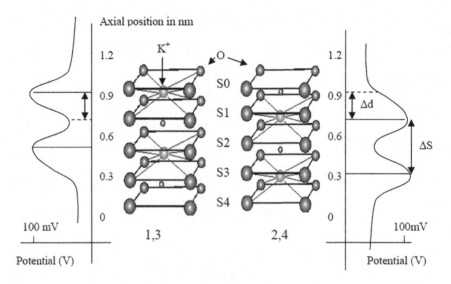

**FIGURE 10.21** Ion trapping within the selectivity filter's two conformational states of K$^+$ channels. As explained earlier, the site S0 is outside and the site S4 toward the intracellular domain of the channel. The confining potentials are modeled as homogeneous harmonics to the left (relating to the 1,3) and to the right side (relating to the 2,4 configuration). During the closed gate state, ions can be modeled as being confined to two moving potentials, shifted along the z-axis by distance $\Delta d$. The "filter" accounts for roughly one-sixth of the total transmembrane length of the channel protein. Adjacent K$^+$ locations are separated by $\Delta S \sim 0.3$ nm, in their S1–S3 and S2–S4 configurations (Bernroider and Roy, 2005).

view of time-dependent perturbation theory. Thus, ion selectivity of a channel protein is partially addressed. Nonselected ions (e.g., $Na^+$ in $K^+$ channels), leading to a separation of energy levels between the preferred filter positions, would still allow state oscillations to occur, but with maximum levels for P in the range of $4V^2/\omega_0^2 < 1$. The effect of the external membrane field/potential, perturbing the local fields within the selectivity pore of the channel, resides within the relative offset between the energy separation of ion channel states and the external field strength. Weak perturbations caused by external field variations can therefore drive the system completely between the degenerate ion channel coordination states 1,3 and 2,4.

A superposition of ion states in the permeation filter of channels that exist at a separation $\Delta S \sim$ 0.3 nm is hypothetically possible. We may "conceptually" close the filter from its environment and describe the state of $K^+$ ions by a spatial superposition. If we consider $|\Psi\rangle$ as the ions state vector for the low-$K^+$ structure and $|k_{i,j}\rangle$ as the ket-vector of either of two conformations 1,3 (i=1,j=3) and 2,4 (i=2, j=4), we get the following general equation of quantum mechanics:

$$|\Psi\rangle = \alpha|k_{1,3}\rangle + \beta|k_{2,4}\rangle$$

For two state conformations, where potassium ion transits,

$$|\alpha|^2 + |\beta|^2 = 1$$

In case of symmetric energy landscapes for both 1,3 and 2,4 conformations (Figure 10.19), we theoretically expect to see equal probability for the potassium ions to occupy either states, hence,

$$|\alpha|^2 + |\beta|^2 = 0.5$$

### Uncertainty of Ion's Position at a Given Time: Is the Quantum Uncertainty Principle Justified in Ion Channels?

The Heisenberg uncertainty principle in quantum mechanics states that it is not possible to know simultaneously (at the same time) the exact geometrical location (position) and momentum of a particle. That is, the more exactly the position is determined, the less known is the momentum, and vice versa. As we predict, using the explanations in this chapter based on various studies quoted here, that the states of ions $K^+$ in KcsA channel may be addressed quantum mechanically, it is expected that the quantum mechanical wave function that represents the ion's movement inside the channel also satisfies other related principles of quantum mechanics. Uncertainty principle is such a fundamental concept that may be satisfied in case of $K^+$ channels at certain physical conditions.

Summhammer et al. (2018) recently reported some results based on their simulations. They monitored the $K^+$ ion during the transition from site S4 to site S3 in the SF of the KcsA channel (see Figure 10.21). The simulation included the corresponding carbonyl groups of Thr74, Thr75, and Val76. In their comparison of a classical and a quantum mechanical (QM) calculation of the motion, they found various results demonstrating fundamental quantum mechanical parameters. Figure 10.22 shows the transition behavior between S4 and S3. Though the classical ensemble splits after around 0.8 ps (middle), the QM distribution goes beyond the barrier to S3 almost completely (right).

Figure 10.23 shows the time-dependent probabilities to find an ion in conformation S3 when the ion was implanted into the conformation S4. At the initial velocity of 300 m/s, the classical particles are observed not to cross to S3. Most classical particles with velocity of 900 m/s are in S3 after 0.5 ps. However, eventually about 45% return to S4 due to oxygen charge-derived forces (the virtual spring that returns these ions to their equilibrium positions with vibrations around 3 THz). The QM wave exhibits a small, but remaining probability (<10%) of returning to S4. Figure 10.24 demonstrates the deviation of Tyr75 carbonyl oxygens from their equilibrium positions as $K^+$ passes by.

Here, the focus was on comparison between the classical standard MD setting and the QM version under the same interaction potentials between all constituents and the same initial position and

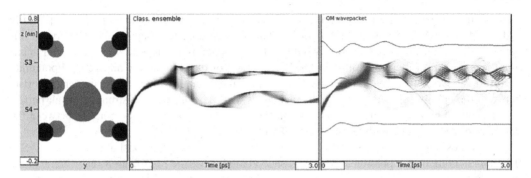

**FIGURE 10.22**  Transition behavior between S4 and S3 (left insert) for a classical ensemble (middle) and the simulated QM wave packet (right), with shades of black and blue coding normalized probability densities for location and time. Red lines (right) are the z-coordinates of carbonyl oxygens. (For color figure see eBook figure.)

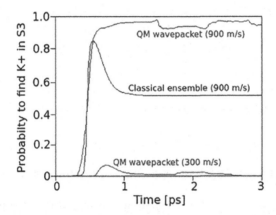

**FIGURE 10.23**  Time-dependent probabilities to find an ion in S3, when the ion was implanted into S4 with different mean onset velocities (900 m/s blue top for the QM wave packet, black for the classical ensemble) and at 300 m/s for the QM wave packet (with some probability <0.1 to cross over to S4).

velocity distribution of the ion. This yielded comparable results under situations where the $K^+$ ion is coordinated at a specific site (e.g., S4) and oscillates within this site due to its thermal energy. Even at the initially relatively low kinetic energy levels of a $K^+$ ion, the QM wave shows a nonvanishing probability that the ion could make a transition to site S3 in the filter (as seen around <1 ps after onset). This observation is interesting as it occurs within just one picosecond (i.e., well within the expected decoherence time due to thermal noise from protein backbone atoms transmitted to carbonyl atoms).

The observations render an important role for decoherence and the quantum to classical transition. Decoherence of the ion's wave function happens after about 1 ps, a time when almost all of its probability distribution has penetrated the barrier, leads to classical behavior, which can be seen to avoid the return to the previous location. So, decoherence actually "guides" the particle into one direction in the filter, and an oscillation between quantum and classical states cooperate in a directed transport through the potential landscape of the filter.

## 10.5  CONCLUDING REMARKS

In all these example studies, specific biological environments have been considered to address the channel conductance and inherent energetics. However, a question may arise regarding

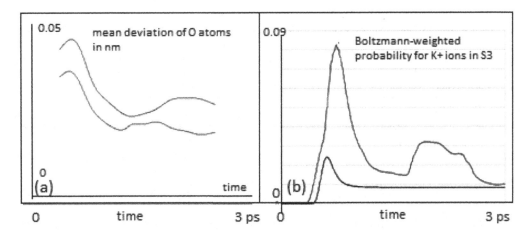

**FIGURE 10.24**   (a) Mean deviation of Tyr75 carbonyl oxygens from their equilibrium positions, while a $K^+$ is moving past from location S4 to S3 (in nm). The classical particles are in black, QM wave packets in blue; (b) Probability for a $K^+$ ion to be found in S3, setting out from S4 with mean velocities between 100 m/s and 900 m/s, weighted according a Boltzmann velocity distribution at 310 K (blue QM, black classical). (For color figure see eBook figure.)

understanding the biological environments correctly and choice of related physical parameters associated with the channel. While doing theoretical research to address and understand ion channel conductance and energetics we may use freedom of choosing parameters from a wide range of possible values, most of which are hypothetical. We may consider distribution of charges inside channel in hydrated or anhydrate environments. We may consider the channel interior as a region equivalent to that free from any biological entity that participate in qualifying the physiological condition, consisting of charge distribution, closed or connected to the physiological environment. We may hypothetically withdraw water molecules from the channel and consider the dynamic (or conditionally static, depending on the electrostatic interaction-based localized pinning effects) presence of two $K^+$s inside the channel, where ions occupy identical energy states in 1,3 and 2,4 configurations (see Figures 10.17a and 10.18). In that case, we may hypothesize that charges inside the channel may experience QM tunneling between two energy states with equal probability being in either state. This quantum entanglement condition can be theoretically addressed through numerical computations on energetics of ions in both of these configurations, 1,3 and 2,4. For addressing the energetics in both states and calculation of free energy difference (expected to approach to 0 according to Figure 10.19) between them can easily be made using screened Coulomb interactions among charges inside the channel (Ashrafuzzaman and Tuszynski, 2012; 2013). We are working to provide a theoretical explanation combining classical and QM calculations and numerical computations. A thorough understanding on the pattern of the energy landscapes inside channels will be made. The classical versus QM analysis of the $K^+$ ion states are going to be revisited using some novel techniques that we are going to develop.

## REFERENCES

Ashrafuzzaman, M., Tseng, C., Duszyk, M., & Tuszynski, J. A. (2012). Chemotherapy drugs form ion pores in membranes due to physical interactions with lipids. *Chemical Biology & Drug Design, 80*(6), 992–1002. doi:10.1111/cbdd.12060

Ashrafuzzaman, M., & Tuszynski, J. (2012). Regulation of channel function due to coupling with a lipid bilayer. *Journal of Computational Theoretical NanoScience, 9*(4), 564–570. doi:10.1166/jctn.2012.2062

Ashrafuzzaman, M., & Tuszynski, J. A. (2013). *Membrane Biophysics*. Berlin: Springer. doi:10.1007/978-3-642-16105-6

Bell, R. (1980). *The Tunnel Effect in Chemistry*. London: Chapman & Hall.

Bernroider, G., & Roy, S. (2005, May 23). Quantum entanglement of K⁺ ions, multiple channel states, and the role of noise in the brain. Retrieved December 07, 2020, from https://www.spiedigitallibrary.org/conference-proceedings-of-spie/5841/1/Quantum-entanglement-of-K-ions-multiple-channel-states-and-the/10.1117/12.609227.full

Brixner, T., Stenger, J., Vaswani, H. M., Cho, M., Blankenship, R. E., & Fleming, G. R. (2005). Two-dimensional spectroscopy of electronic couplings in photosynthesis. *Nature, 434*(7033), 625–628. doi:10.1038/nature03429

Calhoun, T. R., Ginsberg, N. S., Schlau-Cohen, G. S., Cheng, Y., Ballottari, M., Bassi, R., & Fleming, G. R. (2009). Quantum coherence enabled determination of the energy landscape in light-harvesting complex II. *The Journal of Physical Chemistry B, 113*(51), 16291–16295. doi:10.1021/jp908300c

Catterall, W. (2014). Structure and function of voltage-gated sodium channels at atomic resolution. *Experimental Physiology, 99*(1). doi:10.1113/expphysiol.2013.071969

Chancey, C. C., George, S. A., & Marshall, P. J. (1992). Calculations of quantum tunnelling between closed and open states of sodium channels. *Journal of Biological Physics, 18*(4), 307–321. doi:10.1007/bf00419427

Collini, E., Wong, C. Y., Wilk, K. E., Curmi, P. M., Brumer, P., & Scholes, G. D. (2010). Coherently wired light-harvesting in photosynthetic marine algae at ambient temperature. *Nature, 463*(7281), 644–647. doi:10.1038/nature08811

Conti, F., & Stühmer, W. (1989). Quantal charge redistributions accompanying the structural transitions of sodium channels. *European Biophysics Journal, 17*(2), 53–59. doi:10.1007/BF00257102

Díaz-Franulic, I., Sepúlveda, R. V., Navarro-Quezada, N., González-Nilo, F., & Naranjo, D. (2015). Pore dimensions and the role of occupancy in unitary conductance of Shaker K channels. *Journal of General Physiology, 146*(2), 133–146. doi:10.1085/jgp.201411353

Doyle, D. A., Cabral, J. M., Pfuetzner, R. A., Kuo, A., Gulbis, J. M., Cohen, S. L., . . . MacKinnon, R. (1998). The structure of the potassium channel: Molecular basis of K⁺ conduction and selectivity. *Science, 280*(5360), 69–77. doi:10.1126/science.280.5360.69

Engel, G. S., Calhoun, T. R., Read, E. L., Ahn, T., Mančal, T., Cheng, Y., . . . Fleming, G. R. (2007). Evidence for wavelike energy transfer through quantum coherence in photosynthetic systems. *Nature, 446*(7137), 782–786. doi:10.1038/nature05678

Greeff, N. G., & Forster, I. C. (1991). The quantal gating charge of sodium channel inactivation. *European Biophysics Journal, 20*(3), 165–176. doi:10.1007/bf01561139

Guidoni, L., & Carloni, P. (2002). Potassium permeation through the KcsA channel: A density functional study. *Biochimica Et Biophysica Acta (BBA) - Biomembranes, 1563*(1–2), 1–6. doi:10.1016/s0005-2736(02)00349-8

Guy, H. (1990). Models of voltage- and transmitter-activated membrane channels based on their amino acid sequences. In C. A. Pastemak (Ed.), *Monovalent Cations in Biological Systems*. Boca Raton, FL: CRC Press.

Gwan, J., & Baumgaertner, A. (2007). Cooperative transport in a potassium ion channel. *The Journal of Chemical Physics, 127*(4), 045103. doi:10.1063/1.2756531

Hildner, R., Brinks, D., Nieder, J. B., Cogdell, R. J., & Hulst, N. F. (2013). Quantum coherent energy transfer over varying pathways in single light-harvesting complexes. *Science, 340*(6139), 1448–1451. doi:10.1126/science.1235820

Jortner, J. (1976). Temperature dependent activation energy for electron transfer between biological molecules. *The Journal of Chemical Physics, 64*(12), 4860–4867. doi:10.1063/1.432142

Klinman, J. (1989). Quantum mechanical effects in enzyme-catalysed hydrogen transfer reactions. *Trends in Biochemical Sciences, 14*, 368–373. doi:10.1016/0968-0004(89)90010-8

Lee, H., Cheng, Y., & Fleming, G. R. (2007). Coherence dynamics in photosynthesis: Protein protection of excitonic coherence. *Science, 316*(5830), 1462–1465. doi:10.1126/science.1142188

Long, S. B., Campbell, E., & Mackinnon, R. (2005). Crystal structure of a mammalian voltage-dependent shaker family K⁺ channel. *Science, 309*(5736), 897–903. doi:10.1126/science.1116269

March, N. D., Prado, S. D., & Brunnet, L. G. (2018). Coulomb interaction rules timescales in potassium ion channel tunneling. *Journal of Physics: Condensed Matter, 30*(25), 255101. doi:10.1088/1361-648x/aac40b

Moldenhauer, H., Díaz-Franulic, I., González-Nilo, F., & Naranjo, D. (2016). Effective pore size and radius of capture for K⁺ ions in K-channels. *Scientific Reports, 6*(1). doi:10.1038/srep19893

Morais-Cabral, J. H., Zhou, Y., & Mackinnon, R. (2001). Energetic optimization of ion conduction rate by the K⁺ selectivity filter. *Nature, 414*(6859), 37–42. doi:10.1038/35102000

Parson, W. W., & Warshel, A. (2004). A density-matrix model of photosynthetic electron transfer with microscopically estimated vibrational relaxation times. *Chemical Physics, 296*(2–3), 201–216. doi:10.1016/j.chemphys.2003.10.006

Romero, E., Augulis, R., Novoderezhkin, V. I., Ferretti, M., Thieme, J., Zigmantas, D., & Grondelle, R. V. (2014). Quantum coherence in photosynthesis for efficient solar-energy conversion. *Nature Physics, 10*(9), 676–682. doi:10.1038/nphys3017

Salari, V., Moradi, N., Sajadi, M., Fazileh, F., & Shahbazi, F. (2015). Quantum decoherence time scales for ionic superposition states in ion channels. *Physical Review E, 91*(3). doi:10.1103/physreve.91.032704

Salari, V., Naeij, H., & Shafiee, A. (2017). Quantum interference and selectivity through biological ion channels. *Scientific Reports, 7*(1). doi:10.1038/srep41625

Sato, C., Sato, M., Iwasaki, A., Doi, T., & Engel, A. (1998). The sodium channel has four domains surrounding a central pore. *Journal of Structural Biology, 121*(3), 314–325. doi:10.1006/jsbi.1998.3990

Savikhin, S., Buck, D. R., & Struve, W. S. (1997). Oscillating anisotropies in a bacteriochlorophyll protein: Evidence for quantum beating between exciton levels. *Chemical Physics, 223*(2–3), 303–312. doi:10.1016/s0301-0104(97)00223-1

Schlau-Cohen, G. S., Ishizaki, A., Calhoun, T. R., Ginsberg, N. S., Ballottari, M., Bassi, R., & Fleming, G. R. (2012). Elucidation of the timescales and origins of quantum electronic coherence in LHCII. *Nature Chemistry, 4*(5), 389–395. doi:10.1038/nchem.1303

Stühmer, W., Conti, F., Suzuki, H., Wang, X., Noda, M., Yahagi, N., . . . Numa, S. (1989). Structural parts involved in activation and inactivation of the sodium channel. *Nature, 339*(6226), 597–603. doi:10.1038/339597a0

Summhammer, J., Sulyok, G., & Bernroider, G. (2018). Quantum dynamics and non-local effects behind ion transition states during permeation in membrane channel proteins. *Entropy, 20*(8), 558. doi:10.3390/e20080558

Tegmark, M. (2000). Importance of quantum decoherence in brain processes. *Physical Review E, 61*(4), 4194–4206. doi:10.1103/physreve.61.4194

Vaziri, A., & Plenio, M. B. (2010). Quantum coherence in ion channels: Resonances, transport and verification. *New Journal of Physics, 12*(8), 085001. doi:10.1088/1367-2630/12/8/085001

Vos, M. H., Rappaport, F., Lambry, J., Breton, J., & Martin, J. (1993). Visualization of coherent nuclear motion in a membrane protein by femtosecond spectroscopy. *Nature, 363*(6427), 320–325. doi:10.1038/363320a0

Westenhoff, S., Paleček, D., Edlund, P., Smith, P., & Zigmantas, D. (2012). Coherent picosecond exciton dynamics in a photosynthetic reaction center. *Journal of the American Chemical Society, 134*(40), 16484–16487. doi:10.1021/ja3065478

Zigmantas, D., Read, E. L., Mančal, T., Brixner, T., Gardiner, A. T., & Fleming, G. R. (2006). 2D electronic spectroscopy of the B800–B820 LH3 light-harvesting complex. *Femtochemistry VII*, 372–376. doi:10.1016/b978-044452821-6/50052-6

# Index

Printed in the United States
by Baker & Taylor Publisher Services